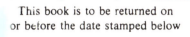
Paddy O'Sullivan
Reading Lists

Paleolimnology and the Reconstruction of Ancient Environments

Paleolimnology and the Reconstruction of Ancient Environments

Paleolimnology Proceedings of the XII INQUA Congress

Edited by

RONALD B. DAVIS

Reprinted from Journal of Paleolimnology,
with the addition of
a more extensive Introduction

Kluwer Academic Publishers

Dordrecht / Boston / London

Library of Congress Cataloging-in-Publication Data

ISBN 0-7923-0571-X

Published by Kluwer Academic Publishers,
P.O. Box 17, 3300 AA Dordrecht, The Netherlands.

Kluwer Academic Publishers incorporates
the publishing programmes of
Martinus Nijhoff, Dr. W. Junk, D. Reidel and MTP Press.

Sold and distributed in the U.S.A. and Canada
by Kluwer Academic Publishers,
101 Philip Drive, Norwell, MA 02061, U.S.A.

In all countries, sold and distributed
by Kluwer Academic Publishers Group,
P.O. Box 322, 3300 AH Dordrecht, The Netherlands.

Printed on acid-free paper

Printed in Belgium

Contents

The scope of Quaternary paleolimnology

Ronald B. Davis
*Department of Botany & Plant Pathology and Institute for Quaternary Studies,
University of Maine, Orono, ME 04469, USA*

Key words: paleolimnology, lake, lake sediment, Quaternary, limnology

Abstract

I describe Quaternary paleolimnology on the basis of a review of abstracts published for meetings of:
(1) American Society of Limnology and Oceanography (ASLO), 10 meetings since 1980; (2) International
Association of Theoretical and Applied Limnology (SIL), 4 since 1977, (3) International Symposia on
Paleolimnology (ISP), 4 since 1967, and (4) International Union for Quaternary Research (INQUA), 5
since 1969. A total of 9538 abstracts were scanned to find 678 with paleolimnological content. A data
base constructed from the 678 contains frequencies of coverage of techniques, variables, themes,
interpretive aspects, and character and geography of study sites. These data indicate that Quaternary
paleolimnology has been a diverse science dealing with many of the same aspects of lakes as neolimnology
but with a longer time perspective. Most frequently studied paleolimnological characteristics were trophic
state, water chemistry (particularly salinity, pH, alkalinity (ANC), micronutrients and oxygen), water
levels, lake morphology, and mixis and other hydrology. Lake biological variables that received greatest
attention were diatoms, pigments, Cladocera, Mallomonadaceae, non-siliceous algae, Ostracoda, and
Mollusca. Most often considered to influence these characteristics and variables were climate;
catchment vegetation, soil, geology, land use and erosion; water chemistry; aerial and non aerial
pollutants; sedimentation; and tectonism. Most frequent chronologic sequences were (1) late-glacial to
present, and (2) modern (*ca.* 0.3 ka to present). Lakes in moist temperate and boreal regions were most
heavily studied.

Of the four series, INQUA covered the longest time scales (to late Tertiary), but emphasized the last
100 ka. INQUA stressed outside-lake geomorphology (as it relates to lake) and lake morphology, physical
forcing functions (e.g., climate and tectonism), hydrologic factors including water levels, paleosalinities,
and reconstruction of paleoclimates. In contrast, SIL and ASLO rarely covered pre-15 ka. Most SIL and
ASLO abstracts dealt with only the most recent ~0.3 ka. Of strong interest to SIL and ASLO were the
effects of catchment vegetation and soils, land uses, and pollutants (e.g., acid deposition) on past lake
chemistry, biology, and trophic conditions. To infer these conditions from sediment contents, frequent
use was made of modern analogues and, starting in the 1970's of microfossil (mostly diatoms) transfer
functions based on calibration data sets. In several respects, ISP subject coverage and approaches were
intermediate between those of INQUA and SIL/ASLO.

Major improvements have been made since the 1960's, particularly in chronology, study of paleoclimate, and application of multivariate statistical techniques for paleoenvironmental inference related to
human impacts on lakes. Further improvements are needed. I recommend (1) increased use of charred
particles, chrysophytes, and non-siliceous algae; (2) use of greater variety of microfossil groups for
quantitative environmental inference, and extension of inference to trophic and other environmental
parameters; (3) application of quantitative inference to fine scale stratigraphic sequences that predate

human effects, to document ranges of natural variation; further study of (4) lake responses to catchment disturbance and succession; (5) effects on lakes of climate, vegetation and soil along elevational gradients; (6) paleolimnology of beaver activity; (7) core replication and representativeness; and (8) increased emphasis on biological taxonomy. Finally, paleolimnology would gain by greater study of lakes in tropical and arctic areas.

General introduction

The major objective of this chapter is to provide a useful and appropriate introduction to this book. I have chosen to do this by describing the science of Quaternary paleolimnology in general, to demonstrate how the book's contents fit within the science. I describe the science as it has been since *ca.* 1970, indicating shortcomings, shifts in emphasis and improvements in the period, and discussing promising avenues for future research.

There have been several reviews of all or major parts of paleolimnology, including Frey (1969, 1974), chapter 6 in Birks & Birks (1980), Binford *et al.* (1983), Engstrom & Wright (1984), Brugam (1984), papers in Gray (1988), and Smol (1989). Those reviews were based on general knowledge of, and experience in paleolimnology by the authors. While this paper also derives from personal knowledge and experience, it differs from prior reviews in being based directly on a data set derived from large numbers of abstracts. These abstracts were published for series of meetings held since the late 1960's. The data set not only allows for an objective assessment of paleolimnology in that period, but also it establishes a firm base-line for assessment of future trends. This approach leads to a non-traditional type of 'review paper'. Instead of summarizing and interpreting large numbers of articles, and listing references to those articles, the paper describes and interprets a data base. The original books of abstracts published for the meetings comprise the primary reference material for the work.

In Part I, the major part of this paper I describe, summarize and discuss the data set. In Part II, I briefly review the paleolimnology content of the XII INQUA Congress, and introduce the reader to the contents of this book of proceedings.

Part I. Survey of paleolimnology abstracts

Introduction

Apart from meetings devoted exclusively to paleolimnology, the two major types of meetings that include substantial numbers of papers on Quaternary paleolimnology are those covering: (1) general Quaternary studies, and (2) limnology. The abstracts published by the leading international societies representing these two areas were reviewed for this study. In addition, abstracts of the major North American limnological (and oceanographic) society (ASLO) were covered because in the past decade it has included many papers and sometimes entire sessions on paleolimnology. Specifically, abstracts from four series of meetings were reviewed: (1) International Symposia on Paleolimnology (ISP), (2) Congresses of the International Union for Quaternary Research (INQUA), (3) Congresses of the International Association of Theoretical and Applied Limnology (SIL), and (4) winter and annual meetings of the American Society of Limnology and Oceanography (ASLO) (Table 1). Paleolimnology papers occasionally have been presented at meetings of many other societies including the Ecological Society of America and the International Association for Ecology, but it was not possible to cover all series of meetings for this paper.

The sample of abstracts was chosen to represent the full scope of Quaternary paleolimnology. On the one hand, INQUA broadly has covered changes in geological, climatological, paleoecological, and archeological aspects of the earth during the Quaternary Period, including consideration of the atmosphere, oceans, lands, rivers/streams, wetlands, and lakes. Paleolimnological studies in that context often have been

Table 1. Numbers of abstracts (all subjects), paleolimnology abstracts*, and percent paleolimnology abstracts published for four series of meetings.

	International Union for Quaternary Research (INQUA), International Congresses				
	1969 France	1973 New Zealand	1977 England	1982 USSR	1987 Canada
All abstracts	457	274	418	654	1086
Paleolimnology abstracts	30	12	25	33	97
Percent	6.6	4.4	6.0	5.1	8.9

	International Symposia on Paleolimnology			
	1967 Hungary	1976 Poland	1981 Finland	1985** Austria
Paleolimnology abstracts	53	76	74	57

	International Association of Theoretical and Applied Limnology (SIL), Congresses			
	1977 Denmark	1980 Japan	1983 France	1987 New Zealand
All abstracts	608	455	780	424
Paleolimnology abstracts	32	18	23	21
Percent	5.3	4.0	2.9	5.0

	American Society of Limnology and Oceanography, Winter (1980) and Annual Meetings									
	1980 ***CAL, WA, TENN	81 WISC	82 N. CAR	83 NFLD	84 BC	85 MINN	86 RI	87 WISC	88 COLO	89 AK
All abstracts	695	312	301	239	339	518	534	352	368	284
Paleolimnology abstracts	10	12	8	1	7	38	11	14	16	10
Percent	1.4	3.8	2.7	0.4	2.1	7.4	2.1	4.0	4.3	3.5

 * See text for definition of 'paleolimnologic abstract'.
 ** Original book of abstracts for 1985 symposium was not available; abstracts of papers in the proceedings were used (Löffler, 1986).
 *** CAL = California; WA = Washington; TENN = Tennessee; WISC = Wisconsin; N. CAR = North Carolina; NFLD = Newfoundland; BC = British Columbia; MINN = Minnesota; RI = Rhode Island; COLO = Colorado.

viewed as only a part of the effort to understand changes in regional landscapes (e.g., tectonics) or changes in atmospheric dynamics on regional or global scales (e.g., glacial/interglacial cycles). In interpreting lake development, emphasis frequently has been placed on exogenous factors. On the other hand, coverage by SIL and ASLO (limnological part) has been more constrained. It has concentrated on the physical, chemical, and biological dynamics of lakes, and on those same aspects of rivers and streams. The terrestrial parts of catchments, and exchanges with the atmosphere have been considered important only to the extent that they affect lakes and streams. Endogenous aspects of lake development have been considered to be important. In the aforementioned respects, the emphases and coverage by ISP have been intermediate between the two 'extremes'.

The data set, and methods

The sample covered only the most recent approximately two decades of paleolimnology, with emphasis on the last decade. I included at least four meetings from each series of meetings to assure that differences in content due to the geographic location of meetings largely would be averaged out. The following number of meetings were covered: INQUA 5, ISP 4, SIL 4, and ASLO 10 (Table 1). I scanned 9358 abstracts to find ones with paleolimnological content. Six-hundred and seventy-eight 'paleolimnology abstracts' were found and studied in detail. The scanning and study took place in the course of ten weeks. Over this period, I tried to be consistent in the classification of abstracts' contents. Undoubtedly, there were subtle shifts in outlook during the period, and certainly my attention to detail was more acute at some times than others, so the data are not perfectly consistent. However, the bold patterns and trends which have emerged from the data cannot be assigned to these possible shortcomings.

A large percentage of the abstracted studies have since been published as full papers. Although I have read many of the papers, the data used for this report were derived only from the abstracts (except for the chapters of this book). Authors attempted to include in their abstracts the most important aspects of their studies. The approach used for this study has the advantage of emphasizing these most important aspects, and establishing a consistent data base for present and future comparisons. The disadvantage is that subsidiary parameters and themes covered in the authors' research or in their full papers undoubtedly were omitted from their abstracts and therefore from these data. Also, some bias has derived from the differing length of abstracts. There has been considerable variation in the length of INQUA and ISP abstracts: generally they have run between 200 and 600 words. On the other hand, the lengths of SIL and ASLO abstracts have been more circumscribed: SIL abstracts have run about 200 words and ASLO abstracts 150-200 words.

Several criteria were used for determining if an abstract was to be counted as paleolimnological. Abstracts were included in the data base only if they contributed to the understanding of lakes in the past. Abstracts primarily on river/stream or wetland environments were counted as paleolimnological only if they included lacustrine sequences. Limnological reconstructions based entirely on historical documents were considered paleolimnological only if they extended more than 50 years before the study (there were very few of these).

A number of INQUA abstracts described long-term Pleistocene continental stratigraphic sequences which included lacustrine units, for purposes of stratigraphic correlation and/or description of regional or global environmental change. Other INQUA abstracts focused entirely on late-glacial through Holocene lake sediments (e.g., many palynologic studies), with the aim of understanding factors outside the lake (e.g., terrestrial vegetation and associated paleoclimate). These abstracts occasionally contained information relevant to understanding the past lake, in which case they qualified as paleolimnological. Certain studies dealing with surface and near-surface sediments were aimed at understanding sediment chemical dynamics and/or exchanges of substances with the water column. Several of these qualified as paleolimnological because the information was useful for taphonomic insights or for modern analogues relevant to paleolimnological sequences. Fortunately, the vast majority of abstracts counted as paleolimnological were clearly so; problematic inclusions occurred in only a small percentage of cases.

The following major categories of information were sought from each paleolimnological abstract: (1) techniques, and classes of information and variables; (2) themes and interpretive aspects; and (3) character and geography of the study site. These major categories included 55, 78, and 74 subcategories, respectively. There was some unavoidable repetition and overlap among subcategories (Table 2). For category 2, no judgements on the validity of authors' interpretations and conclusions were made; the information was

Table 2. Subject coverage of paleolimnology abstracts published for the four series of meetings listed in Table 1, given as mean percent of abstracts per series, and mean of the four means (all). The percentages for this book are based on the full papers.

	INQUA	ISP	SIL	ASLO	ALL	This book
TECHNIQUES, AND CLASSES OF INFORMATION AND VARIABLES						
Dating, correlation, and related techniques:						
Varve dating	2.2	8.0	7.2	2.7	5.04	0
Amino acid recemation	0	0	1.2	0	0.31	0
Fission track	0.2	0	1.2	0	0.36	0
Pb-210, and rarely other radiometric dating	1.8	9.0	12.0	16.6	9.85	0
Stratigraphic markers associated with historical record	4.8	9.2	13.2	4.6	7.97	0
Volcanic as markers, dated non-historically	2.6	0.5	7.0	3.4	3.37	15
Pollen zone correlation	3.6	14.7	0.7	0	4.76	15
Magnetic zone correlation	6.4	4.5	2.5	0	3.35	8
Other and/or unspecified stratigraphic correlation	13.4	7.2	3.5	0.6	6.19	15
Time-series/spectral/Faurier analyses	0.2	0.2	0	0	0.11	0
Indicator-fossils of geologic period(s)	10.6	0.5	0	0	2.77	0
Other techniques:						
Remote sensing	2.6	0	0	1.0	0.90	0
Sediment profiling – seismic/acoustic	2.0	2.2	5.0	1.0	2.56	0
Coring, core subsampling, & related problems	0	0.5	2.2	1.4	1.04	0
Other, including statistics	2.6	14.5	15.2	6.4	9.69	31
Information from outside the present lake or basin of extinct lake, applied to understanding the past lake:						
Paleosoils	5.2	0	0	0	1.30	8
Aeolian deposits incl. sand & loess	8.0	1.7	0	0	2.44	0
Lake shoreline/strandline/terrace	19.4	4.2	0	0	5.91	38
Other geomorphology	26.2	4.2	0	0.6	7.76	31
Trees rings and dendroclimatology	0.8	0.5	0	0	2.12	0
Physical and chemical aspects and contents of sediment:						
Lithology and general description	51.2	41.2	30.7	5.8	32.25	85
Chemistry	24.8	35.0	35.0	46.3	35.27	31
Organic pollutants incl. PAH and soot	0	2.7	1.2	16.8	5.20	0
Stable isotopes	4.4	1.7	3.5	0	2.41	15
Whole lake sediment budget	0.8	0.7	1.5	0	0.76	0
Biological variables including specified biochemicals:						
Diatoms	17.6	36.2	27.2	54.9	34.00	46
Chrysophytes	0.6	6.0	3.0	17.8	6.85	15
Non-siliceous algae	6.0	10.7	7.2	1.2	6.42	23
Fungi	0.4	0	0	0	0.08	8
Pigments	3.8	9.5	13.5	12.1	9.72	23
Lignins & derivatives	0	0.5	1.2	0.3	0.51	0
Pollen/spores, vascular plant & bryophyte	21.4	25.5	16.5	11.9	18.82	54
Plant macrofossils (non-bryophyte)	9.0	8.2	3.5	0	5.19	38
Bryophyte macrofossils	0.6	6.2	0	0	1.71	0
Charred bioparticles incl. charcoal.	1.2	1.2	3.0	3.9	2.34	0
Protozoa (Arcellacea, Testacea, Foraminifera)	2.4	1.5	1.2	0	1.29	0
Porifera	0	1.5	0	0.9	0.60	0
Turbellaria	2.0	1.0	0	0	0.70	8
Rotifera	0	0.5	0	0	1.20	0

Table 2. (continued).

	INQUA	ISP	SIL	ASLO	ALL	This book
Bryozoa (Ectoprocta)	0	1.5	0	0	0.37	0
Ostracoda	12.8	5.7	2.5	1.2	5.49	31
Cladocera	4.8	13.0	13.5	6.8	9.52	23
Other Crustacea; Crustacea-general	3.4	1.0	1.0	0	1.30	15
Mollusca	11.4	8.2	2.2	0	5.47	23
Annelida	0.6	0.5	0.7	0	0.46	0
Chironomidae & Ceratopogonidae	2.8	5.2	5.2	1.7	3.70	15
Chaoborus	0	2.7	1.2	1.2	1.30	0
Trichoptera	0.4	0.2	0	0	0.16	15
Coleoptera	0.4	1.5	0.7	0	0.71	15
Oribatidae	1.6	0.5	0	0	0.52	0
Pisces	1.4	4.0	5.7	0	2.79	8
Amphibia	0	0	0.7	0.	0.19	0
Reptilia	0	0	0.7	0	0.19	0
Birds (waterfowl)	0.6	0	0.7	0	0.34	0
Mammalia	5.4	1.0	0.7	0	1.79	0

THEMES AND INTERPRETIVE ASPECTS

External factors considered in abstract to have caused specified physical/chemical- and general-limnological conditions and/or changes in study lake(s) in the past:

	INQUA	ISP	SIL	ASLO	ALL	This book
Astronomic factors incl. solar cycles	2.8	0.7	1.2	0	1.20	0
Climate (mainly; also glaciers)	49.0	22.7	11.2	7.2	22.55	77
Elevational gradient	1.2	0.7	0	2.2	1.04	15
Marine incursions/transgressions	8.8	0.7	0.7	0	2.57	0
Other aspect of change in sea level	8.0	1.0	1.2	0	2.56	8
Tectonic change incl. earthquake	19.2	10.2	5.0	2.5	9.24	46
Volcanic eruption, ash fall, lava dam	2.0	1.7	2.2	4.9	2.72	0
Hydrothermal activity incl. springs	0	0.2	0	0	0.06	0
River flows: inputs and outputs	18.6	5.0	3.5	2.9	7.50	23
Catchment vegetation, soils, and geological deposits	9.2	13.0	15.0	15.1	13.07	8
Wetland/peatland (w/p) development and/or w/p disturbance in catchment &/or lake basin	2.6	3.2	2.2	1.2	2.32	0
Fire in catchment	0.2	0	0.7	1.4	1.04	0
Beavers	0	0	1.2	0	0.31	0
Anthropogenic – land use incl. mining	9.2	20.5	20.2	15.6	16.39	38
Anthropogenic – pollution (non-aerial)	2.0	13.7	19.0	19.5	13.56	8
Anthropogenic – air pollution incl. acid	1.0	6.0	10.5	26.2	10.92	8

Factors considered in abstract to have caused specified biological conditions and/or changes in the study lake(s) in the past:

	INQUA	ISP	SIL	ASLO	ALL	This book
Climate (mainly; also glaciers)	11.0	16.2	14.7	5.2	11.80	77
Elevational gradient	0.2	1.2	0	3.0	1.11	8
Catchment vegetation, soils, and geological deposits	2.8	10.0	10.0	11.2	8.50	15
Wetland/peatland in catchment	0.8	1.2	1.0	1.5	1.14	0
Fire in catchment	0	0	3.7	2.6	1.59	0
Human damming	0.6	1.2	0.7	1.4	1.00	0
Dredging	0	0	1.2	1.4	0.66	0
Change in lake morphology	3.0	0	0	1.2	1.05	0
Lake ice cover	0	0	1.2	0	0.31	0
Sedimentation incl. volcanic ash fall	2.4	6.7	2.5	7.8	4.86	23

Table 2. (continued).

	INQUA	ISP	SIL	ASLO	ALL	This book
Water chemistry	15.6	22.5	14.5	42.8	23.85	54
Food availability in lake	0	1.0	2.5	0	0.87	0
Grazing and predation in lake	0	2.7	0	1.4	1.04	0
Competition in lake	0	1.0	1.2	0	0.56	0
Macrophyte substrates in lake	0	0.5	0	0	0.12	0
Isolation of lake (re. evolution)	3.2	0.5	0	0	0.92	0
Land bridges (re. evolution)	0.2	0	1.2	0	0.36	8
Paleolimnological aspects and factors for which interpretations were given without consideration to their causal effects:						
Atmospheric deposition (*natural*, other than pollen & spores)	0.8	0.2	0	0	0.20	0.25
Catchment vegetation & soil incl. erosion	10.0	17.7	12.5	14.9	13.79	15
Fires	0.6	1.0	0.7	1.9	1.06	0
Allochthonous vs autochthonous input	1.2	3.7	0.7	0.6	1.57	8
Lake morphology & lake connections	30.8	9.7	7.7	2.1	12.60	8
Ice-block and thermokarst processes	3.0	2.2	0	0	1.31	8
Earthquakes	0.6	0	0	1.0	0.40	0
Light conditions	2.4	1.0	2.2	1.2	1.71	0
Ice cover	1.6	0	1.2	0	0.71	0
Seasonality incl. ice cover period	5.4	3.5	3.2	0.7	3.21	0
Mixis and thermal stratification	4.4	6.0	5.0	4.0	4.85	23
Lake water levels	41.0	19.5	5.5	2.3	17.07	69
Other hydrology	22.4	10.5	2.7	4.2	9.96	38
Water salinity and conductance	23.2	13.0	2.7	5.3	11.06	23
Water pH, acidification, & alkalinity	2.8	16.2	9.0	37.8	16.46	8
Water dissolved organic carbon	0	0	1.2	0	0.31	0
Other water chemistry	8.8	21.7	19.5	39.8	22.46	23
Stability of ecosystem	0.6	0.5	1.2	0	0.59	0
Trophic state	10.8	32.0	26.2	26.6	23.91	46
Diversity and species richness	3.4	6.2	3.2	2.2	3.77	0
Community reconstruction	4.2	5.7	3.5	1.5	3.74	38
Littoral vs offshore inputs	1.6	6.0	2.5	3.4	3.37	8
Successional (autogenic) responses	0.4	3.0	1.5	1.7	1.65	8
Migrations, dispersal, & colonization	3.8	1.5	2.7	0	2.01	8
Biological response times and lags	0.4	1.2	0.7	0	0.60	15
Evolution/speciation & endemism	7.0	3.7	4.7	0	3.87	8
Applied paleolimnology:						
Reconstruction of sea levels	0.6	2.7	1.2	0	1.15	0
Reconstruction of climate	30.8	12.7	10.7	3.1	14.35	38
Impact on lake of land use	9.8	19.2	24.5	14.4	16.99	23
Impact on lake of non-aerial pollutants	2.0	14.2	22.7	19.7	14.67	8
Impact on lake of air pollutants including acids	1.0	7.5	12.0	27.1	11.90	8
Impact on lake of anthropogenic alteration of lake level (damming & draining)	0	2.2	0	1.4	0.91	0
Prospecting for economic deposits	0	1.0	0	0	0.25	0
Uses of economic deposits	0.6	0.5	1.2	0	0.59	0
Biogeography incl. island biogeography:	5.8	4.2	2.5	0	3.14	8
Modern analogues and transfer functions:	11.8	19.7	28.5	28.5	22.14	31

8

Table 2. (continued).

	INQUA	ISP	SIL	ASLO	ALL	This book
Sediment physical dynamics:	20.4	24.2	28.0	12.9	21.39	0
Taphonomy, actuopaleontology, and formation of chemical deposits:	5.4	10.5	7.5	12.9	9.07	8
Wetland stages of development & hydroseres:	8.6	9.0	1.2	0	4.71	0
Interpretive difficulties:						
Dating (incl. old C & bioturbation)	4.8	5.0	1.0	3.6	3.60	38
Taxonomic (biological)	0	0.5	0	0	1.25	0
Past effects of lake on external factors/conditions:						
Climate	0.8	0	0	0	0.20	8
Rivers	0.6	0	0	0	0.15	8
Shoreline modification by in-lake forces:						
By lake ice	0.4	0	0	0	0.10	0
By waves and currents	4.4	0.5	0	0	1.22	0

CHARACTER AND GEOGRAPHY OF THE STUDY SITE

	INQUA	ISP	SIL	ASLO	ALL	This book
Origin of study lake:						
Glacial incl. ice block depression	32.6	52.5	43.2	47.0	43.84	54
Permafrost	2.2	0	0	0	0.55	0
Tectonic – inland	34.4	15.2	19.5	11.4	20.14	38
Tectonic – marine/isostatic	3.8	6.2	2.2	0	3.07	8
Coastal (shoreline) processes	0.8	0.2	1.2	0	0.57	8
Volcanic – crater/maar & lava dam	3.2	4.5	6.2	0	3.49	0
Meteor	0.2	0.2	0	0	0.11	0
Solution	3.8	6.2	1.2	3.2	3.62	8
Riverine	2.0	1.7	2.7	0	1.62	8
Plunge pool	0	0	1.2	0	0.31	0
Land-slide	2.2	0.2	0	0	0.61	0
Deflation and dune dam	6.0	1.2	0	0	1.81	0
Telmatic	0.2	0	1.0	0	0.30	8
Anthropogenic (impoundment)	0	1.7	1.5	12.2	3.86	8
Period(s) represented:						
Ancient sediment						
Pre-Pliocene Tertiary	6.0	5.2	3.2	0	3.62	0
Pliocene	8.0	3.0	9.7	0	5.19	8
Pleistocene – early	4.4	1.2	1.2	0	1.72	0
Pleistocene – middle	5.8	1.7	5.2	0	3.20	8
Pleistocene – late pre- ~100 ka	8.4	3.2	2.0	0.6	3.56	8
Pleistocene unspecified, and all Quaternary except last ~100 ka	12.2	1.5	1.5	0	3.80	0
Early-late Wisc., ~100–~15 ka	31.6	7.7	6.2	2.1	11.92	15
More recent sediment						
Late-glacial/Holocene transition	13.8	1.0	2.5	3.0	5.07	15
Late-glacial to present	39.4	30.5	10.5	9.4	22.45	54

9

Table 2. (continued).

	INQUA	ISP	SIL	ASLO	ALL	This book
Early Holocene to present	6.6	7.0	2.0	6.2	5.45	0
Mid-late Holocene to present	5.8	7.2	9.2	6.4	7.17	8
'Modern' ($\gtrsim$0.3 ka)	13.6	37.2	58.5	70.5	44.96	0
Extinct lake (proglacial, pulvial, etc.):	34.6	7.5	0.7	0	10.71	31
Climate around site:						
Moist/wet arctic and antarctic	8.6	1.7	2.5	0	3.21	0
Semi-arid/arid arctic and antarctic	1.6	1.0	4.0	0	1.65	0
Moist temperate including boreal	28.0	60.7	69.5	67.6	56.46	62
Semi-arid/arid temperate-boreal	21.8	5.0	3.0	3.2	8.25	15
Moist subtropical	1.4	3.0	6.5	3.7	3.65	0
Semi-arid/arid subtropical	1.6	3.0	0	0	1.15	0
Moist tropical	7.0	3.0	0.7	0.9	2.91	0
Semi-arid/arid tropical	9.8	4.7	0.7	2.4	4.42	0
Rheic type now: – endo	17.0	10.7	1.5	4.2	8.36	15
– exo	46.0	63.5	75.5	83.8	67.20	69
Salinity now: – saline	20.6	11.5	2.5	5.3	9.97	15
– fresh	44.6	66.0	63.0	82.3	63.97	69
Geographic location:						
Greenland and Iceland	0	0.7	2.5	0	0.81	0
Europe (Ural Mts. & westward; Caspian Sea & westward; north of Caucasus Mts.; north of Turkey)						
Scandinavia	2.2	19.2	13.5	3.6	9.64	0
North continental & central	5.8	11.5	3.2	0	5.14	15
East, incl. European-USSR	14.6	24.7	4.0	0	10.84	8
South (touch/near Mediterranean, &/or Black, &/or Caspian Seas)	10.6	4.2	4.7	0	4.90	0
United Kingdom	2.0	6.0	6.2	1.2	3.86	0
Asia except Asia Minor						
USSR north of 50° N, except all USSR east of 90° E	2.6	4.2	1.5	0	2.09	0
China	1.0	0	0	0	0.25	0
Japan	2.2	1.5	20.0	0	5.92	8
Southeast & south Asia, west to Persian Gulf and including Iran (non-USSR)	0.6	0	0	0	0.15	0
USSR: S of 50° N, W of 90° E	6.8	5.0	0.7	0	3.14	0
Asia Minor (south of Caucasus Mts. and Black Sea, west of Iran, to west coast of Mediterranean Sea and east coast of Red Sea)	1.4	2.0	0.7	0	1.04	0
New Guinea	2.4	0	0	0	0.60	0
Australia and Tasmania	6.2	1.2	0	0	1.86	0
New Zealand	0.2	1.2	2.2	0	0.92	0
Antarctica	1.6	0	1.2	0	0.71	0
Africa						
North of 15° N	4.4	1.2	0.7	0	1.60	0
15° N to 10° S (tropical)	7.6	2.7	0.7	3.0	3.52	0
South of 10° S	0.2	0	0	0	0.05	0

Table 2. (continued).

	INQUA	ISP	SIL	ASLO	ALL	This book
South America						
Colombia	0.2	0	0	0	0.05	0
Venezuela	0.8	0.2	0	0	0.26	0
Brazil	0.2	0	0	0	0.05	0
Ecuador incl. Galapagos	0	0.7	0	0	0.19	0
Bolivia	0.8	0.5	0	0.6	0.47	0
Argentina	0.4	0	0	0	0.10	0
North and Central America						
Caribbean/Antilles	0.4	0	0	1.5	0.47	0
Central America and Mexico	4.8	1.7	0.7	1.2	2.12	0
Southwest US	2.6	1.5	1.7	8.5	3.59	15
Northwest US & S-west Canada	5.6	1.0	3.2	7.9	4.44	8
Canada north of 55° N	2.0	0.2	1.2	1.2	1.17	0
Alaska	0.4	0	1.5	0	0.47	0
N-east US, S-east Canada	4.0	4.2	11.7	31.4	13.75	8
Laurentian Great Lakes region of US and Canada including south-central Canada	10.4	7.5	13.2	28.8	14.99	15
Mid-west US	1.6	0	0	2.8	1.10	0
South-central & southeast US	1.2	2.0	1.2	5.7	2.36	0
Total papers	2889	260	2267	3942	9358	13
Total paleolimnology papers	197	260	94	127	678	13
(percent)	6.8	100	4.1	3.2	–	100

classified and entered into the data base in all cases. For most abstracts there were multiple entries for at least one of the aforementioned three major categories.

Results

The 678 paleolimnology abstracts were well distributed among the four series of meetings (Table 1). Apart from ISP, INQUA was strongest in its average percentage of paleolimnology abstracts (6.8 percent of total abstracts, from 1969 to 1987). ASLO was weakest (3.2 percent, from 1980 to 1989). The following results refer to the paleolimnology abstracts only: the term 'abstract', used alone will mean paleolimnology abstract; the term 'scientist(s)' (e.g., 'INQUA scientists') will refer only to the persons presenting paleolimnology papers.

There was variation in the number and percent paleolimnology abstracts at different meetings of the same series (Table 1). The amount of paleolimnological research, and the number of practicing paleolimnologists vary from one region to another. Most obvious high points were the 1985 ASLO annual meeting at Minneapolis and the 1987 INQUA Congress at Ottawa. Special efforts were made for those conferences to organize paleolimnological symposia.

The abstracts covered wide varieties of types of information, variables, techniques, themes and interpretations, and site characteristics (Table 2). The complete data, including tabulations for individual meetings are too voluminous to reproduce here, although trends based on each series of meetings will be described. Copies of the complete data are available from the author. Apart from trends based on the original data, the following text largely is based on Table 2.

(1) Techniques, and classes of information and variables

Dating, correlation, and other techniques
The abstracts covered a wide range of approaches to dating and correlation. Carbon-14 was a frequently used dating technique, reflecting the emphasis on the period since 50 ka. However, C-14 was not included in Table 2 because *direct* reference to it in abstracts was very low compared to its actual frequency of use (as indicated by reading many of the full papers). While the abstracts often gave dates in the range of C-14, I could not always correctly assume that C-14 was actually used. Indirect dating methods, such as correlation with C-14 dated sequences elsewhere for pollen zones, volcanic ash layers, and magnetic signatures were used in many cases. For sediments older than the effective range of C-14 dating, classical stratigraphic techniques including indicator fossils were often used. Such geological approaches largely were confined to INQUA abstracts.

The emphasis in the 1970's and 1980's on analysis of short cores, particularly for reconstructing sequences of lake pollution and lake responses to pollution, has led to the frequent use of Pb-210. This radiometric dating technique is limited in effectiveness to the past *ca.* 125 years. To extend dating beyond that limit, and to the beginning of the Industrial Revolution and just before, stratigraphic markers of historically dated events have been used, for example, landscape disturbances of known date as indicated in sediment by pollen, charcoal, and geochemical markers. Markers of events in the past few decades (e.g., Cs-137 marker for atomic bomb testing in the atmosphere), and of events in much earlier periods (e.g., volcanic ash markers of major eruptions/ash falls such as Mazama ash, dated elsewhere by C-14) also have been used. The most accurate dating technique, varve counting was used at many of the sites where varves were present. Techniques suitable for dating 'modern' ($\gtrsim 0.3$ ka) sediments were most frequently employed by SIL and ASLO scientists.

Through the 1970's and 1980's the frequency of use of magnetic correlation and Pb-210 dating increased markedly. Also increasingly used in that period were: (1) multivariate statistical techniques for environmental calibration of microfossils, and (2) freeze coring for obtaining detailed stratigraphic resolution in upper sediments, particularly in varved sediments.

The technical approaches of ISP scientists were generally intermediate between those of INQUA and SIL/ASLO. However, the ISP group made particularly heavy use of pollen zone correlation for dating. That usage reflects a high frequency of study of European sites, and a strong interest in the late-glacial through Holocene period — encompassing the classical European pollen zonation. The almost exclusive emphasis by ASLO scientists on modern sequences resulted in the most limited variety of dating techniques used (Table 2).

Classes of information and variables
Information from outside an extant lake or from the basin of an extinct lake can be useful for understanding the past lake, for example from paleosoils, aeolian deposits, strandlines and other geomorphological features. Strandlines have been used to indicate past, higher lake levels. Lake levels are highly relevant to the interpretation of paleoclimates. As a result of their stronger geological and climatological interests, INQUA scientists most frequently have studied outside-the-lake physical features.

The geological emphasis of INQUA also was reflected by the frequent description of lithology. 'Lithology and general description' (Table 2) included percent water, loss on ignition, structure (varves, turbidites, other laminae, concretions and crusts, fecal pellets, granulometry, etc.), and sediment typology.

Sediment chemistry had the reverse distribution among the abstracts. ASLO scientists have had the strongest interest in relating sediment chemistry to water chemistry and to pollution history. They have given much attention to sediment content of heavy metals. Recently, sediment content of polycyclic aromatic hydrocarbons (PAH in Table 2) — which originate largely from fossil fuel combustion, has received their increased attention.

At least 30 different biological variables were mentioned in the abstracts. Certain of the variables have been widely studied by paleolimnologists regardless of their scientific orientation, most notably diatoms and pollen. Somewhat less frequently used have been 'non-siliceous algae' (including Characeae, *Pediastrum*, *Botryococcus*, and photosynthetic bacteria), plant and bacterial pigments, charred bioparticles, Ostracoda, Cladocera, and Chironomidae. Biological variables that have seen increased use in the past *ca.* 20 years include charred bioparticles and Chrysophyta.

Biological variables that have been more widely studied by INQUA than SIL/ASLO scientists include pollen and spores, plant macrofossils, Protozoa, Turbellaria, Ostracoda, Crustacea other than Cladocera and Ostracoda, Mollusca, Oribatidae, and Mammalia. Ostracoda and Mollusca have been especially useful for stratigraphic study and environmental inference based on pre-Holocene sediments. The more biological/limnological orientation of SIL and ASLO scientists has resulted in greater interests in Chrysophyta, plant and bacterial pigments, and charred bioparticles. Starting in the late 1970's, sedimentary remains of diatoms, and in the 1980's also Chrysophytes have been of especial interest to ASLO scientists as indicators of lake acidification resulting from anthropogenic acid deposition.

For many of the biological variables, ISP had an intermediate frequency between INQUA and SIL/ASLO. The ISP abstracts contained the strongest overall variety of biological variables, reflecting a broad and diverse representation at ISP meetings, despite the relatively small number of participants. However, as discussed later, diversity at ISP did not extend to geographic representation.

(2) *Themes and interpretive aspects of abstracts*

The abstracts contained many different themes and interpretive aspects. These are grouped into categories (Table 2) which are summarized below.

External factors considered in abstract to have caused specified physical/chemical- and general-limnological conditions and/or changes in study lake(s) in the past

Of the 16 factors under this heading (Table 2), the most frequently cited were (a) climate (mainly; also glaciers), (b) catchment vegetation, soils and geological deposits (including biogeochemistry and natural disturbances), and (c) several anthropogenic factors including land use, non-aerial pollution, and air pollution. Tectonism and river flows were somewhat less frequently cited. From the late 1960s into the 1980s, interests in the paleolimnological effects of air pollutants (particularly acid deposition) and volcanic eruptions have increased.

More frequently cited in INQUA than SIL/ASLO abstracts as having caused past limnological conditions or changes were astronomic factors, climate and glaciers, marine incursions, other sea level changes, tectonism, and river flows. The reverse was true for effects of volcanic eruptions; catchment vegetation, soils and geological deposits; beavers; and anthropogenic factors. ISP abstracts were intermediate in their percentages for most of these factors.

Factors considered in abstract to have caused specified biological conditions and/or changes in study lake(s) in the past

Of the 17 factors under this heading, the most frequently cited were water chemistry (including salinity and mineral nutrients), climate (mainly; also glaciers), and vegetation and soil of catchment. No other factors came close in their scores. Comparisons of SIL/ASLO to INQUA indicate higher SIL/ASLO scores for various effects of the catchment, and for in-lake effects of dredging, ice cover, sedimentation including volcanic ash fall, and biological interactions. The data also indicate a high interest of ASLO participants in lake water chemistry. Climate (mainly; also glaciers) scored surprisingly high for ISP and SIL as a cause of biological conditions/changes. Factors claimed in abstracts to affect 'general-limnological conditions' but not specified biological conditions (c.f. previous two paragraphs), including tectonism, elevation, and river flows, obviously have biological effects.

Paleolimnological aspects and factors for which interpretations were given without consideration to their causal effects

Among these 26 aspects and factors (Table 2), the following were most frequently interpreted in abstracts: trophic state; water pH; water salinity and conductance; other water chemistry including oxygen and micronutrients; lake water levels; catchment vegetation and soil, including erosion; and lake morphology and lake connections. Mixis and thermal stratification, including formation and destruction of meromixis, and other hydrological factors were somewhat less frequently interpreted. The original data again indicated that in the 1970's and 1980's interest in atmospheric acid deposition increased. On the other hand, interest in biological diversity and its possible relationship to ecosystem stability has decreased.

Trophic state has been a 'classic' theme of paleolimnologists. That interest was considerably buoyed in the 1960's and 1970's by public concern over cultural eutrophication of lakes (Likens, 1972). The numbers of paleolimnological studies of cultural eutrophication have declined since the mid-1970's. A few of the abstracts tallied under trophic state dealt with the development of dystrophy. Studies of water pH, including alkalinity (acid neutralizing capacity) and lake acidification have partly replaced cultural eutrophication as a major interest in the 1980's, but trophic state has remained a major theme of paleolimnology.

Physical and chemical aspects and factors relating to long-term Quaternary change were of greatest interest to INQUA participants, including lake water levels and other hydrology, lake morphology and connections, and water salinity. 'Other hydrology' (Table 2) includes currents and waves, groundwater inputs, flood inputs, and water budgets. Salinity was often considered in relation to lake levels and climate, or in relation to marine incursions. Of greatest interest to SIL/ASLO participants were: pH, acidification and alkalinity; other water chemistry; trophic state; and catchment vegetation and soils. ISP frequencies were intermediate for most of the aforementioned aspects. However, ISP stood out for its most frequent references to: catchment vegetation and soils; comparisons and distinctions between allochthonous and autochthonous inputs; and between littoral and offshore inputs to deep water sediments; community reconstruction (including aquatic macrophytes and invertebrates); biological diversity; and successional (autogenic) responses to sudden change including anthropogenic disturbance.

Applied paleolimnology

Quaternary geologists and paleoclimatologists have made heavy use of paleolimnology for their reconstructions. Paleoclimate reconstruction for continental areas has been a strong component of INQUA abstracts. That research has most often been based on studies of paleo-lake levels, paleo-salinities, and stable isotopes. Most basins used for the work have been in semi-arid and arid regions, and most of these have been endorheic for much of their history.

The other major area of application of paleolimnology has been elucidation of past impacts of people on lakes. The paleolimnological approach has been useful because direct limnological data on impacts such as cultural eutrophication and acidification are especially sparse for the period prior to 1960. Paleolimnological studies increasingly have improved our understanding of the timing, nature and degree, and geographic extent of these impacts, as recently emphasized by Smol (1989). This has been particularly true for the impact of acid deposition.

Paleolimnology also has been useful for studying human impacts that are not so heavily concentrated in recent decades, particularly the impacts on lakes of land uses in catchments. These land uses have included logging and forest clearance, agriculture and irrigation, urban development, and diffuse-source non-aerial pollution associated with some of those activities. Certain of these studies in Europe and the tropics have dealt with impacts as old as mid-Holocene.

Biogeography including island biogeography
Abstracts on biogeography of lake biota have variously included issues on evolution and/or theoretical ecology — including consideration of lakes as 'islands'. These subjects were largely confined to INQUA and ISP, in line with the relatively long-term chronologies covered.

Modern analogues and transfer functions
Modern analogues were referred to by many authors for interpreting sediment deposits and fossil assemblages. In most cases modern analogues were used in a qualitative manner; in fewer cases, but increasingly in the 1980's transfer coefficients based on modern analogues were applied to microfossils (e.g., diatoms) for quantitative inference of past environmental conditions. A small percentage of the 1980's abstracts dealt with the construction of calibration data sets from (1) microfossils in surface sediments and (2) contemporary water chemistry in the same lakes (usually 30-70 of them) in a region. From the relationships between 1 and 2, calibration equations containing transfer functions (coefficients) were derived. Such formal approaches were most frequently reported by SIL and ASLO scientists whose aims were the reconstruction of human impacts on lakes.

Sediment physical dynamics have been of interest to paleolimnologists because of effects on the sedimentary record. Resuspension of sediment by water movements and by bioturbation can hamper paleolimnological study by reducing temporal resolution of the record and destroying varves. On the other hand, if mixing of sediment has been limited in depth (= time) it can be advantageous by integrating short-term temporal variation (which is not always of interest), thereby reducing the need for closely-spaced samples. Subaqueous erosion, when sporadic (e.g., turbidity flows) can complicate sediment chronologies and interpretations. Many abstracts included interpretations of paleolimnological results in terms of sediment physical dynamics; a few reported on modern studies of the dynamics themselves as they relate to paleolimnology.

Taphonomy, actuopaleontology, and formation of chemical deposits
Understanding the processes of formation of sedimentary deposits and their contents can be helpful for paleolimnological interpretation. Preservation (and decomposition) of organic remains including pigments, diagenesis, chemical mobilities in sediment, development of lithological character including concretions and crusts are all involved in the formation of lacustrine deposits. In many abstracts these processes were part of paleolimnological interpretations; a very few abstracts reported on actual studies of the processes in relation to paleolimnology.

Wetland stages of development and hydroseres
A number of abstracts included description of wetland stages of development (only when a lacustrine stage was also included was an abstract tallied as paleolimnological). Open water lacustrine stages usually preceded wetland stages, but in a few instances the reverse was true. Coverage of wetland phases was most frequent for INQUA and ISP abstracts; SIL and ASLO abstracts were more strictly limnological.

(3) *Character and geography of the study site*

Much has been learned about the nature of paleolimnology by classifying information on study sites. Site information has been classified, as follows: origin(s) of study lake, period(s) represented, climate around study site, and geographic location.

Origin of study lake
Many of the study lakes had multiple modes of origin; all modes given in abstracts were tallied. The relative frequencies of lake origin types reflected, only in part, the worldwide abundance of types. Glacial origins lead the list. Inland tectonic lakes were studied about half as frequently. Less often studied, but still among the most often were:

tectonic – marine/isostatic, volcanic – crater/maar and lava dam, solution, and anthropogenic lakes. The heavy INQUA emphasis on inland tectonic lakes, many of them extinct and located in arid and semi-arid regions, has reflected interests in long stretches of Quaternary time (tectonic lakes tend to be very old) and reconstruction of paleo-climates (tectonic lakes are/were often located in intermontane basins that have been especially sensitive to moisture supply throughout time). The emphasis in ASLO abstracts on anthropo-genic lakes has reflected that group's emphases on the recent period and on human impacts. ASLO's coverage of lake types was much more limited in variety than that of the other groups.

Periods represented
The terms 'ancient sediment' and 'ancient period' in this paper signify periods older than the most recent late-glacial period, namely older than ~ 15 ka. When more than one ancient period was covered by the same sedimentary sequence, the abstract was tallied for each of the periods. In some abstracts, Pleistocene chronologies were broadly stated or vague, in which cases the ab-stracts were tallied under the composite category 'Pleistocene unspecified, and all Quaternary except last ~ 100 ka' (Table 2). The periods listed under 'more recent sediment' (post- ~ 15 ka) are overlapping; sequences were not tallied under more than one of those periods. However, abstracts tallied for one of the two periods that include the late-glacial (Table 2) could also have been tallied for one or more of the ancient periods.

While this paper deals with Quaternary paleo-limnology, the list (Table 2) starts with the Tertiary because a number of Quaternary studies were directed at sedimentary sequences starting in that period. The majority of abstracts covered sediments younger than 15 ka, but a substantial percentage (12%) covered part or all of the last major glacial cycle, ~ 100- ~ 15 ka (the Wisconsin Period of North America). Sequences starting in the late-glacial period (~ 15- ~ 10 ka) and ex-tending to the present were the object of study in 22 percent of the abstracts. The most commonly

studied (44%) lacustrine sequence was 'modern', i.e. less than 0.3 ka.

Ancient sediment (pre- ~ 15 ka) was most fre-quently covered by INQUA abstracts, less fre-quently by ISP and SIL abstracts, and least (by far) frequently by ASLO abstracts. The same re-lationships applied to the study of extinct lakes. In contrast, the modern period was covered least by INQUA abstracts, and most by ASLO abstracts.

Climate around study site
For an extant lake, the present climate was tallied, for an extinct lake the predominant climate during existence of the lake was tallied if it was given or could be inferred from the abstract. Climate is a strong determinant of limnological character. Lakes in moist temperate and boreal regions have been most frequently studied. A weak second in frequency was lakes in semi-arid and arid temperate and boreal climates. Most of these were in the INQUA series. That series also was strongest for lakes in moist arctic and antarctic climates, and for lakes in moist to arid tropical climates. The relatively high frequencies of moist subtropical climates for SIL and ASLO largely derive from studies in Florida which increased in the 1980's. Overall, INQUA had the most balanced climatic coverage. The reverse can be said for ASLO.

Rheic and salinity types
Saline lakes in endorheic basins have been much less frequently studied than freshwater lakes in exorheic basins. There is, or course, a much greater worldwide abundance of the fresh/exor-heic type, which accounts, in large part for the greater study of them. INQUA's relatively strong coverage of the saline/endorheic type is correlated with its coverage of arid regions, and relates to interests in paleoclimates and glacial cycles.

Geographic location
The regions of most frequent study for the three international series of abstracts taken together, in order of frequency were: (1) Europe including Scandinavia and United Kingdom; (2) North America, especially northeast United States,

southeast Canada, and the Laurentian Great Lakes region of both nations; (3) Asia, especially USSR and Japan; and (4) Africa, especially equatorial and northern parts. ASLO, a North American organization is not directly comparable. Apart from equatorial Africa, studies in the tropics have been scarce. Alaska and northern Canada also were sparsely studied. There were no abstracts on southeast Asia and Indonesia.

The original data included large variations in geographic coverage relating to locations of meetings (locations given in Table 1). This relationship probably derived from the often-prohibitive expense of distant travel, scarcity in some nations of foreign exchange, and, at times from travel prohibitions related to international power politics. Four examples follow. (1) The largest number of INQUA abstracts on eastern European and USSR lakes were published in 1982 for the Moscow meeting; the smallest number in 1973 for Christchurch. (2) The largest number of ISP abstracts on Scandinavian lakes were published in 1981 for the Joensuu meeting; the smallest number in 1967 for Tihany. (3) The largest number of SIL abstracts on Japanese lakes were published in 1980 for the Kyoto meeting; the smallest number in 1977 for Copenhagen. (4) The largest number of ASLO abstracts on northwest US and southwest Canada were published in 1984 for the Vancouver meeting; no abstracts on that region were published for several meetings elsewhere in North America.

Abstracts on African lakes peaked in the 1970's. There is not yet an indication of recovery from the low frequency of the 1980's. Abstracts on People's Republic of China (PRC) lakes first appeared at the 1982 INQUA Congress (one abstract), and again (two abstracts) at the 1987 Congress. This may signal a trend, but abstracts on PRC lakes have not yet appeared in the other series. Abstracts on Caribbean lakes and on southeastern US lakes increased in the 1980's, largely due to the activities of E.S. Deevey's group at University of Florida. INQUA and SIL abstracts on southwestern US lakes increased in the 1980's, in part due to expanding interest in paleoclimates.

International coverage by INQUA was most far reaching and balanced. Most notable was INQUA's relatively strong coverage of equatorial and southern hemisphere areas. ISP coverage was most strongly skewed toward Europe.

(4) *Summary of results*

Quaternary paleolimnology is a diverse science dealing with many of the same aspects of lakes as neolimnology but with a longer time perspective. The following limnological characteristics were most frequently studied: trophic state, water chemistry (particularly salinity, pH, alkalinity (ANC), micronutrients and oxygen), water levels, lake morphology, and mixis and other hydrology. Lake biological variables that received the most attention were diatoms, pigments, Cladocera Chrysophytes, non-siliceous algae, Ostracoda, and Mollusca. Most frequently considered to influence these limnological characteristics and biological groups were climate; catchment vegetation, soil, geology, land use and erosion; water chemistry; sedimentation; aerial and non-aerial pollutants; and tectonism. Other important subjects of paleolimnological study were biogeography, modern analogues (including formulation and application of transfer functions), sediment physical dynamics, taphonomy, and wetland stages of development. Most frequently studied chronological sequences were (1) late-glacial ($\sim$ 15-10 ka) to present, and (2) modern ($\sim$ 0.3 ka to present). Lakes in moist temperate and boreal regions were most heavily studied.

Papers with paleolimnological content regularly have been a part of INQUA Congresses. Of the four series, INQUA has generally covered the longest time scales — in some studies back to the late Tertiary. But even INQUA mainly has concentrated on the most recent 100 ka, with greatest emphasis on the last 15 ka. The INQUA approach has been relatively geological: lots of lithological description, study of outside-the-lake geomorphology (as it bears on lake history) and lake morphology, emphasis on physical forcing functions (e.g., climate and tectonism), interest in hydrologic factors (including water levels), paleosalinities, and reconstruction of paleoclimates.

SIL and ASLO scientists have less frequently reported on deposits older than 15 ka. More than 50 percent of SIL and ASLO studies were limited to the most recent ~0.3 ka. Changes in trophic state have been a strong interest. Many SIL and ASLO studies were directed toward understanding the effects of catchment vegetation and soils on lake chemistry, biology, and trophic state. Emphasis has been placed on anthropogenic effects including land uses and pollutants, and on sediment chemical and microfossil indicators of those effects. Use of modern analogues has been strong, and development of calibration data sets by SIL and ASLO scientists has been increasing. In many respects ISP approaches and subject coverage have been intermediate between INQUA and SIL/ASLO.

Discussion

Improvements since the 1960's
In the past two decades, paleolimnology has undergone several improvements. Uses of carbon-14 and Pb-210 dating have substantially increased, leading to greatly improved chronologies. In the 1980's, accelerator (AMS) C-14 dating, feasible for milligram quantities of carbon, has permitted the dating of terrestrial plant macrofossils from lake sediments, thereby avoiding the 'old carbon effect' associated with whole sediment containing remains of aquatic biota. Paleomagnetism, a powerful correlation technique has also seen greater use in this period. Multivariate statistical techniques developed since ~1970 (e.g., canonical correspondence analysis) (ter Braak & Prentice, 1988; ter Braak & van Dam, 1989) not only have improved our understanding of modern relationships between organisms and their environments, but also have led to refined calibrations of microfossils for paleoenvironmental inference (Smol, 1989). Mallomonad chrysophytes have begun to supplement diatoms as indicators of past water chemistry (Smol, 1986).

Paleolimnological studies of lake pH, first applied in the 1970's to the problem of anthropogenic acidification of lakes, have substantially contributed to understanding the chronology and degree of acidification. At the same time, these well financed studies have introduced improvements in technique and in outlook, leading to increased use of paleolimnology for formulation and testing of hypotheses and for corroboration of limnological and biogeochemical models (e.g., Wright et al., 1986).

Underexploited variables and techniques
Several variables have been underexploited relative to their potential usefulness. Only a few will be mentioned here. First, charcoal analysis which is simple to carry out yet quite revealing in regard to paleoclimates, catchment influences, and human impacts is worthy of much wider use. Fire frequency increases when climate becomes drier (Clark, 1988). Charcoal (and fire) analysis can contribute to understanding natural and anthropogenic dynamics of vegetation and soils in lake catchments (Tolonen, 1986; Patterson et al., 1987), and the impact of these dynamics on lakes. Second, the relatively new technique of soot analysis (Griffin & Goldberg, 1979), can provide stratigraphies paralleling fossil fuel usage, and these can serve as chronostratigraphic tools (Renberg & Wik, 1984). Both charcoal and soot analyses have become more numerous since the 1960's, but still are omitted from many studies where they would be useful.

Third, while the use of siliceous scales of mallomonad chrysophytes as paleolimnological indicators has increased in recent years (Smol, 1986), further increase would be productive. The other commonly used group of siliceous algae, diatoms, tends to be scarce or absent in the plankton of unproductive acidic (pH $\gtrsim 5.5$) lakes (Charles & Smol, 1988). Mallomonad chrysophytes are strictly planktonic, and certain taxa are relatively abundant in such lakes. Other mallomonad taxa abound in more eutrophic and higher pH lakes (Kristiansen, 1986; Siver & Hamer, 1989), where they have been least used for paleolimnological reconstruction.

Fourth, while calibration equations for paleo-limnological inference have been developed for certain environmental parameters and applied in limited contexts, there is room for considerable further development and application. Equations derived from diatom and chrysophyte calibration data sets have been used for pH, alkalinity, and Al reconstructions in short-core studies (Davis, 1987; Charles & Smol, 1988). The same approach has been used for dissolved organic matter (DOC) reconstruction in Norwegian lakes (Davis et al. 1985; Davis, 1987). More recently, diatom calibration equations for DOC have been developed for North American lakes ('PIRLA II' project: D. Charles, pers. comm.). Calibration of fossil groups in addition to diatoms and chryso-phytes, and applications of the approach to additional aspects of lake water chemistry would help to put paleolimnology on a firmer quantitative basis.

Fifth, the remains of non-siliceous algae have potential for much wider use. These remains are abundant in many lake sediments, and can be helpful for understanding past limnological conditions (van Geel, 1986). The severe sediment processing techniques used for siliceous algae and pollen destroy the remains of most non-siliceous algae. Study of these remains on parallel series of slides prepared by gentler techniques (Cronberg, 1986) could significantly augment paleolimnological interpretations.

Underexploited themes and objectives
Several infrequently appearing themes are ripe for further inquiry, and potentially can yield a rich harvest of paleolimnological insights. First, limnological responses to sudden destruction of catchment vegetation and soil organic matter, as by forest fire, are well-suited to fine-resolution paleolimnologic study, particularly in lakes with varved sediments. Fire and other agents of destruction (e.g., logging) of catchment ecosystems can drastically alter soil nutrient conditions, patterns of leaching and erosion, and amounts and contents of runoff to lakes (Likens & Bormann, 1979). Disturbance events are followed by vegetational succession and soil development.

Potential physico-chemical and biological responses of lakes to these changes can be detected from several sedimentary variables. Boucherle et al. (1986) investigated lake responses to the rapid decline of *Tsuga canadensis* in catchment forest. Improved calibration equations for indicators (e.g., chrysophytes) of relevant water column parameters (e.g., chlorophyll *a*) would greatly advance this research. Many intriguing hypotheses can be formulated around the general question: do lake responses track changes in the catchment?

Second, much can still be learned about lake trophic state. Approaches used in the 1960's and 1970's for paleolimnological study of trophic state (Davis & Norton, 1978) are no longer adequate. Efforts were afoot in the 1970's to place paleolimnologic inference of trophic state on a more quantitative basis, for example, diatom calibrations for epilimnetic total phosphorus (an index of trophic state) and conductance were developed by Bailey (1978) and Bailey & Davis (1978). These efforts were sidetracked by acid deposition. Recent signs of return to the trophic question have appeared in the form of improved approaches for calibrating diatoms as trophic state indicators (Agbeti, 1987; Christie, 1988). Smol (1989) has recently discussed this promising trend in greater detail.

Third, it would be most helpful for interpretation of quantitative paleolimnological reconstructions of modern ($\gtrsim 0.3$ ka) human impacts on lake characteristics (e.g., acidity, trophic state, and other) to have comparably quantitative information on earlier natural variation of the same characteristics. The importance of a particular finding of anthropogenic change will depend in part on whether the degree (and rate) of change is within or beyond the range of natural variation. In the 1980's almost all of the applications of calibration equations have been to modern ($\gtrsim 0.3$ ka) sedimentary sequences for understanding human impacts on lakes. While human impacts including acidification and cultural eutrophication are worthy of further paleolimnological attention, it is crucial at this point to extend the new methods to longer-term non-anthropogenic sequences. In addition to standard resolution study of natural long-term chemical change and trophic development (as in Whitehead *et al.*, this volume), particular attention should be payed to fine-scale stratigraphic resolution to document ranges and rates of natural variation most relevant to the time scales of human impacts. Studies of lake responses to catchment disturbance and succession, discussed earlier, are relevant to this effort.

Fourth, study of series of lakes with similar basin morphology and geology along elevational gradients (see Whitehead *et al.*, this volume) can increase understanding of limnological responses to the climatic, vegetational, and pedological gradients associated with elevation. These gradients, along short elevational distances, essentially parallel those over longer latitudinal distances and can conveniently serve as latitudinal analogues.

Fifth, the limnological/hydrological effects of beavers have barely been touched on by paleolimnologists. It may be possible to investigate limnological/hydroseral responses to beaver activity by coordinated studies of modern limnology and paleolimnology (sediment and peat sequences) in series of beaver impoundments of differing age including sites that have succeeded to wetlands. I have observed many steam-valley fens that appear to be old beaver flowages (linear series of impoundments). How old are they? Beaver dams, degraded and largely overgrown, were often discernable. A valuable supplement to dating the sediment and peat deposits would be the dating of the beaver dams themselves. How can this be done?

The population of beavers in North America has been greatly reduced by intensive trapping since early colonial time. Considering the present major impact of beavers on the hydrology and ecology of boreal landscapes (Naiman *et al.*, 1986), including many thousands of beaver impoundments and wetlands that originated as impoundments, the impact prior to colonial time must have been immense. Beaver modification of drainages may be cyclic along with fire rotation and the resultant succession that controls availa-

bility of food (e.g., *Populus* spp.) for beaver. Beavers may hasten that succession, and shorten the cycle by reducing populations of early-successional tree species. In many cases, when beavers abandon a flowage due to reduced availability of food, the abandoned dams are destroyed by action of frost and ice, spring freshets and other processes, and there is a return to a narrow stream course surrounded by an upland ecosystem. In other cases, abandoned impoundments gradually fill with sediment and succeed to long-lasting wetlands/peatlands. If food supply and other requirements for beavers in the wetland and surrounding upland again become favorable, the animals may return and flood the wetland. Undoubtedly, there is a great deal of variation in these patterns (Naiman *et al.*, 1986). Can these postulated paleolimnological and hydroseral sequences/cycles be reconstructed from sedimentary sequences?

Sixth, interpretations of biological remains in sediment cores have often failed to consider biological interactions as possible causes of change(s). Grazing, predation and other interactions can have major effects on aquatic communities and the fossil record of them (Nilssen & Sandoy, 1990). By combined study in the same core of remains from (1) phytoplankton (e.g., scaled chrysophytes) (2) herbivorous zooplankton (e.g., certain Cladocera), and (3) predatory zooplankton (e.g., *Chaoborus*) it is possible to infer certain community responses to interactions. For example, Uutala (1989) has shown for Adirondack (NY, USA) lakes that the remains (primarily mandibles) of the predatory *Chaoborus americanus*, a species that is entirely limnetic (does not migrate by day to the sediment as do certain other *Chaoborus* spp.) first appear in Pb-210 dated sediment at a time when fishes are known from the historical record to have been eliminated by acidification (demonstrated from diatom and chrysophyte remains). Further application of such approaches would help to place paleolimnology on a firmer ecological base.

20

Interpretive difficulties

Establishing a proper chronology was the most frequently cited (in abstracts) difficulty, particularly the interpretation of problematic C-14 results. A reading of many of the full papers has indicated that authors frequently have underestimated the chronology problem. Most understated, however, wre difficulties associated with biological taxonomy. It was rare to find reference to this problem in abstracts. Most experienced paleolimnologists recognize the dependence of accurate conclusions on correct taxonomy, yet even the full length papers usually avoided or belittled the problem. In some recent paleolimnological research programs, the problem has received forthright, well-organized and energetic attention (e.g., PIRLA Project: Kingston, 1986). Additional efforts of this type, and increased support for taxonomic study are called for.

A nagging worry of paleolimnologists is the extent to which their single core of mid-lake sediment can be used to interpret conditions and changes in the entire lake. Frey (this volume) deals with the representation of littoral versus offshore biotic remains in mid-lake cores. Recognition that sedimentation processes associated with river/stream inputs, focusing of sediment to the deepest part of the lakes, and finer scale dynamics cause horizontal-spatial variation in sediment character and content, and particulary that transport properties vary for different components of sediment, has led to many questions regarding the 'representativeness' of single mid-lake cores. Recently, these questions have stimulated important studies of replicate mid-lake cores (e.g, Kreis, in press), but a much broader attack on this problem is needed.

Underexploited types and locations of sites

Studies of tropical and arctic lakes have lagged behind studies of temperate and boreal lakes: in particular, tropical lakes of the solution type, and arctic lakes of the permafrost (meltout) type have been understudied. Even in temperate and boreal regions some types have been slighted, most notably riverine and telmatic lakes.

Quaternary paleolimnology has had a strong bias toward study of lakes in moist temperate and boreal regions (56.5% of all studies). This bias is understandable, given the large number of research universities in those regions, the relatively favorable financial base of many of those universities, and the nearby abundance and accessibility of lakes. Expansion of paleolimnological research in other regions should be encouraged. Efforts to broaden the geographic coverage of International Symposia on Paleolimnology, perhaps by development of a travel support program for scientists from 'third world' developing nations, would be a step in the right direction.

Part II. Paleolimnology proceedings of the XII INQUA Congress, and contents of this book

The XII INQUA Congress was held in Ottawa, Canada in July, 1987. The meeting contained an unprecedentedly large number of papers on paleolimnology (Table 1 lists the Congresses back to 1969). Ninety-seven or 8.9 percent of a total of 1086 abstracts had some appreciable paleolimnological content. Four half-day sessions of oral presentations explicitly on paleolimnology took place: (1) Paleolimnology and Associated Paleoenvironments, (2) Paleolimnology and its Application to Reconstruction of Paleoclimate, (3) Paleolimnology: Recent Advances, and (4) The Lake Biwa Record. In addition, a poster session on paleolimnology was held. The first oral session consisted of papers invited by R.B. Davis and H. Löffler; several of these papers are contained in this book. Additional papers in the book come from other paleolimnology sessions. Certain of the papers come from sessions not explicitly labeled as paleolimnological. Fifteen oral and poster sessions entitled 'paleohydrology', 'paleoclimates', and/or 'paleoenvironments', as well as other sessions contained scattered paleolimnology papers.

Contents of book, in relation to paleolimnology in general

I cannot hope to include in one short book of papers all aspects of the field of paleolimnology. Nevertheless, the book does cover a wide range and variety of paleolimnological techniques, variables, topics and concepts Table 2). As expected, the contents are more typical of the geological orientation of INQUA than the more strictly limnological orientation of SIL and ASLO. This is exemplified by (1) the greater than average coverage of sediment description and lithology, old strandlines and geomorphology, former lake levels, 'other hydrologic factors', paleosalinity and paleosalinity indicators (e.g., ostracodes), endorheic basins, extinct lakes, paleoclimatic inferences and effects, tectonic effects, and late-glacial and older sediments, and (2) the scant inclusion of short-core studies that disclose recent human impacts on lakes, and absence of related techniques such as Pb-210 dating. However, the book has stronger biological and limnological emphases than are typical of INQUA, as exemplified by its coverage of (1) siliceous algae and pigments, (2) limnological effects of catchment vegetation, soils, and land use, (3) community and trophic state reconstructions, and (4) biological effects of water chemistry including pH. The preponderance of moist-temperate study sites is more typical of SIL and ASLO than INQUA. North American and, to a lesser extent, continental European authors and study sites dominate the book; unfortunately, the tropics receive no coverage.

Organization of book

While there are varying degrees of overlap in content and approach between the chapters, it has been possible to sort them into six groups, each group representing somewhat different aspects of, and/or approaches to paleolimnology. This first chapter constitutes Part I. The six remaining parts are described below.

Part II deals with pluvial and non-pluvial lake stages, changing water level and salinity, inlet river diversion, abandoned outlets, and relationships of the aforementioned to past climate and tectonism, as illustrated by studies in presently arid, intermontane basins of western United States. This predominantly geological approach is illustrated by two papers: *Chapter 2* by C.G. Oviatt on late-glacial and Holocene lake fluctuations in Sevier Lake basin, Utah, and *Chapter 3* by J.P. Bradbury, R.M. Forester, and R.S. Thompson on late Quaternary changes in the Walker Lake basin, Nevada. *Chapter 2* primarily is based on geomorphic and stratigraphic studies of lacustrine and alluvial deposits outside the present Sevier Lake, including old shorelines. Paleolimnological events in the last 14 ka are explained in terms of tectonism and climatic models. *Chapter 3* makes heavier use of biological evidence (diatoms, ostracodes, and pollen) in mid-lake sediment cores for the purpose of reconstruction of physical and chemical changes in a lake basin in the last 40 ka. Major emphasis is places on inlet river diversion, first away from the lake, and much later back to the lake, to illustrate a major point: straightforward interpretation of lake stages in terms of paleoclimate alone can be misleading.

Part III covers past glacial meltwater inflows and changing lake levels and climates, inferred from lacustrine and associated deposits in and around the Laurentian Great Lakes. *Chapter 4* by D.F.M. Lewis and T.W. Anderson describes episodes of major glacial meltwater inflow from upper proglacial to lower 'non-glacial' great lakes, and associated changes in lake water temperature and levels, and lake surface configurations, lake connections and drainage ways, and lake-influenced climates (as inferred from pollen) in late-glacial and early Holocene time. The authors suggest that the cold climate event at 11.0-10.5 ka near the lower lakes was due to the meltwater influx, and did not involve a major air mass shift related to the Younger Dryas episode. *Chapter 5* by F.M.G. McCarthy and J.H. McAndrews deals with short-term changes in rates of water level rise along the north shore of lake Ontario from 4.2 to 2.0 ka, as inferred from sediment (including pollen and plant macrofossils) in a small embayment/lake, and correlates episodes of higher rates with periods of cool, wet climate. *Chapter 5* makes heavier use of biological evidence, but like *Chapter 4* its primary objective is the reconstruction and explanation of past physical changes in lakes.

Part IV deals with difficult problems and controversies in paleolimnology. *Chapter 6* by B. Ammann describes two sets of synchronous changes in biological variables in Swiss lakes and landscapes, the first set around 12.5 ka and the second around 10 ka. The problem is that the sedimentary interval covering each set has a series of closely spaced C-14 AMS dates which are virtually identical, suggesting that each set of changes was virtually instantaneous. Ammann postulates that the constant dates are artifacts of periods of reduced atmospheric C-14 resulting from increased ocean circulation coinciding with each interval. *Chapter 7* by D. Frey is a review and discussion of the interpretive problems arising from the differing contributions of littoral versus offshore communities to the fossil contents of cores of offshore sediment. He argues that in thermally stratified lakes the littoral and offshore areas may have very different developmental trajectories, and therefore the remains from each part of the lake should be considered separately, but that separate interpretation for the littoral zone from an offshore core is difficult.

Chapter 8a by I.R. Walker & R.W. Mathewes describes the late-glacial and postglacial sequences of chironomid remains in two British Columbia lakes. As in their earlier paper (Walker & Mathewes, 1987a), the authors interpret the rapid change in chironomid fauna around 10 ka to have been caused by climatic warming. Following the 1987 paper, the interpretation was disputed in a commentary by Warner & Hann (1987) — which elicited a rejoinder by Walker and Mathewes (1987b). Similarly, in this volume the climatic interpretation is questioned, this time by W.F. Warwick (*Chapter 8b*), and followed by a response by Walker & Mathewes (*Chapter 8c*). This debate is important: (1) to Quaternary scientists in general because of their interest in environmental reconstruction from biological evidence, especially the reconstruction of paleoclimates, and (2) to Quaternary paleoecologists in particular because it demonstrates the great relevance of understanding the present functioning of ecosystems and ecology of organisms for interpreting past ecological changes.

Part V covers studies of lake development since late-glacial time in two, moist north-temperate areas. *Chapter 9* gives the results of a multidisciplinary project by D.R. Whitehead, D.F. Charles, S.T. Jackson, J.P. Smol, and D.R. Engstrom on changes in water chemistry and trophic state in three lakes along an elevational gradient in New York State, USA. The research adds to the growing realization that many lakes in glaciated regions were more productive in late-glacial and early Holocene time than they have been at any time since. Emphases are placed on the effects of catchment vegetation and soils on nutrient fluxes, pH and associated limnological changes. *Chapter 10*, by H. Züllig briefly summarizes the use of carotenoid pigments in paleolimnology, and then goes on to review studies of short cores and long cores from Swiss lakes. By means of highly specific pigment analyses and other variables, Züllig is able to interpret the cores in terms of oxygen conditions in bottom waters, formation and destruction of meromixis and change in trophic state, and to demonstrate the sensitivity of the lakes to natural and human influences.

Part VI covers classifications, variables, and techniques. In *Chapter 11*, L. Saarse presents a classification of Estonian lake basins based on their origins, a lithological classification of Estonian lake sediments, and a geographic summary of those aspects of the republic's lakes. The classifications are based on large data bases and wide experience by the author. The paper provides a useful overview for limnologists and paleolimnologists, and for students in these fields. *Chapter 12* by N.E. Williams reviews the use of larval caddisfly (Trichoptera) remains from lacustrine and fluvial habitats for reconstruction of past environments. These remains, rarely found in the offshore, deep-water cores so commonly studied, are sometimes abundant in littoral and stream deposits. Environmental inferences from caddisfly remains can include '....type of water body, water depth, current speed, substrate, aquatic vegetation and surrounding terrestrial vegetation'. Williams' ongoing work will facilitate future paleoecological use of caddisfly remains. *Chapter 13* by Roy Thompson & R.M. Clark describes an improved method of sequence slotting, a numerical technique for core comparison and matching. While such approaches have been commonly used by marine stratigraphers, they can also be helpful to paleolimnologists for sparsely dated sequences, and for corroboration of dates (which may be in error) in core matching.

Part VI contains only *Chapter 14* by S. Takahashi, dealing with origins, speciation and endemism of fishes and invertebrates in Lake Biwa, Japan. That paper is unusual for these proceedings in covering an evolutionary theme. It is included as an example of a type of study occasionally reported by INQUA participants, but very rarely by ISP, SIL and ASLO scientists whose focus has generally been on time scales too short for metazoan speciation to occur.

Acknowledgements

D.S. Anderson, T-O Rhodes, and Fengsheng Hu helped in a variety of ways in the preparation of this chapter. Book preparation was greatly facilitiated by the secretarial assistance of D. Wilbur of the University of Maine, Department of Botany and Plant Pathology. J. Smol provided major editorial input and facilitated publication. H. Löffler arranged the initial reviews of three of the chapters. Without the reviewers this book would not have been possible; their names are: T.W. Anderson, P.J. Barnett, R.W. Battarbee, L. Benson, R.B. Brugam, T.L. Crisman, D.L. Danielopol, P. DeDeckker, E.S. Deevey, D.R. Engstrom, W.R. Ferrand, O.S. Flint, J.R. Glew, W. Hofmann, P.F. Karrow, J.W. King, I.L. Kornfield, D.A. Livingstone, R.W. Mathewes, A.R. McCune, E.H. Muller, S.A. Norton, R. Schmidt, F.A. Street-Perrott, E.B. Swain, W.F. Warwick & T. Webb.

References

Agbeti, M. D., 1987. The use of diatom assemblages to determine paleotrophic changes in lakes. M.Sc. Thesis, Brock Univ., St. Catherines, Ontario, Canada: 140 pp.
Bailey, J. H., 1978. Quantitative comparison of limnological characteristics and diatom surface sediment assemblages in nineteen Maine lakes. M.S. Thesis, University of Maine, Orono, USA: 45 pp.
Bailey, J. H. & R. B. Davis, 1978. Quantitative comparison of lake water quality and surface sediment diatom assemblages in Maine, USA lakes. Verh. int. Ver. Limnol. 20: 551.
Binford, M. W., E. S. Deevey & T. L. Crisman, 1983. Paleolimnology: an historical perspective on lacustrine ecosystems. Ann. Rev. Ecol. Syst. 14: 255–286.
Birks, H. J. B. & H. H. Birks, 1980. Quaternary palaeoecology. E. Arnold, London, 289 pp.
Boucherle, M. M., J. P. Smol, T. C. Oliver, S. R. Brown & R. McNeely, 1986. Limnologic consequences of the decline in hemlock 4800 years ago in three Southern Ontario lakes. Hydrobiol. 143: 217–225.
Brugam, R. B., 1983. Holocene paleolimnology. In H. E. Wright, Jr. (ed.), Late-Quaternary Environments of the United States. Univ, Minnesota Press, Minneapolis: 208–221.
Charles, D. F. & J. P. Smol, 1988. New methods for using diatoms and chrysophytes to infer past pH of low-alkalinity lakes. Limnol. Oceanogr. 33: 1451–1462.
Christie, C. E., 1988. Surficial diatom assemblages as indicators of lake trophic status. M.Sc. Thesis, Queen's Univ., Kingston, Ontario, Canada: 124 pp.
Clark, J. S., 1988. Effect of climate change on fire regimes in northwestern Minnesota. Nature 334: 233–235.
Cronberg, G., 1986. Blue-green algae, green algae and chrysophyceae in sediments. In B. E. Berglund (ed.), Handbook of Holocene Palaeoecology and Palaeohydrology, J. Wiley & Sons, Chichester, UK: 507–526.
Davis, R. B., 1987. Paleolimnological diatom studies of acidification of lakes by acid rain: an application of Quaternary science. Quat. Sci. Rev. 6: 147–163.
Davis, R. B., D. S. Anderson & F. Berge, 1985. Paleolimnological evidence that lake acidification is accompanied by loss of organic matter. Nature 316: 436–438.
Davis, R. B. & S. A. Norton, 1978. Paleolimnologic studies of human impact on lakes in the United States, with emphasis on recent research in New England. Pol. Arch. Hydrobiol. 25: 99–115.
Engstrom, D. R. & H. E. Wright, Jr., 1984. Chemical stratigraphy of lake sediments as a record of environmental change. In E. Y. Haworth & J. W. G. Lund (eds.), Lake Sediments and Environmental History. Univ. Minnesota Press, Minneapolis: 11–67.
Frey, D. G., 1969. The rationale of paleolimnology. Mitt. int. Ver. Limnol. 17: 7–18.

Frey, D. G., 1974. Paleolimnology. Mitt. int. Ver. Limnol. 20: 95–123.

Gray, J. (ed.), 1988. Paleolimnology: aspects of freshwater paleoecology and biogeography. Elsevier, Amsterdam: 678 pp.

Griffin, J. J. & E. D. Goldberg, 1979. Morphologies and origin of elementary carbon in the environment. Science 206: 563–565.

Kingston, J., 1986. Diatom analysis – basic protocol. In D. F. Charles & D R. Whitehead (eds.), Paleoecological Investigation of Recent Lake Acidification, Methods and Project Description. Electric Power Research Institute, Palo Alto, California, Rpt. EA-4906: 6–1 – 6–11.

Kreis, R. G., Jr. Sources and estimates of variability associated with paleolimnological analyses of PIRLA sediment cores. J. Paleolim. In press.

Kristiansen, J., 1986. Silica-scale bearing chrysophytes as environmental indicators. Br. phycol. J. 21: 425–436.

Likens, G. E. & F. H. Bormann, 1979. The role of watershed and airshed in lake metabolism. Arch. Hydrobiol. 13: 195–211.

Löffler, H. (ed.), 1986. Paleolimnology. Hydrobiol. 143: 431 pp.

Naiman, R. J., J. M. Melillo & J. E. Hobbie, 1986. Ecosystem alteration of boreal forest streams by beaver (*Castor canadensis*). Ecology 67: 1254–1269.

Nilssen, J. P. & S. Sandoy, 1990. Cladocera, predation and lake acidification. In R. W. Battarbee, I. Renberg, J. F. Talling & J. Mason (eds.), Palaeolimnology and lake acidification. Phil. Trans. R. Soc., Ser. B, London. (in press).

Patterson, W. A., III, K. J. Edwards & D. J. Maguire, 1987. Microscopic charcoal as a fossil indicator of fire. Quat. Sci. Rev. 6: 3–23.

Renberg, I. & M. Wik, 1984. Dating recent lake sediments by soot particle counting. Verh. int. Ver. Limnol. 22: 712–718.

Siver, P. A. & J. S. Hamer, 1989. Multivariate statistical analysis of the factors controlling the distribution of scaled chrosophytes. Limnol. Oceanogr. 34: 368–381.

Smol, J. P., 1986. Chrysophycean microfossils and indicators of lakewater pH. In J. P. Smol *et al.* (eds.), Diatoms and Lake Acidity. W. Junk Publ., The Hague, Neth.: 275–287.

Smol, J. P., 1989. Paleolimnology — recent advances and future challenges. In R. De Bernardi (ed.), Scientific Perspectives in Theoretical and Applied Limnology. CNDR, PALLANZA. (in press).

ter Braak, C. J. F. & I. C. Prentice, 1988. A theory of gradient analysis. Adv. ecol. Res. 18: 271–317.

ter Braak, C. J. F. & H. van Dam, 1989. Inferring pH from diatoms: a comparison of old and new calibration methods. Hydrobiol. 178: 209–223.

Tolonen, K., 1986. Charred particle analysis. In B. E. Berglund (ed.), Handbook of Holocene Palaeoecology and Palaeohydrology, J. Wiley & Sons, Chichester, UK: 485–496.

Uutala, A. J., 1989. Historical fish population status inferred from *Chaoborus* stratigraphy. Vth Int. Symp. Palaeolim. Abstracts, Cumbria, U.K.: 107.

van Geel, B., 1986. Application of fungal and algal remains and other microfossils in palynological analyses. In B. E. Berglund (ed.), Handbook of Holocene Palaeoecology and Palaeohydrology, J. Wiley & Sons, Chichester, UK: 497–526.

Walker, I. R. & R. W. Mathewes, 1987a. Chironomidae (Diptera) and postglacial climate at Marion Lake, British Columbia, Canada. Quat. Res. 27: 89–102.

Walker, I. R. & R. W Mathewes, 1987b. Chironomids, lake trophic status, and climate. Quat. Res. 28: 431–437.

Warner, B. G. & B. J. Hann, 1987. Aquatic invertebrates as paleoclimatic indicators. Quat. Res. 28: 427–430.

Wright, R. F., B. J. Cosby, G. M. Hornberger & J. N. Galloway, 1986. Comparison of paleolimnological with MAGIC model reconstructions of water acidification. Wat. Air Soil Pollut. 30: 367–380.

Late Pleistocene and Holocene lake fluctuations in the Sevier Lake Basin, Utah, USA

Charles G. Oviatt
Department of Geology, Kansas State University, Manhattan, KS 66506, USA

Key words: paleolimnology, Sevier Lake, Great Basin, western United States, Lake Bonneville

Abstract

Sevier Lake is the modern lake in the topographically closed Sevier Lake basin, and is fed primarily by the Sevier River. During the last 12 000 years, the Beaver River also was a major tributary to the lake. Lake Bonneville occupied the Sevier Desert until late in its regressive phase when it dropped to the Old River Bed threshold, which is the low point on the drainage divide between the Sevier Lake basin and the Great Salt Lake basin. Lake Gunnison, a shallow freshwater lake at 1390 m in the Sevier Desert, overflowed continuously from about 12 000 to 10 000 yr B.P., into the saline lake in the Great Salt Lake basin, which continued to contract. This contrast in hydrologic histories between the two basins may have been caused by a northward shift of monsoon circulation into the Sevier Lake basin, but not as far north as the Great Salt Lake basin. Increased summer precipitation and cloudiness could have kept the Sevier Lake basin relatively wet.

By shortly after 10 000 yr B.P. Lake Gunnison had stopped overflowing and the Sevier and Beaver Rivers had begun depositing fine-grained alluvium across the lake bed. Sevier Lake remained at an altitude below 1381 m during the early and middle Holocene. Between 3000 and 2000 yr B.P. the lake expanded slightly to an altitude of about 1382.3 m. A second expansion, probably in the last 500 years, culminated at about 1379.8 m. In the mid 1800s the lake had a surface altitude of 1379.5 m. Sevier Lake was essentially dry (1376 m) from 1880 until 1982. In 1984–1985 the lake expanded to a 20th-century high of 1378.9 m in response to abnormally high snow-melt runoff in the Sevier River. The late Holocene high stands of Sevier Lake were most likely related to increased precipitation derived from westerly air masses.

Introduction

This paper summarizes current knowledge about the late Pleistocene and Holocene history of Sevier Lake and its predecessors in western Utah. The conclusions are based on field observations of the geomorphology and stratigraphy of lacustrine and alluvial deposits in the Sevier Desert. In addition, a paleoclimatic hypothesis is presented to explain the apparent lack of synchroneity between lakes in the Sevier Lake basin and the Great Salt Lake basin during the period from about 12 000 to 10 000 yr B.P. The Holocene records of the two basins seem to have been similar.

Sevier Lake is a shallow, ephemeral saline lake in one of the closed basins of the Basin and Range physiographic province in western Utah (Fig. 1). During historic high stands Sevier Lake has had a surface area of about 400 km² and has been fed

Originally published in
Journal of Paleolimnology **1**: 9–21.

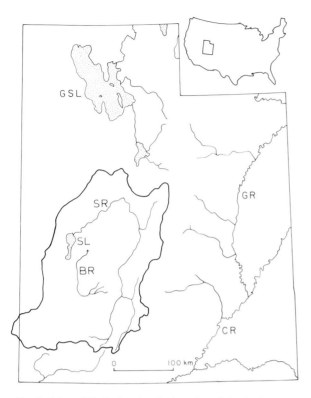

Fig. 1. Map of Utah showing the location of the Sevier Lake basin (heavy line). GSL = Great Salt Lake; SR = Sevier River; BR = Beaver River; SL = Sevier Lake; CR = Colorado River; GR = Green River.

primarily by surface inflow from the Sevier River. Prior to the establishment of irrigation in the drainage basin in the mid 1800s, the Beaver River also contributed a minor amount of water to the lake, and was an important tributary to the late Pleistocene expanded lakes.

The Sevier Lake drainage basin is approximately 42 000 km² in area, and includes part of the high plateaus of central and southern Utah, and the Tushar and Pavant Ranges. Altitudes in the drainage basin range from over 3600 m in the Tushar Mountains to 1376 m on the floor of Sevier Lake playa. Precipitation in the lower parts of the basin is about 150 mm, but increases to over 1000 mm in the mountains. Mean annual lake evaporation on the basin floor is about 1500 mm. Mean annual temperatures range from 10 °C on the basin floor to about 4 °C at high altitudes in the mountains.

The Sevier Lake basin is bounded on the north by the Great Salt Lake basin, from which it is separated by a low divide – the Old River Bed threshold – presently at an altitude of 1400 m (Fig. 2). Both basins were inundated by Lake Bonneville during the late Wisconsin. However, the two basins became separated after about 12 000 yr B.P. during the regression of Lake Bonneville.

Previous work on the paleolimnological history of Sevier Lake has been cursory and of a reconnaissance nature. Most notable is the work of Gilbert (1890), who visited the lake in 1872 and 1880, and whose monumental work on Lake Bonneville was published as U.S. Geological Survey Monograph 1. However, Gilbert had little to say about the post-Bonneville history of Sevier Lake. More recently Whelan (1969) and Hampton (1978) have studied the saline minerals of Sevier Lake playa. Archaeological investigations by Simms and Isgreen (1984) and Simms (1985) have contributed to our understanding of the Holocene prehistory of the basin, and have generated some valuable radiocarbon dates.

Morrison (1965, 1966) and Varnes and Van Horn (1984), among others, have suggested that the Bonneville basin was innundated to a level almost as high as the Provo Shoreline (about 1460 m in the Sevier Desert) shortly after 10 000 yr B.P. This lake is represented stratigraphically by what Morrison (1965) referred to as the Draper Formation. However, numerous independent radiocarbon dates, amino acid analyses, regional mapping, and stratigraphic data from throughout the Bonneville basin, including those reported in this paper, do not support the concept of a Draper lake cycle. See Scott *et al.* (1983) for a critical review of the Draper Formation and lake cycle. See also Miller *et al.* (1980), Spencer *et al.* (1984), Currey & Oviatt (1985), Oviatt & McCoy (1986), and Oviatt (1987a, 1987b) for further data and interpretations.

The chronology of late Pleistocene and Holocene lake fluctuations in the Sevier Lake basin is summarized in Fig. 3 and in Table 1. Radiocarbon dates depicted in Fig. 3 are listed in Table 2. All the radiocarbon dates presented here

Table 1. Explanation of lettered features on Fig. 3.

Letter	Explanation[a]
A	Regression of Lake Bonneville from the Provo Shoreline
B	Postulated unmodified Old River Bed threshold altitude (1390 m)
C	Lake Gunnison overflowing at an altitude of 1390 m; age determined by radiocarbon dates 6, 7, and 8 (Table 2)
D	Sevier Lake confined to levels below this altitude (1381 m) as shown by post-Lake Gunnison soil buried by late Holocene beach gravel (see Fig. 5)
E	Rise of Sevier Lake to 1382.3 m; age determined by radiocarbon date 1 on gastropod shells in Sevier River alluvium that is graded to a former high stand of the lake at 1382.3 m
F	Rise of Sevier Lake to 1379.8 m; age inferred from regional paleoclimatic patterns (see Currey & James 1982; McKenzie & Eberli 1985)
G	Sevier Lake in A.D. 1872 at 1379.5 m; age inferred from observations of G. K. Gilbert (1890), and from ax-cut driftwood forming a strandline at 1379.5 m, which must have been deposited after White settlement of the area in the mid-1800s
H	Late 20th century rise of Sevier Lake to 1378.9 m in A.D. 1984–1985

[a] Altitudes of prehistoric and historic lake levels were determined by handleveling from benchmarks to the crests of beach ridges surrounding Sevier Lake.

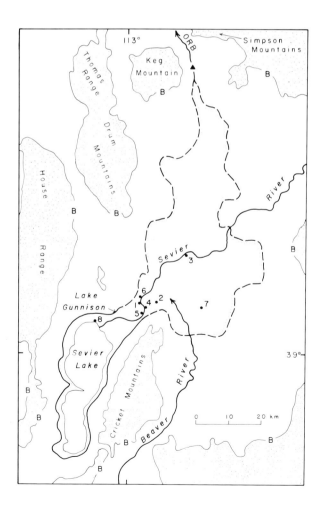

Fig. 2. Map showing part of the Sevier Lake basin. The line labelled B marks the Bonneville Shoreline of Lake Bonneville for reference (modified from Currey, 1982). Numbers 1–7 indicate radiocarbon localities (Table 2). Locality 8 is discussed in the text. The solid triangle marks the Old River Bed threshold. ORB = Old River Bed.

are on shells or disseminated organic carbon in sediments. Neither of these two types of materials are ideal for radiocarbon dating; but the results are stratigraphicaly consistent, and in the absence of more reliable samples, they allow for a first approximation of a lake-level chronology. Geomorphic reconnaissance studies in the basin beginning in 1979 (Currey, 1980, 1982; Isgreen, 1986; Oviatt, 1987a) are the primary sources of data for the chronology presented in this paper.

Lake fluctuations

Lake Bonneville
Shorelines and deposits of Lake Bonneville are widespread in the Sevier Lake basin. Lake Bonneville transgressed into the Sevier Lake basin between about 21000 and 20000 yr B.P.,

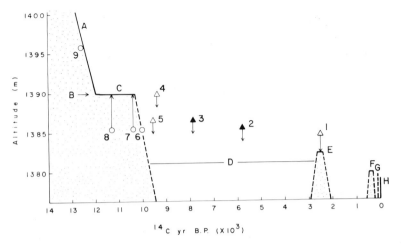

Fig. 3. Lake Gunnison and Sevier Lake time-altitude diagram for the last 14 000 yr. Open circles indicate radiocarbon dates on shells from lacustrine deposits; numbers are keyed to radiocarbon dates in Table 2. The vertical arrows on dates 7 and 8 indicate that the dates are on deposits of Lake Gunnison, which formed beaches and spits at an altitude of 1390 m. Date 6 is on regressive shorezone sand of Lake Gunnison. Open triangles indicate radiocarbon dates on shells from alluvium in the Sevier Desert. Solid triangles indicate radiocarbon dates on disseminated organic matter in alluvial deposits in the Sevier Desert. The downward-pointing arrows indicate that in all cases Sevier Lake was below the altitude of the dated alluvial sample at the time the sample was deposited. Refer to Table 1 for an explanation of the lettered features on the diagram.

and flooded the basin for approximately 8000 years, a period that included the time of development of the Bonneville Shoreline and the Provo Shoreline (Currey & Oviatt, 1985). Lake Bonneville regressed to the level of the Old River Bed threshold between about 13 000 and 12 000 yr B.P. (Currey & Oviatt, 1985; Thompson *et al.*, 1987). A radiocarbon date of 12 490 ± 130 yr B.P. on *Anodonta* shells from regressive-phase shorezone deposits in the Sevier Desert provides some local control on the timing (date 9, Table 2; Isgreen, 1986). Based on regional mapping of shorelines and deposits (Currey, 1982; Oviatt, 1987a), Lake Bonneville regressed to the level of the Old River Bed threshold, and the Sevier Lake basin became separate from the Great Salt Lake basin, shortly after this date.

Lake Gunnison

With the segregation of Lake Bonneville into two separate lakes, the lake in the Sevier Desert continued to overflow into the Great Salt Lake basin along the Old River Bed, an abandoned river channel cut in the deposits of Lake Bonneville

(Gilbert, 1890). The overflowing freshwater lake in the Sevier Lake basin has been referred to as Lake Gunnison (Currey & James, 1982: 34; Oviatt, 1987). Evidence for Lake Gunnison consists of geomorphic, stratigraphic, and radiocarbon data as discussed below.

The Old River Bed threshold is a key geomorphic control on lakes in the Sevier Lake basin and therefore is described briefly here. The threshold is formed by alluvial fans and Quaternary lacustrine deposits on the sloping piedmonts of the Simpson Mountains and Keg Mountain, which converge to form a broad saddle about 90 km north of Sevier Lake (Fig. 2). Water discharging from the Sevier Lake basin carved a shallow trench along the lowest part of the saddle.

The modern altitude of the threshold is about 1400 m, which for two reasons is higher than its late Pleistocene altitude during the period of overflow. First, the threshold area has rebounded isostatically a greater amount than has the deepest part of the basin in the vicinity of modern Sevier Lake. The differential isostatic rebound was caused by the removal of the Lake Bonneville water load, which was greatest in the northern part of the Lake Bonneville basin (Gilbert, 1890; Currey, 1982). Differential isostatic rebound between the Old River Bed threshold and the deepest part of the Sevier Lake basin may be as great as 4.5 m (Currey, 1982; Oviatt, 1987). Second, Holocene alluvial fans have partially filled in the threshold since overflow ceased. Auger holes in the Holocene alluvial fill at the threshold indicate that the alluvium extends downward to an altitude of at least 1396 m, and therefore, the rebounded altitude of the late Pleistocene threshold is 1396 m or somewhat lower.

A prominent shoreline surrounding Sevier Lake consists of erosional scarps, and gravel beaches and spits having a crestal altitude of 1390 m at all mappable points around the shore of the lake. One spit at this shoreline on the northeast shore of the lake is about 10 km long and thus represents a significant period of relatively stable water level and wave activity. A stable water level is most likely produced by over-flowing at an outlet, and therefore Currey (1982) has suggested that the prominent shoreline at 1390 m around Sevier Lake was formed during the last major period of overflow at the Old River Bed threshold (Lake Gunnison). Gilbert (1890: 183) predicted the existence of such a lake, but did not identify its shoreline.

Fine-grained deposits of Lake Gunnison are exposed at only a few localities in cutbanks along the Sevier River (localities 4, 5, and 6, Fig. 2). A radiocarbon date of 10 360 ± 225 yr B.P. on *Anodonta* shells (date 7, Table 2) collected from lagoon deposits that interfinger with spit gravel provides a date on the spit itself, and therefore on the age of the lake at a specific, geomorphically determined level. The spit has a crestal altitude of 1390 m, and is one segment of the Lake Gunnison Shoreline as mapped around the perimeter of Sevier Lake (Currey, 1982). The altitude of Lake Gunnison must have been slightly less than 1390 m during the formation of the spit, but the surface altitude of Lake Gunnison is regarded as 1390 m for simplicity of discussion.

Two other radiocarbon dates on Lake Gunnison deposits are listed in Table 2. Both are

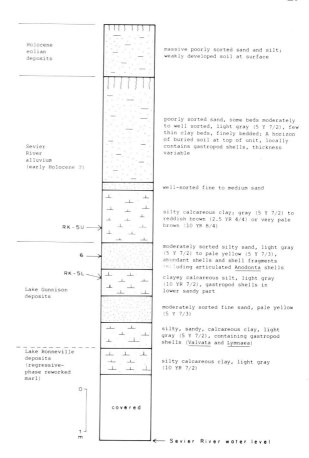

Fig. 4. Measured section in Lake Gunnison and post-Lake Gunnison deposits at locality 4 (see Fig. 2), exposed on the east bank of the Sevier River. Number 6 refers to radiocarbon date in Table 2. RK-5U and RK-5L are ostracode samples discussed in text.

on *Anodonta* shells. A measured section at the collection locality of sample number 6 is illustrated in Fig. 4. An important observation at this locality is that the marly sediments at the base of the section, which are interpreted as fine-grained lake-bottom sediments deposited and partly reworked during the regressive phase of Lake Bonneville, are conformable with the overlying sediments deposited in Lake Gunnison. Similar stratigraphic relationships exist at all localities where the contact between Lake Bonneville and Lake Gunnison deposits has been observed, including localities 5 and 6 (Fig. 2), and in a stream cut through a Lake Gunnison beach at the south end of Sevier Lake. This conformable relationship suggests that the Sevier Lake basin did

Table 2. Radiocarbon dates referred to on Fig. 3.

No.	[14]C date[a]	Dated materials[d]	Altitude[e]	Locality[f]	Lab No.	Reference[g]	Depositional setting
1	2560 ± 75[b] 2300 ± 70[c]	gastropod shells[h]	1385	1	Beta-17882	A	alluvial mud underlying terrace of Sevier River graded to late Holocene beach at 1382.3 m
2	5930 ± 220	disseminated organic carbon	1386	2	Beta-8011	B	alluvial mud above level of lake
3	7930 ± 110	disseminated organic carbon	1387	2	Beta-12988	C	alluvial mud above level of lake
4	9345 ± 160[b] 9110 ± 150[c]	gastropod shells[h]	1390	3	Beta-17878	A	alluvial mud above level of lake
5	9570 ± 430[b] 9340 ± 420[c]	gastropod shells[h]	1387	2	Beta-12987	C	alluvial mud above level of lake
6	10070 ± 130[b] 9760 ± 130[c]	*Anodonta* shells	1385	4	Beta-19455	A	lacustrine sand overlying Lake Gunnison carbonate mud
7	10360 ± 225[c]	*Anodonta* shells	1385	5	GX-6776	D	lacustrine mud that interfingers with Lake Gunnison spit gravel
8	11270 ± 110[b] 11000 ± 100[c]	*Anodonta* shells	1385	6	Beta-17883	A	lacustrine sand at distal end of Lake Gunnison spit
9	12490 ± 130[b]	*Anodonta* shells	1396	7	Beta-8348	E	lacustrine sand and gravel in Lake Bonneville spit

[a] radiocarbon dates in yr B.P.
[b] ^{13}C-adjusted date on gastropod shells.
[c] non-adjusted radiocarbon date on shells.
[d] All shells used in dating were fresh-appearing and unweathered. *Anodonta* shells retained their original pearly luster.
[e] meters above sea level.
[f] refer to Fig. 2.
[g] A = Oviatt (1987); B = Simms and Isgreen (1984); C = Simms (1985); D = Currey (1980); E = Isgreen (1986).
[h] *Lymnaea, Helisoma,* & *Physa.*

Table 3. Ostracodes in samples from locality 4[a]

RK-5U

    Limnocythere ceriotuberosa abundant, well preserved
    Candona patzcuaro? sensu Delorme
    Candona caudata
    Physocypria globula

RK-5L

    Candona patzcuaro? sensu Delorme
    Candona caudata
    Limnocythere ceriotuberosa some reworked
    Cytherissa lacustris generally reworked
    Cytheromorpha sp.
    Cyprideis beaconensis some reworked

[a] Identifications by R. M. Forester, U.S. Geol. Survey, Denver (1987, pers. commun.).

not dry out following the regression of Lake Bonneville and prior to the inception of Lake Gunnison. Rather, the basin experienced continuing overflowing conditions. This contrasts with the lake in the hydrologically closed Great Salt Lake basin during the same time period, which regressed to very low levels (Currey & Oviatt, 1985).

Two samples of sediments containing ostracodes were collected at locality 4 (Figs. 2 & 4, Table 3). R. M. Forester (1987, pers. commun.) suggests that the ostracodes in sample RK-5L (Fig. 4) usually live in bicarbonate-depleted water and that many of them appear to be reworked, whereas the ostracodes in sample RK-5U usually live in bicarbonate-enriched water. From field interpretations of the stratigraphy at locality 4,

RK-5L was deposited in Lake Gunnison while it was overflowing and wave activity was strong nearby. RK-5U was deposited during the regression of Lake Gunnison when wave activity was decreased and sediment was derived from an alluvial soruce, the Beaver or Sevier Rivers. Bicarbonate-enriched water, as indicated by RK-5U, could have been introduced by the Beaver River, which from mapping evidence is known to have emptied into the lake near locality 4. The modern Beaver River, upstream near Minersville, is enriched in bicarbonate (Hahl & Mundorff 1968, Table 1). These interpretations must be regarded as preliminary because they are based on limited data.

Lake Gunnison was probably not over 15 m deep at its deepest point during the period of overflow. However, its surface area was approximately 2000 km², and it covered much of the floor of the Sevier Desert as a shallow lake and deltaic-marsh system. The deepest water and most pronounced wave activity were centered in the vicinity of modern Sevier Lake.

Early and middle Holocene
No lacustrine deposits of early or middle Holocene age have been identified in the Sevier Lake basin. Four radiocarbon dates on Sevier River alluvium (dates 2, 3, 4, & 5, Table 2) show that the lake remained below the level of Lake Gunnison, and therefore did not overflow during the early and middle Holocene. Differential isostatic rebound of the Old River Bed threshold would have been proceeding rapidly during the early Holocene, and this uplift of the threshold, when combined with a probable climatic change after about 10 000 yr B.P., would have prevented extensive lake expansions.

Stratigraphic evidence for the upper limits of early and middle Holocene lakes consists of a buried soil exposed at locality 8 (Fig. 2). At this locality, beach gravel of late Holocene age (E in Fig. 1, Table 1) overlies the buried soil, which is developed in sediments interpreted as Lake Bonneville beach deposits and fine-grained pre-Bonneville lake beds (Fig. 5). The soil consists of an oxidized Bw horizon (7.5 yr 6/4) with weak to

Table 4. Climatic regions of Mitchell (1976) depicted in Fig. 7.

Region	Winter	Summer
II	frequent intrusion of Pacific air masses	dominated by warm, relatively dry air masses
V	infrequent intrusion of Pacific air masses	dominated by warm, relatively dry air masses
VI	infrequent intrusion of Pacific air masses	summer rainy season due to monsson air from Gulfs of Mexico and California

moderate subangular blocky to prismatic structure (terminology after Soil Survey Staff, 1975 and Guthrie & Witty, 1982). A Bk horizon below the Bw horizon includes flecks and small nodules of calcium carbonate, and some carbonate on ped faces and as rootlet fillings (stage I morphology after Machette 1985). Surface soils in this area

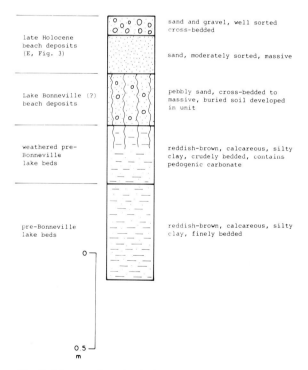

Fig. 5. Measured section in lacustrine deposits at locality 8 (see Fig. 2), exposed in a stream cut on the north shore of Sevier Lake. The altitude of the top of the section is about 1381 m.

having a similar degree of development are developed in Lake Bonneville deposits, and in alluvium of early Holocene age. Only a thin vesicular A horizon in fine-grained eolian deposits has been locally developed in the late Holocene beach gravel.

The buried soil at locality 8 therefore represents a relatively long period of development following the regression of Lake Gunnison and prior to burial by beach gravel of late Holocene age. Therefore, the soil indicates that the lake remained below an altitude of 1381 m during the early and middle Holocene. Fluctuations below 1381 m during this time period are probable, but evidence for them has not been found.

Late Holocene

During the late Holocene Sevier Lake expanded and contracted a number of times and deposited prominent beach ridges during its expanded phases. The two most prominent late Holocene beach ridges can be mapped around the perimeter of Sevier Lake and have crestal altitudes of 1382.3 m and 1379.8 m (Fig. 6).

The higher and older of the two major beach ridges (A in Fig. 6) has been dated by its geomorphic relationship with dated alluvium along the Sevier River upstream from the lake. A well-defined terrace underlain by fine-grained alluvium can be mapped for many kilometers upstream from Sevier Lake along the Sevier River where the river is entrenched into Lake Gunnison and early Holocene alluvial deposits. The terrace has a low gradient (0.4 m/km) and is graded to about the level of the higher late Holocene beach at 1382.3 m. The beach and the terrace are not directly connected because of the complicated geomorphic relationships created by the delta of the Sevier River and some eolian sand dunes. However, the terrace flattens, and has an altitude of 1383 m within 5 km of the lake shore.

The sediments underlying the terrace are mostly silty clay, are flatlying and finely bedded, and do not contain typical fluvial sedimentary textures or structures, such as point-bar sands, lenses of sand or gravel, ripple-laminations, or cross-bedding. Many layers contain organic matter and abundant mollusk shells, including the genera *Anodonta*, *Helisoma*, *Gyraulus*, *Lymnaea*, and *Physa*. Therefore, I interpret the deposits underlying the terrace as representing sedimentation in shallow non-saline water, with a high

Fig. 6. Stereo pair showing the two major late Holocene beaches on the north shore of Sevier Lake. A = higher beach; B = lower beach. Photos taken on June 27, 1987.

suspended-sediment load and abundant organics. The geomorphic setting and the morphometry of the deposit suggest that the river and the lake were interacting to produce an estuary along the lower reach of the Sevier River.

Gastropods collected from one of the organic-rich layers of fine-grained fill below the terrace at locality 1 (Fig. 2) yielded a radiocarbon date of 2560 ± 75 yr B.P. (date 1, Table 2). I interpret this date as indicating the time of deposition of the higher late Holocene beach at Sevier Lake. Great Salt Lake also experienced a high stand at about this time (Currey & James, 1982: 40; McKenzie & Eberli, 1985).

The lower prominent beach (B in Fig. 6) has not yet been dated. It is younger than the higher beach as shown by the relative degree of eolian silt cover on its surface (Fig. 6). The silt is derived from the Sevier Lake playa during periods when the lake is dry. The more extensive silt cover on the higher beach suggests that there was a relatively long period of dryness following its deposition and prior to the deposition of the lower beach. Although it is likely that an estuary formed along the lower Sevier River while the lower beach was being formed, such deposits have not yet been recognized. Therefore, the age of the lower beach must be inferred.

Because the record of late Holocene lake fluctuations of the Great Salt Lake is apparently similar to the dated parts of the Sevier Lake record, it is used here to infer the age of the lower prominent beach at Sevier Lake. The Great Salt Lake experienced a number of expansions between about A.D. 1400 and A.D. 1850 that have been documented by studies of the shoreline record (Currey & James, 1982: 40–41), and the oxygen-isotope record in sediment cores from the floor of the lake (McKenzie & Eberli, 1985). I infer, therefore, that the lower prominent beach at Sevier Lake was produced during a lake expansion, or series of expansions, between about A.D. 1400 and 1850.

Historic record
In 1872 G. K. Gilbert visited Sevier Lake and estimated that it was about 4.5 m deep (3080.5 m), and observed that the water was saline (Gilbert, 1890: 224–228). Gilbert revisited the lake in 1880, and in the intervening eight years, the lake had dried almost completely, thus creating a salt playa. The observed decline in the lake may have been caused partly by the increasing use of water for irrigation upstream along the Sevier and Beaver Rivers, but a change in climate was probably also involved. In 1873 the Great Salt Lake reached its highest recorded level of the 19th century, and by 1880 it had dropped about 1.5 m (Gilbert, 1890; Arrow, 1984). As with Sevier Lake, some of that drop was caused by irrigation, but much of the decline was caused by a decrease in precipitation in the basin (Arnow,

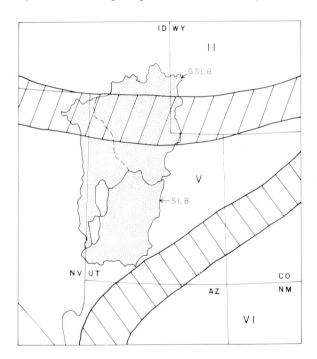

Fig. 7. Map showing the Sevier Lake drainage basin (SLB), the Great Salt Lake drainage basin (GSLB), and climatic regions (Roman numerals) suggested by Mitchell (1976: Fig. 3). Boundary zones between climatic regions are shown by lined areas. The modern Great Salt Lake receives 92% of its surface-water inflow from major rivers draining the Wasatch and Uinta Mountains in the eastern part of the drainage basin (Arnow 1984). The area west of the dashed line contributes no surface water to modern Great Salt Lake and probably contributed no, or little, water in the late Pleistocene. However, the area west of the dashed line probably contributed a significant volume of surface water to Lake Bonneville at the Gilbert stage.

1984). It is likely, therefore, that the Sevier Lake decline was also caused by a precipitation decrease.

A strandline consisting of driftwood logs along the north shore of Sevier Lake in the vicinity of locality 8 (Fig. 2) probably dates from the high stand of 1872 that Gilbert observed. Most of the logs are juniper, and many of them have distinct ax-cuts on their ends. Therefore, the driftwood strandline, at an altitude of 1379.5 m, had to have been deposited after the region was settled by ax-wielding emigrants (after about 1850), but before 1880, by which time the lake was dry. The logs were carried to Sevier Lake by large floods in the Sevier River then redistributed on the shoreline by waves.

From 1880 until about 1982 Sevier Lake was dry, with the exception of thin films of water that accumulated on the playa surface in extremely wet years during that period (Whelan, 1969). In 1982 the lake began to rise, and by 1985 it had reached an altitude of 1378.9 m and a depth of about 3 m in response to the melting of heavy snow that had accumulated in the mountains. Large floods of snow-melt runoff were generated on the Sevier River, especially in 1982–1984. The Great Salt Lake also rose dramatically during the same time period and created widespread flood damage in that basin (Kay & Diaz, 1985). Both lakes responded rapidly to greater-than-average winter precipitation, possibly associated with the El Niño/Southern Oscillation, which was unusually strong in 1982–1983 (Rasmussen & Wallace, 1983; Canby, 1984; Cane, 1986).

Paleoclimatic significance of lake fluctuations

Several conclusions about paleoclimate in this region during the last 12 000 years can be drawn by comparing the records of lake fluctuations in the Sevier Lake basin with those of the Great Salt Lake basin. The interpretations presented below must be regarded as preliminary because of the relatively poor dating control on the beaches around Sevier Lake. However, they are presented here as testable hypotheses that can be accepted or rejected as further evidence becomes available.

During the period from about 12 000 to 10 000 yr B.P. Lake Gunnison continuously overflowed from the Sevier Lake basin and therefore had a positive water balance. But during this same period, Lake Bonneville dropped to levels lower than modern levels in the Great Salt Lake basin, and thick beds of mirabilite were deposited in the north arm of the lake (Eardley, 1962; Spencer et al., 1984; Currey & Oviatt, 1985). An unconformity marked by mud cracks, marsh sediments, and oxidized sediment at many localities below the Gilbert Shoreline in the Great Salt Lake basin records this period of low lake levels in that basin (Miller et al., 1980; D. R. Currey, 1986, pers. commun.). The Gilbert Shoreline represents a lake transgression about 10 500 yr B.P. to about 15 m higher than the historic average level of Great Salt Lake (Currey & Oviatt, 1985). Therefore, Lake Gunnison was partly out-of-phase with the lake in the Great Salt Lake basin.

In contrast to the late Pleistocene part of the record, the middle and late Holocene high-lake stands in the Sevier Lake basin were apparently in-phase with those in the Great Salt Lake basin. Therefore, during the Holocene the lake basins were responding to similar climatic forcing, but between 12 000 and 10 000 yr B.P. they were apparently in different climatic regions.

The model of climatic regions of Mitchell (1976) may be used to present a hypothesis to explain the observed patterns of lake fluctuations in the two basins (Fig. 7; Table 4). In this hypothesis I assume that Mitchell's climatic regions, which are based on modern climatic data, apply in a general way to past climates in the western United States, and that the boundaries of those regions can shift as climate changes.

Based on these assumptions, the differing hydrology of the two basins during Lake Gunnison time may be explained by a northward shift of the 'Arizona monsoon' climatic region VI (Fig. 7; Table 4). A northward shift of the summer monsoon climatic region could have kept the Sevier Lake basin relatively wet and under summer cloud cover, while the Great Salt Lake basin dried out under the influence of climatic region V. This interpretation is consistent with that of Spaulding

& Graumlich (1986), who suggest on the basis of independent paleobotanical evidence that the monsoon circulation was established in postglacial time in the American Southwest about 12 000 yr B.P. Forester & Markgraf (1984), Davis (1986, 1987), Davis & Sellers (1987), and Forester (1987) have arrived at similar conclusions. However, other authors (including Markgraf & Scott, 1981; Van Devender, 1987; L. V. Benson, 1987, pers. commun., R. S. Thompson, 1987, pers. commun.) have interpreted pollen, pack-rat midden, and paleohydrologic data as indicating that the monsoon circulation was not operating in the southwestern United States during the period 12 000 to 10 000 yr B.P. Therefore, although a southern moisture source (monsoon) best explains the contrasting paleohydrologic records of Lake Gunnison and the Great Salt Lake, this hypothesis must be tested by further independent evidence before it can be confidently accepted.

The development of the Gilbert Shoreline (about 11 000 to 10 000 yr B.P.), and part of the period of overflow from the Sevier Lake basin, appear to be coincident with the European Younger Dryas event, which has now been recognized in stratigraphic sequences in Atlantic Canada (Mott et al., 1986). But, it is not known at this point whether there was a causal link between high lake levels in the eastern Great Basin, and climatic cooling in Europe and Atlantic Canada.

For most of the Holocene, the boundaries of the climatic regions of the western United States must have been similar to the modern conditions. During the early and middle Holocene, the monsoon (region VI) probably did not shift as far northward as it had between 12 000 and 10 000 yr B.P. because both Great Salt Lake (Currey & James, 1982) and Sevier Lake were relatively low during these periods. Both lake basins were probably dominated by climatic region V during periods when lake levels were low. High lake levels in the late Holocene may have been caused by precipitation increases derived from western air masses, some of which may have been associated with the El Niño/ Southern Oscillation (L. V. Benson, 1987, pers.

commun.). The El Niño/Southern Oscillation developed as a coherent ocean/atmosphere circulation pattern after 5000 yr B.P. (Rollins et al., 1986).

Acknowledgements

This study was partially funded by the Utah Geological and Mineral Survey and the U.S. Geological Survey as part of a Cooperative Geologic Mapping (COGEOMAP) project in 1986. Scott Oviatt and Ward Taylor helped in the field. Discussions with Don Currey, Owen Davis, Rick Forester, Marilyn Isgreen, Steve Simms, Geoff Spaulding, and Bob Thompson have been valuable in developing the ideas expressed in this paper. Rick Forester identified and interpreted ostracodes. I am grateful to Larry Benson, Ron Davis, Paul Kay, and an anonymous reviewer for constructive review comments.

References

Arnow, T., 1984. Water-level and water-quality changes in Great Salt Lake, Utah, 1847–1983. U.S. Geol. Survey Circ. 913: 22 pp.

Canby, T. Y., 1984. El Niño's ill wind. Natl. Geog. 165: 144–183.

Cane, M. A., 1986. El Niño. Ann. Rev. Earth Planet. Sci. 14: 43–70.

Currey, D. R., 1980. Radiocarbon dates and their stratigraphic implications from selected localities in Utah and Wyoming. Encyclia, J. Utah Acad. Sci. Arts Lett. 57: 110–115.

Currey, D. R., 1982. Lake Bonneville: Selected features of relevance to neotectonic analysis. U.S. Geol. Survey Open-File Rept. 82-1070: 30 pp.

Currey, D. R. & S. R. James, 1982. Paleoenvironments of the northeastern Great Basin and northeastern Basin rim region: A review of geological and biological evidence. Soc. Am. Arch. Pap. 2: 27–52.

Currey, D. R. & C. G. Oviatt, 1985. Durations, average rates, and probable causes of Lake Bonneville expansions, stillstands, and contractions during the last deep-lake cycle, 32 000 to 10 000 years ago. In P. A. Kay & H. F. Diaz (eds.), Problems of and prospects for predicting Great Salt Lake levels: Papers from a conference held in Salt Lake City, March 26–28, 1985. Center for Public Affairs and Administration, Univ. Utah: 9–24.

Davis, O. K., 1986. A late glacial pluvial maximum, the history of the Arizona monsoon, and the astronomical

36

theory of climatic change. Am. Quat. Asso. (AMQUA) Prog. Abs.: 127.

Davis, O. K., 1987. Late Quaternary global temperature change versus seasonal precipitation change in the western U.S. Geol. Soc. Am. Abs. Prog. 19: 636.

Davis, O. K. & W. D. Sellers, 1987. Contrasting climatic histories for western North America during the early Holocene. Curr. Res. Pleist. 4: 87–89.

Eardley, A. J., 1962. Glauber's salt bed west of Promontory Point, Great Salt Lake. Utah Geol. Min. Survey Spec. Stud. 1: 12 pp.

Forester, R. M., 1987. Late Quaternary paleoclimatic records from lacustrine ostracodes. In W. F. Ruddiman & H. E. Wright, Jr. (eds.), North America and adjacent oceans during the last deglaciation. Geol. Soc. Am., Geol. N. Am. K-3: 261–276.

Forester, R. M. & V. Markgraf, 1984. Late Pleistocene and Holocene seasonal climatic records from lacustrine ostracode assemblages and regional (pollen) vegetational patterns in southwestern U.S.A. Am. Quat. Asso. (AMQUA) Prog. Abs.: 43–45.

Gilbert, G. K., 1890. Lake Bonneville. U.S. Geol. Survey Mono. 1: 438 pp.

Guthrie, R. L. & J. E. Witty, 1982. New designations for soil horizons and layers and the new Soil Survey Manual. Soil Sci. Soc. Am. J. 46: 443–444.

Hahl, D. C. & J. C. Mundorff, 1968. An appraisal of the quality of surface water in the Sevier Lake basin, Utah, 1964. Utah Dept. Nat. Res. Tech. Pub. 19: 41 pp.

Hampton, D. A., 1978. Geochemistry of the saline and carbonate minerals of Sevier Lake playa, Millard County, Utah. M.S. thesis, Univ. Utah, Salt Lake City: 75 pp.

Isgreen, M. C., 1986. Holocene environments in the Sevier and Escalante Desert basins, Utah: A synthesis of Holocene environments in the Great Basin. M.S. thesis, Univ. Utah, Salt Lake City: 134 pp.

Kay, P. A. & H. F. Diaz (eds.), 1985. Problems of and prospects for predicting Great Salt Lake levels. Papers from a conference held in Salt Lake City, March 26–28, 1985. Center for Public Affairs and Administration, Univ. Utah: 309 pp.

Machette, M. N., 1985. Calcic soils of the southwestern United States. In D. L. Weide (ed.), Soils and Quaternary geology of the southwestern United States. Geol. Soc. Am. Spec. Pap. 203: 1–21.

Markgraf, V. & L. Scott, 1981. Lower timberline in central Colorado during the past 15 000 yr. Geology 9: 231–234.

McKenzie, J. A. & G. P. Eberli, 1985. Late Holocene lake-level fluctuations of the Great Salt Lake (Utah) as defined from oxygen-isotope and carbonate contents of cored sediments. In P. A. Kay & H. F. Diaz (eds.), Problems of and prospects for predicting Great Salt Lake levels. Papers from a conference held in Salt Lake City, March 26–28, 1985. Center for Public Affairs and Administration, Univ. Utah: 25–39.

Miller, R. D., R. Van Horn, W. E. Scott & R. M. Forester, 1980. Radiocarbon date supports concept of continuous low levels of Lake Bonneville since 11 000 yr B.P. Geol. Soc. Am. Abs. with Prog. 12: 297–298.

Mitchell, V. L., 1976. The regionalization of climate in the western United States. J. Appl. Metero. 15: 920–927.

Morrison, R. B., 1965. Lake Bonneville: Quaternary stratigraphy of eastern Jordan Valley, south of Salt Lake City, Utah. U.S. Geol. Survey Prof. Pap. 477: 80 pp.

Morrison, R. B., 1966. Predecessors of Great Salt Lake. Utah Geol. Soc. Guidebook Geol. Utah 20: 77–104.

Mott, R. J., D. R. Grant, R. Stea & S. Occhietti, 1986. Late-glacial climatic oscillation in Atlantic Canada equivalent to the Allerød/Younger Dryas event. Nature 323: 247–250.

Oviatt, C. G., 1987a. Quaternary geology of part of the Sevier Desert, Millard County, Utah. Utah Geol. Min. Survey Open-File Rept. 106: 120 pp.

Oviatt, C. G., 1987b. Lake Bonneveille stratigraphy at the Old River Bed, Utah. Am. J. Sci. 287: 383–398.

Oviatt, C. G. & W. D. McCoy, 1986. New radiocarbon and amino acid age constraints on Holocene expansions of Great Salt Lake, Utah. Am. Quat. Asso. (AMQUA) Prog. Abs.: 103.

Rasmussen, E. M. & J. M. Wallace, 1983. Meteorological aspects of the El Niño/Southern Oscillation. Science 222: 1195–1202.

Rollins, H. B., J. B. Richardson III & D. H. Sandweiss, 1986. The birth of El Niño: Geoarchaeological evidence and implications. Geoarchaeology 1: 3–15.

Scott, W. E., W. D. McCoy, R. R. Shroba & M. Rubin, 1983. Reinterpretation of the exposed record of the last two cycles of Lake Bonneville, western Untied States. Quat. Res. 20: 261–285.

Simms, S. R., 1985. Radiocarbon dates from 42MD300, Sevier Desert, western Utah. Weber State College, Archaeo. Tech. Prog. Rept.: 9 pp.

Simms, S. R. & M. C. Isgreen, 1984. Archaeological excavations in the Sevier and Escalante Deserts, Utah. Univ. Utah Archaeo. Center Rept. Investig. 83–12: 446 pp.

Soil Survey Staff, 1975. Soil taxonomy. U.S. Dept. Agric. Agric. Handbook 436: 754 pp.

Spaulding, W. G. & L. J. Graumlich, 1986. The last pluvial climatic episodes in the deserts of southwestern North America. Nature 320: 441–444.

Spencer, R. J., M. J. Baedecker, H. P. Eugster, R. M. Forester, M. B. Goldhaber, B. F. Jones, K. Kelts, J. McKenzie, D. B. Madsen, S. L. Rettig, M. Rubin & S. J. Bowser, 1984. Great Salt Lake and precursors, Utah: The last 30 000 years. Contrib. Min. Pet. 86: 321–334.

Thompson, R. S., L. J. Toolin & R. J. Spencer, 1987. Radiocarbon dating of Pleistocene lake sediments in the Great Basin by accelerator mass spectrometry (AMS). Geol. Soc. Am. Abs. Prog. 19: 868.

Whelan, J. A., 1969. Subsurface brines and soluble salts of subsurface sediments, Sevier Lake, Millard County, Utah. Utah Geol. Min. Survey Spec. Stud. 30: 13 pp.

Van Devender, T. R., 1987. Holocene vegetation and climate in the Puerto Blanco Mountains, southwestern Arizona. Quat. Res. 27: 51–72.

Varnes, D. J. & R. Van Horn, 1984. Surficial geologic map of the Qak City area, Millard County, Utah. U.S. Geol. Survey Open-File Rept. 84–115.

Late Quaternary paleolimnology of Walker Lake, Nevada

J. Platt Bradbury, R. M. Forester & R. S. Thompson
MS 919, U.S. Geological Survey, Denver, CO 80225, USA

Key words: paleolimnology, river diversion, climate change, pollen, diatoms, ostracodes, brine shrimp

Abstract

Diatoms, crustaceans, and pollen from sediment cores, in conjunction with dated shoreline tufas provide evidence for lake level and environmental fluctuations of Walker Lake in the late Quaternary. Large and rapid changes of lake chemistry and level apparently resulted from variations in the course and discharge of the Walker River. Paleolimnological evidence suggests that the basin contained a relatively deep and slightly saline to freshwater lake before *ca.* 30 000 years B.P. During the subsequent drawdown, the Walker River apparently shifted its course and flowed northward into the Carson Sink. As a result, Walker Lake shallowed and became saline. During the full glacial, cooler climates with more effective moisture supported a shallow brine lake in the basin even without the Walker River. As glacial climates waned after 15 000 years ago, Walker Lake became a playa. The Walker River returned to its basin 4700 years ago, filling it with fresh water in a few decades. Thereafter, salinity and depth increased as evaporation concentrated inflowing water, until by 3000 years ago Walker Lake was nearly 90 m deep, according to dated shoreline tufas. Lake levels fluctuated throughout this interval in response to variations in Sierra Nevada precipitation and local evaporation. A drought in the Sierras between 2400 and 2000 years ago reduced Walker Lake to a shallow, brine lake. Climate-controlled refilling of the lake beginning 2000 years ago required about one millennium to bring Walker lake near its historic level.

Through time, lake basins in the complex Lake Lahontan system, fill and desiccate in response to climatic, tectonic and geomorphic events. Detailed, multidisciplinary paleolimnologic records from related subbasins are required to separate these processes before lake level history can be reliably used to interpret paleoclimatology.

Introduction

Walker Lake occupies the southernmost basin of Pleistocene Lake Lahontan (Fig. 2), which may have reached its maximum level *ca.* 14 000 years ago (Benson & Thompson, 1987). At its greatest extent, Lake Lahontan reached an altitude of about 1335 m and covered an area of 22 780 km^2 (Benson & Mifflin, 1986). Lake Lahontan was the product of late Pleistocene climatic and hydrologic conditions, but the relative importance of parameters such as precipitation, evaporation,

Originally published in
Journal of Paleolimnology **1**: 249–267.

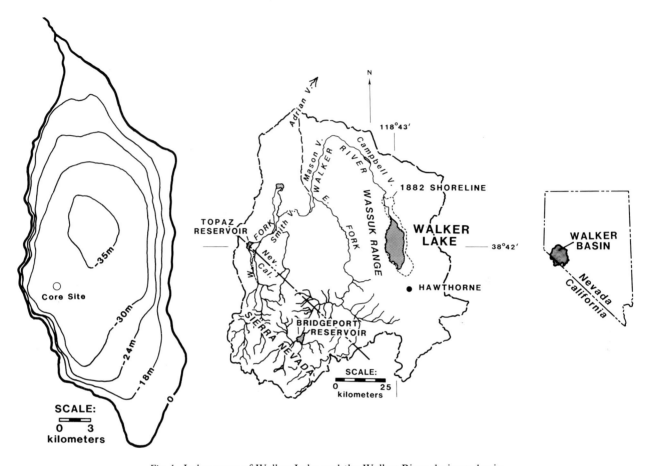

Fig. 1. Index maps of Walker Lake and the Walker River drainage basin.

temperature and runoff is difficult to evaluate as the lake could have resulted from more than one combination of factors (e.g. Benson, 1981).

The chronology of central Lake Lahontan, based on radiometrically dated shoreline tufas and packrat middens, was recently refined (Benson & Thompson, 1987; Thompson *et al.*, 1986). Nevertheless, despite the additional number of dates, shoreline features cannot provide a complete, detailed limnological history because they lack chronostratigraphic continuity. However, accurately dated *in situ* shoreline tufas can provide unambiguous lake level information that is often unobtainable from profundal sediments. Ideally, both paleolimnologic approaches should be utilized to arrive at the most complete history as well as to identify problems with either method.

In this study, diatoms, ostracodes, brine shrimp, and pollen from sediment cores have been used to interpret the paleolimnological history of Walker Lake for the last 30 000 years. This history provides a means of evaluating the Walker Lake shoreline chronologies.

Study area

Walker Lake lies at the western margin of the Great Basin in west-central Nevada in a graben formed by normal faulting as a result of mid- to late Cenozoic crustal extension. Faulting continued through the Holocene (Dempsy, 1987) and seismic profiles show major deformation of lacustrine deposits in Walker Lake between 7000 and 9000 years ago (Benson, 1988).

Walker Lake is an endorheic, warm monomictic, saline (10 g/l), Na-SO$_4$-CO$_3$-Cl lake with a maximum depth of 35 m. It is oriented approxi-

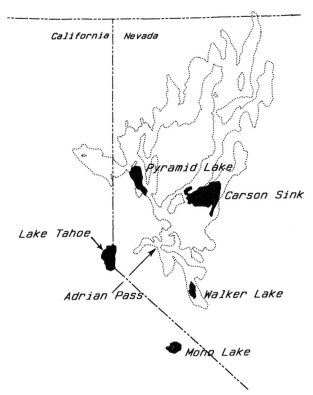

Fig. 2. Index map of the Lahontan Basin showing the maximum extent of Lake Lahontan (dotted line) and extant lakes.

mately north-south with a maximum length of 25 km, a width of 9 km and a surface area of approximately 150 km^2 (Fig. 1). The lake is supported by the Walker River whose discharge is largely derived from snowmelt in the Sierra Nevada. Without this river input, a dry playa would be present in the Walker basin under present climatic conditions (Rush, 1970).

Modern-day Walker Lake level varies between about 1206 and 1212 m elevation in response to variability in discharge from the Walker River (Benson, 1988). The east and west tributaries of the Walker River begin in the Sierra Nevada at elevations above 3000 m and drain an area of about 10 360 km^2. This discharge provides at least 80 percent of the total inflow and much of the clastic sediment to Walker Lake (Rush, 1970).

Walker Lake lies in the rainshadow east of the Wassuk Range and the more distant Sierra Nevada. Annual average precipitation is low (10 cm, Cooper & Koch, 1984), while annual

evaporation is about 120 cm (Benson, 1981). The melting of the Sierran snowpack during May and June produces the greatest discharge to the Walker River. Storage of this discharge in Bridgeport and Topaz reservoirs and use for irrigation during the last 60 years has reduced the inflow to the lake by 60 percent (Benson & Leach, 1979), causing the lake level to fall and salinity to increase (Rush, 1970). The use of Walker River water for irrigation below the major reservoirs also results in an increase in its salinity due to evapotranspiration. Some of this water returns to the Walker River resulting in an increase in the river's salinity in inverse proportion to the river's discharge (Benson & Leach, 1979).

The Walker Lake epilimnion cools and the lake destratifies during the autumn. Typically, circulation takes place after November and lasts until April or early May (Koch *et al.*, 1979). During this period, the bottom waters become oxygenated and nutrients are distributed throughout the lake. The lake stratifies in May and by mid summer the hypolimnion becomes anoxic. Stratification persists despite winds that may produce large clockwise currents in the epilimnion (Koch *et al.*, 1979).

The salinity of Walker Lake when first recorded (1882 A.D.) was about 2560 mg/l (Rush, 1970). In 1977 the salinity of the lake was over 10 000 mg/l total dissolved solids (TDS), owing to the reduction of inflow in the Walker River and high evaporation rates (Cooper & Koch, 1984). By comparison, nearby Pyramid Lake is presently about half the salinity of Walker Lake.

Today, the phytoplankton of Walker Lake is composed of bluegreen algae (chiefly *Nodularia spumigena*), and diatoms (Cooper & Koch, 1984; Koch *et al.*, 1979). Maximum diatom concentrations are low and occur in the winter months when nutrients are circulated to the epilimnion. The few analyses available, from 1975–1977, show cell concentrations that are always below 300 cells per ml and generally in the range of 20–40 cells per ml (Koch *et al.*, 1979; Cooper & Koch, 1984). However, planktonic diatoms are abundant and well preserved in sediment cores from the profundal zone.

Although different planktonic diatoms have different ecological requirements, in general, high diatom abundance relates to increased nutrient loadings from the Walker River and recycling of nutrients from the hypolimnion during circulation. *Stephanodiscus excentricus* and *Cyclotella quillensis* characterize the winter and early spring diatom blooms in both Walker and Pyramid Lakes today while *Chaetoceros elmorei* generally dominates in the late summer and fall (Cooper & Koch, 1984; Hamilton-Galat & Galat, 1983; Galat *et al.*, 1981). In Pyramid Lake, *Stephanodiscus excentricus* and *Chaetoceros elmorei* are replaced by bluegreen algae at higher salinities (Lockheed Ocean Science Labs., 1982) and the relative scarcity of diatoms in Walker Lake today may be due to its rising salinity.

Ostracodes are small (*ca.* 1 mm), largely benthonic crustaceans with a bivalved calcareous carapace. They are common in the lake today and abundant in lake sediment cores and outcrops. The ostracode fauna provides information about environmental parameters near the sediment/water interface such as temperature, salinity, solute composition, and oxygen content (Forester, 1983, 1985, 1986, 1987). *Limnocythere cerio-tuberosa* is the only abundant ostracode living on the lake sediment surface today. This species appears to require seasonally cold water and moderate to high alkalinity, and it tolerates a range of salinity from about 200 to 30 000 mg/l (Forester, 1986, 1987). *Candona caudata* lived in the lake prior to the drawdown due consumptive water use for agriculture. *Candona caudata* lives in seasonally cold, fresh water with salinities up to about 3 g/l in Canada (Delorme, 1978) and throughout the Great Basin (Forester, unpublished data).

Methods

Coring and core sampling

The upper 50 m of sediment in Walker Lake was sampled in the summer of 1984 by coring from a 12- by 15- by 1.5-m barge constructed of steel cassions anchored in the lake 2.6 km east and 0.2 km north of the northeast corner of section 6, T. 9 N, R. 28 E., in a water depth of 33.9 m (Fig. 1). A 3-m surface sediment core was taken with a piston-fitted, 5-cm inside diameter, clear polycarbonate tube. Additional core segments extending to a depth of 12 m were taken with a 5-cm diameter square-rod Livingstone piston sampler. Aluminum drill rod, 3.5 cm in diameter with a weight of 4.8 kg per 3.3 m length was used to drive the samplers into the sediment. The plastic tube and Livingstone samplers were raised and lowered through 11-cm inside diameter steel casing to appropriate depths by a mechanical winch. These core segments are collectively known as core 8. A truck mounted, rotary, wire-line drilling rig (Mobile B-61) was used to raise deeper, 7-cm diameter cores from the same location. These cores were obtained with drilling fluid in 1.5- to 3-m drives, and are collectively known as core 4. Another core, core 5, was taken 7 m to the south of core 4 to partially fill gaps in recovery of core 4 (Benson, 1988).

The piston cores were extruded on the barge and wrapped in plastic sheeting and aluminum foil. The surface core tube was cut at the sediment-water interface to allow overlying water to slowly drain away, and then capped and later extruded and sampled in the laboratory. After scraping the core surface clean, samples for microfossil analysis 2 cm thick were cut from the core at approximately 10-cm intervals. The wire-line cores were longitudinally sectioned and sampled at approximately 1-m intervals for ostracode and pollen analysis. Diatoms were analyzed from smear slides prepared from adjacent, 20-cm, longitudinal scrapings of the cleaned core surface. Generally, one sample every 40 cm was examined for diatoms.

Microfossil analysis

Diatom samples from core 8 were disaggregated in distilled water, and allowed to settle overnight. The supernatant water was decented and the remaining water-saturated sediment mixed. A volumetric aliquot of this water-saturated sediment was resuspended in distilled water and settled onto a coverslip, and dried. The coverslip was mounted in Hyrax resin with a refractive index of 1.65. At least 400 diatoms or two full (17 mm) transects were counted to determine the number of diatoms on the coverslip. Manipulation of aliquot volume and dilution factors allowed determination of the concentration of diatoms in 1 cc of the settled, disaggregated, water-saturated sediment (e.g. Battarbee, 1986). Raw sediment smear slides from adjacent, 20-cm sections of core 4 were mounted in lakeside cement. Diatoms were tabulated along transects to determine the number of diatoms per mm of the count transect. This technique is not precise, but gives a semi-quantitative estimate of diatom concentration averaged over a 20-cm section of the core.

Samples for ostracode analysis of roughly 40 cm^3 (core 8) and of known weight (core 4) were covered with hot water, $NaHCO_3$, and commercial Calgon and then frozen to disaggregate the sediment. The sediment was thawed and washed over a 100-mesh (150 μm) sieve and the residue was air-dried. All valves were size-sorted through a nested set of 3-inch diameter sieves, and the adults counted to determine the number of valves per sample. The percentage of each taxon was calculated from the counts, except for samples with very low ostracode abundance.

Pollen analysis followed the standard procedures for extraction and quantitative enumeration (Gray, 1965). Pollen types of local and (or) limnological importance, algae, and fungi were selected for presentation, if abundant, to provide evidence on the paleolimnology of Walker Lake and the vegetative character of its immediate surroundings.

Results

Chronology: Cores 4 and 8

Organic and (or) carbonate fractions from six horizons in core 8 and 15 horizons in core 4 were

radiocarbon dated (Table 1). The dates from core 8 indicate a maximum age of about 4700 years B.P. and an average sediment accumulation rate of 0.2–0.3 cm/year. A date/depth curve based on all dates (Fig. 3A) provides interpolated age estimates for intermediate, undated parts of the core, and forms the basis of the core 8 chronology of this study.

Radiocarbon dates for core 4 (Table 1, Fig. 3B) are generally consistent below 10 m depth, but

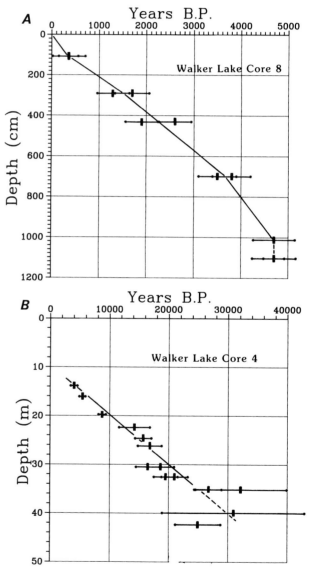

Fig. 3. Radiocarbon ages vs. sample depths for Walker Lake cores 8 (A) and 4 (B). Data from Yang (1988) and Benson (1988).

Table 1. Radiocarbon dates for cores of Walker Lake sediment (Yang, 1988). Sample depths taken from Benson (1988).

	Depth (m)	Age (uncalibrated years B.P.)		
		Organic fraction	Carbonate fraction	Lab number
Surface core	1.00–1.15	360 ± 170	350 ± 100	DE-366
Piston core	2.83–2.99	1300 ± 170	1700 ± 180	DE-367
(12 m)	4.23–4.39	2600 ± 170	1900 ± 170	DE-372
	6.92–7.07	3500 ± 200	3800 ± 200	DE-370
	10.00–10.24	4700 ± 220	4700 ± 220	DE-371
	10.90–11.15	4700 ± 110	4700 ± 230	DE-369
Wireline core 4				
	3.65–3.74	1600 ± 190		DE-350
	7.20–7.28	550 ± 150		DE-351
	10.19–10.26	600 ± 160		DE-352
	13.80–13.87	4000 ± 280		DE-353
	16.04–16.17		5400 ± 250	DE-354
	19.65–19.89		8700 ± 310	DE-368
	22.35–22.45	14 200 ± 1300		DE-355
	24.60–24.67	15 700 ± 680		DE-356
	26.16–26.24	16 800 ± 1000		DE-357
	30.41–30.62	18 700 ± 1100	16 500 ± 1000	DE-358
	32.48–32.62	21 000 ± 1100	19 500 ± 1000	DE-359
	35.09–35.18	32 100 ± 3900	26 700 ± 1100	DE-360
	39.91–39.98	30 900 ± 6000		DE-361
	42.36–42.43	> 36 100	24 900 ± 1900	DE-362
	44.81–44.88	> 33 700		DE-363

show considerable scatter. The carbonate dates tend to be younger than organic dates with the diffrence becoming greater for the older pairs. A date at 3.7 m (1610 ± 185 years B.P.) and near modern dates from 7.2 and 10.2 m depth are probably due to contamination during rotary coring in soft, water-saturated sediments (Benson, 1988). Core 8 was taken because of this problem. Sediments below 12 m are stiffer and less subject to mixing by rotary coring. Nonetheless, some contamination may have accompanied the sediment deformation observed in core 4 and could mean that some of these dates are too young (e.g. Benson, 1988). Despite variable recovery and disturbance during coring, individual dates and date pairs form a generally coherent group that can be used to approximate the age of the sediment in Walker Lake between 13 and 40 m depth. Because there is no obvious way of discriminating between carbonate and organic dates for accuracy, a best- (logically) fit line based on all finite dates below 13 m is used for the chronology of core 4 in this study, with the recognition that other techniques may ultimately become more appropriate for determining the age of these deposits. Sedimentation rates in the best controlled part of this interval (13.8–32.5 m) are about 0.1 cm/year. The slower sedimentation rates for core 4 may result from compaction of the sediments and (or) deposition in shallow lake or playa environments during late Pleistocene.

Biostratigraphy: Core 4

Diatoms are comparatively common between 50 and 35 m, where the assemblage is composed of planktonic species that typically live in moderately saline, alkaline water. The most common species include *Stephanodiscus excentricus,* and *Surirella nevadensis* (Fig. 4), but concentration fluctuates sharply, implying variable limnological conditions.

Walker Lake Core 4

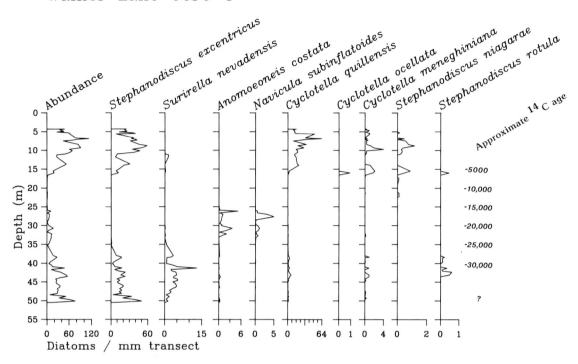

Fig. 4. Stratigraphy of selected diatom taxa from Walker Lake core 4.

Abundant *Surirella nevadensis* indicates salinities in the range of 2500 to 3000 mg/l, and possibly high, river-supported lake levels if historical analogs in Pyramid Lake apply (Bradbury, 1987). *Surirella nevadensis* abundances exhibit a close correspondence with those of *Limnocythere bradburyi* (Fig. 5). This ostracode is not kown from lakes in this region today, and may require warmer water or different circulation patterns than presently occur in Walker Lake (*e.g.* Forester, 1985). It is probable that some nutrient- or turbulence-related aspect of a past higher and perhaps warmer Walker Lake, in addition to salinity, accounts for the presence of *Surirella nevadensis*. *Stephanodiscus excentricus* probably reflects phosphorus and nitrate nutrient fluxes during the period of winter circulation and (or) spring influx of the Walker River (Hamilton-Galat & Galat, 1983).

Ostracode abundance is also high in the interval between 35 and 50 m and the assemblage is composed largely of *Limnocythere ceriotuberosa* and *L. bradburyi,* two taxa that are not known to

live together today (Forester, 1985; 1987). These taxa commonly occurred together throughout the Great Basin in the past (Forester, 1987, unpublished data). Their co-occurrence probably results from a combination of seasonal conditions that involves water temperatures cold enough to meet the physiological needs of *L. ceriotuberosa* and warm enough to allow *L. bradburyi* to both survive winter and reach reproductive maturity in the summer. A lake of moderate depth that circulated in summer could meet the needs of both taxa. The presence of both limnocytherans and of *Candona* sp., together with the absence of other taxa, implies a fluctuating salinity in the probable range 750 to 5000 mg/l, with wider ranges being possible. The actual upper salinity tolerance ranges of these taxa is probably related to the composition of the water and not simply to its salinity (Forester, 1986). If the absence of *C. caudata* is chemically controlled, as is likely, then the salinity probably remained above 2000 mg/l as suggested by the diatoms.

Diatom concentrations fall between 40 and

46

Walker Lake core 4

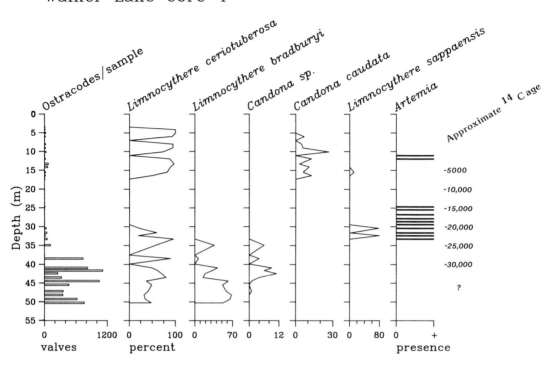

Fig. 5. Ostracode and *Artemia* pellet stratigraphy of Walker Lake core 4.

35 m, and remain low and fluctuating until 26 m (Fig. 4). Between 26 and 35 m, the diatom assemblage is dominated by benthonic species, *Anomoeoneis costata,* both cysts and vegetative cells, and *Navicula subinflatoides.* These taxa are tolerant of both high salinity and large or rapid fluctuations in osmotic pressure caused by variable salinities (Cholnoky, 1968). The cysts of *A. costata,* an unusually tolerant and widespread diatom, suggest that conditions were often too saline for even it to prosper. These cysts, as well as *N. subinflatoides,* can be found in the shallow water, marginal habitats of Mono Lake, California, where salinities can exceed 100 g/l. The ostracode *Limnocythere sappaensis* is also present in this interval. Its presence implies a highly saline, alkaline-enriched water (Forester, 1983. This species commonly co-occurs with the diatom *Anomoeoneis costata,* but probably doesn't live in salinities much above 60 g/l (Forester, 1986). Brine shrimp *(Artemia)* pellets occur frequently (Fig. 5), terrestrial fungal spores become

common, and Cyperaceae and *Sarcobatus* pollen increase to comparatively high percentages (Fig. 6) at this time, respectively indicating marshes and salt flats.

These microfossil assemblages indicate that between approximately 25 000 and about 16 000 years ago (core 4, 35 to 26 m, Fig. 3), Walker Lake was a shallow, saline, and marsh-fringed (Cyperaceae) lake surrounded by salt flats supporting *Sarcobatus* vegetation. At times, especially around 19 000 years ago (*ca.* 27 m), salinities may have been as high as Mono Lake today, and the limnological environment was probably characterized by dramatic fluctuations in salinity and in the area of open water.

Diatoms and pollen are absent or only present at low concentrations between 16 and 26 m. Occasional reworked frustules of *Aulacosira* from nearby Miocene diatomites are present at some levels, whereas other samples contain rare individuals of freshwater benthonic and planktonic diatoms. The abundance of the freshwater taxa is

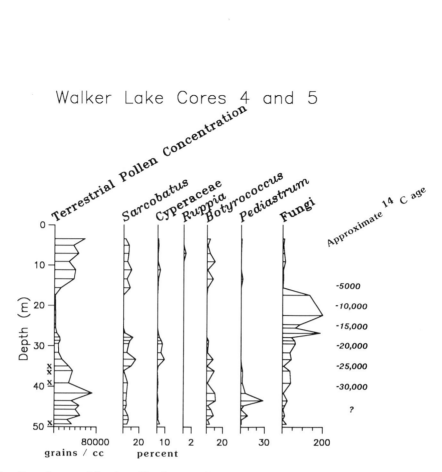

Fig. 6. Selected pollen, algae, and fungi profiles from Walker Lake cores 4 and 5. Levels marked with (x) are from core 5.

so low that they cannot be convincingly related to any local aquatic environment. Ostracodes are also essentially absent between these levels, and as with the diatoms, contain very rare occurrences of freshwater taxa that may have been derived from younger sediments during coring (Fig. 5).

Radiocarbon dates in this interval indicate that sediment deposition occurred between 5400 and about 16 800 years ago (Table 1), although sedimentation may have been discontinuous throughout this time. Pollen analyses (Fig. 6) show very high counts of terrestrial fungal spores and hyphae between 26 and 16 m and highly degraded pollen. High C/N atomic ratios (> 30) imply a dominance of terrestrial organic matter, and highly saline pore fluids in this interval, suggest the former presence of evaporites (Benson & Thompson, 1987). Taken together, these date suggest the absence of a lake in the Walker basin during this period.

Samples between 16 and 4.3 m represent sediments deposited during the last 5000 years

(Table 1), but sections of this interval show evidence of sediment mixing during coring. Nevertheless, diatoms and ostracodes from this interval (Figs. 4 and 5) can be correlated to apparently equivalent events and even *Artemia* pellets occur at the expected horizon. However, core 8 provides an undisturbed, *ca.* 12-m long section from the same site as core 4, and is therefore a more reliable record of the Holocene in Walker Lake.

Biostratigraphy: core 8

Like core 4, the sediments of core 8 are largely composed of fine silt, silt- and sand-sized diatoms, authigenic grains of $CaCO_3$, and lesser amounts of organic detritus. Ostracodes also represent a large part of the sand-sized fraction. Volcanic ashes are present at several horizons in the core and except for an ash from the Inyo Craters (257 cm), all ashes came from the Mono Craters (Bradbury, 1987).

48

Core 8 provides an undisturbed paleo-limnologic record (11.70 m) of Walker Lake for the last 5000 years, and the 10-cm sampling interval for diatoms and ostracodes provides information about limnological changes on a scale of about 50 years. The diatom stratigraphy (Fig. 7) shows significant short-term variation in abundance of all taxa. Most of these fluctuations are constrained by several adjacent analyses. Similar fluctuations occur in grain size, ostracodes, algae and pollen.

The bottom of core 8 (1170 cm) is a dark grey, stiff, silty to sandy sediment with low diatom and ostracode abundances. Euryhaline *Navicula* species, (principally *N. subinflatoides*) are the only common diatoms, and the only ostracode is *Limnocythere sappaensis*. Both groups indicate shallow, highly saline water. *Navicula subinflatoides* currently lives in Mono Lake, California, where the water has a salinity of about 72 g/l. The presence of *Artemia* pellets (Fig. 8) and traces of *Ruppia* pollen at one level (Fig. 9) together with large percentages of *Sarcobatus* pollen presumably

derived from surrounding salt flats, also imply high salinity. This interval (1170–1100 cm) was a period when Walker Lake was little more than a saline wet-playa, although the predominantly freshwater alga *Botryococcus* (Fig. 9) implies at least short periods of fresher water.

Planktonic diatom abundance abruptly increases at 1075 cm signaling the sudden deepening of the lake at that time. *Cyclotella meneghiniana* and *Chaetoceros elmorei* appear first (Fig. 7) indicating eutrophic, moderately saline water. These taxa are followed in succession by concentration spikes of *Stephanodiscus excentricus*, *Stephanodiscus rotula*, *Stephanodiscus niagarae,* and *Cyclotella ocellata* (Fig. 7) indicating a progression from somewhat saline to essentially fresh water. The transition from the initial filling of Walker Lake (1075 cm) to completely freshwater conditions (1036 cm) was rapid according to the vertical slope of the date/depth curve through this interval (Fig. 3A, Table 1). This diatom assemblage occurs in small percentages at about 16 m in core 4 suggesting a downward displacement of

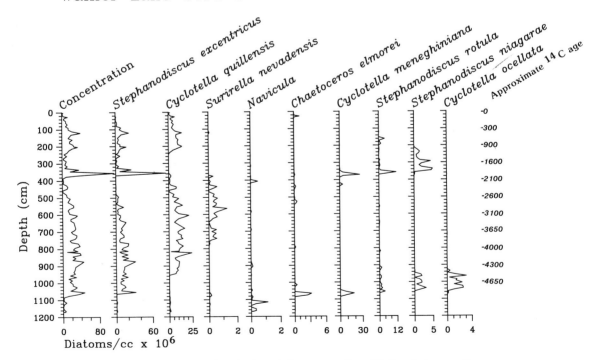

Fig. 7. Stratigraphy of selected diatom taxa from Walker Lake core 8.

Walker Lake Core 8

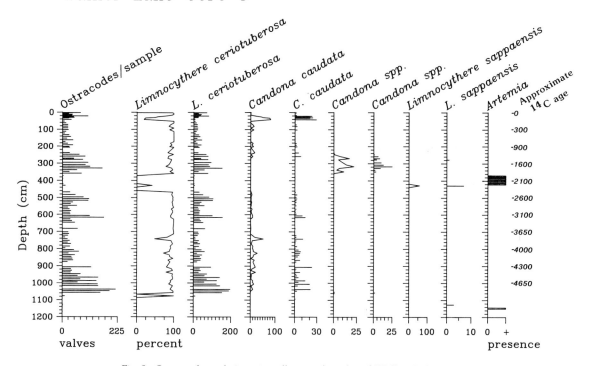

Fig. 8. Ostracode and *Artemia* pellet stratigraphy of Walker Lake core 8.

sediments in core 4 of about 5 m relative to core 8 (Benson, 1988).

Cyclotella ocellata typically occurs in the epilimnion of cool, oligotrophic, freshwater lakes. Its nearest recorded occurrence to Walker Lake is in Lake Tahoe, California (Mahood *et al.*, 1984). Assuming it was not transported to Walker Lake from upstream habitats, its occurrence implies substantially different limnological conditions than exist today.

The ostracode stratigraphy documents the same paleolimnological changes. Ostracode abundance is very low below 1080 cm (Fig. 8), and the occasional, abraded valves were probably transported to the site. The presence of *Artemia* pellets implies salinities above the tolerance of the few ostracodes present and also suggests that these ostracodes do not represent a life assemblage.

The euryhaline ostracode, *Limnocythere cerio-tuberosa,* becomes common at 1077 cm indicating the presence of a saline, perennial water body. *Candona caudata,* which documents the freshen-

ing of the lake, is rare at 1047 cm and becomes common at 1038 cm. The large number of valves of *L. ceriotuberosa* and *C. caudata* in the filling interval implies high ostracode production and oxygenated bottom environments for much of the year. Moderate water depths and low algal productivity suggested by *Cyclotella ocellata* may both have contributed to oxygenated conditions.

Stephanodiscus niagarae and *Cyclotella ocellata* are replaced by *Cyclotella quillensis* at 955 cm, probably reflecting increased salinity from evaporation. *Stephanodiscus excentricus* remains high and variable (Fig. 7). Concentrations of diatoms fluctuate although there is a general trend of decreasing diatom concentration above 900 cm and especially above 600 cm. Salinities around 3000 mg/l would permit limited productivity of *Candona caudata* in the lake, although some part of its distribution between 955 and 700 cm might be due to transport from the Walker River. The prominent drop in abundance of *Limnocythere ceriotuberosa* between 1000 and 900 cm may have been related to a loss of dissolved oxygen at the

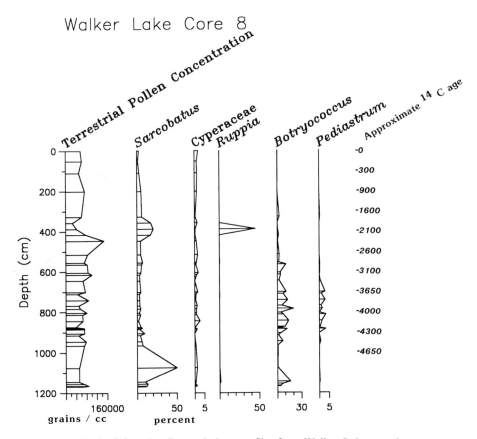

Fig. 9. Selected pollen and algae profiles from Walker Lake core 8.

sediment water interface during seasonal stratification. Ostracode productivity would then have been restricted to periods of circulation when oxygen at the sediment/water interface is sufficient for their survival.

Surirella nevadensis co-occurs with *Stephanodiscus excentricus* and *Cyclotella quillensis* from 741 to 437 cm (Fig. 7). The occurrence of this assemblage in Pyramid Lake during the 1920's (salinity = 3486 mg.l, Hanna & Grant, 1931) implies that Walker Lake was intermittently deeper and fresher than today between 3600 and 2600 years B.P.

The ostracodes in the *Surirella nevadensis* interval vary in abundance. When both *Limnocythere ceriotuberosa* and *Candona caudata* increase, lake salinity is probably falling and oxygenated environments expanding over the lake bottom. When *L. ceriotuberosa* increases but *C. caudata*

declines, lake salinity may be rising, limiting the productivity of *Candona caudata*.

Diatom and ostracode abundance varies around very low values in the interval from 465 cm to 382 cm (Fig. 7, 8). *Limnocythere sappaensis* is present but rare in this interval whereas brine shrimp pellets and *Ruppia* and *Sarcobatus* pollen are common *(Figs. 8, 9)*. Euryhaline *Navicula* species are present in low but significant numbers at a depth of 407 cm (Fig. 7). Collectively the microfossils suggest that this interval is again a highly saline, wet-playa.

The end of the saline wet-playa environment is distinctively marked by the appearance of planktonic diatoms at 377 cm. The order of diatom appearances is the same as for the older interval in core 8, *Cyclotella meneghiniana* – *Stephanodiscus rotula* – *S. niagarae*, although *Chaetoceros elmorei* is poorly represented, and *Cyclotella ocel-*

lata is absent from this freshwater phase. This phase is distinctive in that *Stephanodiscus niagarae* becomes relatively abundant (347 cm), and persists until 224 cm, where its loss is probably due to increasing salinity. *Stephanodiscus excentricus* and *Cyclotella quillensis* then dominate this assemblage until historic time when diatom productivity drops, possibly due to competition with bluegreen algae and to higher salinities.

The ostracode assemblage is characterized by the reappearance of *Limnocythere ceriotuberosa* at 359 cm followed by the appearance of two undescribed species of *Candona*. *Artemia* pellets are abundant at 379 cm, rare at 373 cm, and absent above that horizon, indicating that the salinity of the lake has fallen. The increase in abundance of *Limnocythere ceriotuberosa*, the presence of *Candona* spp., and the absence of *C. caudata* suggests that the salinity may be seasonally above 3000 mg/l. The general abundance of the ostracodes implies that the bottom water of the lake is at least seasonally oxygenated. The appearance of *C. caudata* at 260 cm may imply that the salinity is below 3000 mg/l or that river discharge is high enough to transport this taxon to the core site.

The low abundance of ostracodes from 240 until 40 cm is probably due to seasonal availability of dissolved oxygen in the profundal zone and implies less turnover in the winter or greater summer productivity than today. The increase in ostracode abundance afterwards, especially of *C. caudata*, suggests the lake has oxygen in the profundal zone for longer periods of time and that the salinity is below 3000 mg/l. The greater abundance of *C. caudata* probably means that it is living in the lake and may also be transported to the lake by high discharge. This filling probably resulted in the historic high stand (1882 A.D.) that was 30 m deeper than, and had about 1/3 the modern salinity of modern Walker Lake (Rush, 1970). The loss of *C. caudata* and the increase in abundance of *Limnocythere ceriotuberosa* reflects both the present rise in salinity and the modern circulation structure of the hypolimnion.

Discussion

A large lake existed in the Walker basin from at least 32000 to *ca.* 25000 years B.P. (Fig. 10). This lake appears to have extended back in time beyond the limit of radiocarbon dates. We show the microfossil evidence for this lake to a depth of 50 m in core 4 and believe that this horizon could be 40000 years old or older. By disregarding radiocarbon dates on carbonate as potentially too young, Benson (1988) estimates that this large lake phase ended *ca.* 40000 years ago. Nevertheless, radiocarbon-dated ostracodes at Kawich Playa, 250 km southeast of Walker Lake, place the end of the last, moist phase there (a perennial alkaline marsh) at 27000 $\pm$ 500 years B.P. (AA-1247), and closer by, at Mono Lake, California, high lake stands are also recorded between 25000 and 32000 years B.P. (*e.g.* Forester, 1987).

The presence of *Limnocythere bradburyi* in this middle Wisconsin lake suggests that different climatic, hydrologic and limnologic conditions existed in the Walker Lake basin than during the full glacial or the Holocene. This ostracode species has not yet been found living in Nevada; its principal distribution is in central Mexico, although it is known from one locality in southern New Mexico. Based on environmental conditions where *L. bradburyi* lives today (Forester, 1985, 1987), it might be able to live in Walker Lake if the lake did not stratify in summer and bottom water temperature reached at least 15 to 20 °C. These conditions might exist if summers were windier and warmer, or if warmer winters, and therefore, warmer hypolimnion temperatures could prevent or retard summer stratification.

The depth of Walker Lake during the middle Wisconsin is not known. Pre-late Wisconsin tufa deposits at elevations from 1260 m to 1317 m (Benson & Thompson, 1987; Lao & Benson, 1988) are logical candidates for the shoreline of this lake. If so, the middle Wisconsin Walker Lake may have occasionally entered Adrian Valley to flow into or unite with the northern subbasins of Lake Lahontan (Fig. 2).

Saline marsh conditions existed from at least

Walker basin by geomorphic or tectonic means (King, 1978) rather than reappearing due to climate change. River switching is supported by the presence of moist, but stable climatic conditions in the Sierra Nevada before and after that time (Adam, 1967) and the existence of a paleo river channel and recently active faults between Mason Valley and Adrian pass which connects the Walker River drainage to the Carson River (King, 1978; Davis, 1982; Morrison & Davis, 1984).

This filling episode resulted in a comparatively large, fresh lake in the Walker basin that became progressively saline by evaporation (Bradbury, 1987). Fluctuations in diatom and ostracode concentrations may record different levels of snowpack in the Sierras and meltwater discharge down the Walker River drainage basin. Once the lake reached or exceeded essentially modern depths (4200 years B.P.), the variations in ostracode and diatom abundances may record fluctuations in the water balance tied to lake level influences on circulation, salinity changes, and river discharge.

Gradually decreasing concentrations of diatoms and percentages of *Botryococcus* (Figs. 7 and 9) and finally abrupt decreases in ostracodes (Fig. 8) and increases in *Sarcobatus* (Fig. 9) document the reduction of Walker Lake to a shallow, saline wet playa. *Ruppia* seeds with attached gymnophores in the interval between 2500 and 2150 years B.P. from a nearby core indicate water depths less than one meter (Bradbury, 1987) and imply that Walker Lake was reduced volumetrically by 98% and areally by 93% from the 1968 size (Rush, 1970; Benson & Mifflin, 1986). The reduction of Walker Lake to a saline wet-playa 2500 years ago could have resulted from a diversion of the Walker river to Carson Sink (*e.g.* King, 1978) or to drought in the Sierras. *Ruppia* indicates the presence of saline ponds and therefore some minimal unput from the Walker River; otherwise Walker Lake would be completely dry because evaporation vastly exceeds precipitation (Rush, 1970).

Tufa deposits around Walker Lake and Pyramid Lake (Benson & Thompson, 1987) suggest low levels between 2000 and 3000 years B.P.

(Fig. 11), but because of their temporal discontinuity, they say little about the historical trend of lake level changes. Nevertheless, evidence for a shift to drier climates in nearby parts of the Great Basin, such as at Mono Lake (Stine *et al.*, 1984), Osgood Swamp (Adam, 1967), and the White Mountains (La Marche, 1973) suggests that the low levels of Walker Lake at about the same time reflect regional climate change rather than diversion of the Walker River (Bradbury, 1987). Dry climates were apparently not so severe, however, that the Truckee River entering Pyramid Lake was reduced to a point where the cui-ui, an endemic fish that spawns in the river, became extinct (*e.g.* Benson & Thompson, 1987).

The ostracode and diatom stratigraphies document several events during the subsequent filling of the lake. In contrast to the earlier (about 4700 year B.P.) filling phase, when the successive events occurred very rapidly (identical radiocarbon dates above and below the interval), the filling that began about 2000 years B.P. took place more slowly. The step-like sequence of ostracode and diatom appearances that records a progressive freshening of Walter Lake occurred over a 150-cm stratigraphic interval (approximately 380–230 cm). Extrapolation between averaged radiocarbon dates across this 150-cm interval indicates a span of about 800 years for the filling period (Fig. 3A). The longer filling phase may have resulted from slowly increasing flow of the Walker River under climatic control. Higher sedimentation rates for this interval are also possible, however, especially if the radiocarbon dates on carbonate (Table 1) are considered more accurate reflections of the chronology of this interval. In that case, the filling phase would have taken over two centuries. This rate of filling is still too slow to be ascribed to rediversion of the Walker River back into its basin, but would correspond to a more rapid climatic change.

This filling episode terminated with the development of a moderately deep but fluctuating lake. Decreasing diatom concentrations after 350 years ago may relate to drier climates and lower lake stages of the 17th and 18th Centuries (Rush, 1970). Increases in some diatoms and ostracodes

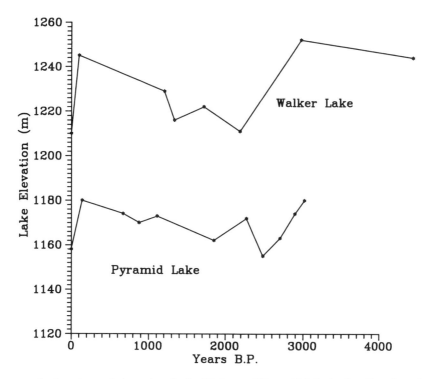

Fig. 11. Late Holocene tufa elevation and chronology in the Walker and Pyramid Lake basins. Data from Benson & Thompson (1987).

at the very top of core 8 probably correlate with the historic high stand observed by Russell in 1882 A.D. (Rush, 1970).

Cores from Walker Lake have recorded deep, high-stand lakes in both the Holocene and in the middle Wisconsin. The apparent absence of any significant body of water in Walker Lake between the early full glacial and the middle Holocene complicates the use of this basin for paleoclimatic reconstruction. Nevertheless, the presence of a saline marsh and the absence of the Walker River during the full glacial indicates that *in situ* precipitation minimally exceeded evaporation at that time. By 15 000 years B.P., assuming that the radiocarbon chronology is satisfactory, the local hydrologic balance was inadequate to support any kind of lake in the basin and Walker Lake desiccated. With the return of the Walker River to Walker Lake in the mid Holocene, this basin once again could record paleohydrologic changes under climatic control. Investigation of earlier paleolimnologic records from this basin must consider the variable influence of the Walker River under both climatic and geomorphic-tectonic controls.

Acknowledgements

A number of people have unselfishly given their data and insights gathered in the course of the Walker Lake core study. Their information has been especially valuable in developing a holistic paleolimnological history of Walker Lake. Because their contributions are not specifically cited in the text, the authors wish to list these individuals and the area of their contributions with great thanks. L. V. Benson: hydrology, hydrochemistry, pore fluid chemistry; D. L. Galat: limnology; P. A. Meyers: organic geochemistry; A. Sarna-Wojicki: tephrochronology; D. Schaeffer: seismic profiling of bottom sediments; J. Smoot: sedimentology; W. Witte & G. Kukla: magnetic susceptibility; I. C. Yang: radiocarbon chronology; J. C. Yount: sediment texture. Work was preformed for the U.S. Department of Energy, Nevada Nuclear Waste Storage Investigations Project (Interagency Agreement DE-AI08-78ET44802).

References

Adam, D. P., 1967. Late-Pleistocene and Recent palynology of the central Sierra Nevada, California, *in* Wright, H. E. & Cushing, E. J., (eds.), Quaternary Paleoecology: Yale University Press, New Haven and London, p. 275–301.

Battarbee, R. W., 1986. Diatom analysis: *in* Berglund, B. E. (ed.) Handbook of Holocene paleoecology and paleohydrology, John Wiley and Sons Ltd., New York, p. 527–568.

Benson, L. V., 1981. Paleoclimatic significance of lake level fluctuations in the Lahontan Basin: Quat. Res. 16: 390–403.

Benson, L. V., 1986. The sensitivity of evaporation rate to climatic change – results of an energy balance approach: U.S. Geological Survey Water-Resources Investigations Report 86–4148, 40 p.

Benson, L. V., 1988. Premiminary paleolimnologic data for the Walker Lake subbasin, California and Nevada: U.S. Geological Survey, Water Resources Investigations Report 87–4258, 50 p.

Benson, L. V. & D. L. Leach, 1979. Uranium transport in the Walker River Basin, California and Nevada: Journal of Geochemical exploration 11: 227–248.

Benson, L. V. & M. D. Mifflin, 1986. Reconnaissance bathymetry of basins occupied by Pleistocene Lake Lahontan, Nevada and California: U.S. Geological Survey Water-Resources Investigations Report 85–4262, 14 p.

Benson, L. V. & R. S. Thompson, 1987. Lake-level variation in the Lahontan basin for the past 50 000 years: Quat. Res. 28: 69–85.

Bradbury, J. P., 1987. Late Holocene diatom paleolimnology of Walker Lake, Nevada: Archiv fur Hydrobiologie, Supplement 79, Monographische Beitrage 1: 1–27.

Cholnoky, B. J., 1968. Die Okologie der Diatomeen in Binnengewassern: J. Cramer, Lehre, W. Germany, 699 p.

Cooper, J. J. & D. L. Koch, 1984. Limnology of a desertic terminal lake, Walker Lake, Nevada, U.S.A.: Hydrobiologia 118: 275–292.

Davis, J. O., 1982. Bits and pieces: the last 35 000 years in the Lahontan area: *(in)* Madsen, D. B. & J. O. Davis. Man and environment in the Great Basin, Society for American Archaeology, SAA Papers 2: 53–75.

Delorme, L. D., 1978. Distribution of freshwater ostracodes in Lake Erie: Journal of Great Lakes Research 4: 216–220.

Demsey, K. A., 1987. Late Quaternary faulting and tectonic geomorphology along the Wassuk Range, west-central Nevada (abs): Geological Society of America, 1987 Annual Meeting, Abstracts with Programs, p. 640.

Forester, R. M., 1983. Relationship of two ostracode species to solute composition and salinity: Implications for paleohydrochemistry: Geology 11: 435–438.

Forester, R. M., 1985. *Limnocythere bradburyi* n. sp.: a modern ostracode from central Mexico and a possible Quaternary paleoclimatic indicator: Journal of Paleontology 59: 8–20.

Forester, R. M., 1986. Determination of the dissolved anion composition of ancient lakes from fossil ostracodes: Geology 14: 796–798.

Forester, R. M., 1987. Records of deglaciation from lacustrine ostracodes, *in* Ruddiman, W. F. & Wright, H. E., Jr. (eds.), North America and adjacent oceans during the last deglaciation, The Geology of North America: v. K-3, Geological Society of America, Boulder, Colorado, p. 261–276.

Galat, D. L., E. L. Lider, S. Vigg & S. R. Robertson, 1981. Limnology of a large, deep, North American terminal lake, Pyramid Lake, Nevada, USA: Hydrobiologia 82: 281–317.

Gray, J., 1965. Palynological techniques: *in* Kummel, B. & Raup, D. (eds.), Handbook of paleontological techniques, W. H. Freeman & Co., San Francisco, p. 471–481.

Hamilton-Galat, K. & D. L. Galat, 1983. Seasonal variation of nutrients, organic carbon, ATP, and microbial standing crops in a vertical profile of Pyramid Lake, Nevada. Hydrobiologia 105: 27–43.

Hanna, G. D. & W. M. Grant, 1931. Diatoms of Pyramid Lake, Nevada: Transactions of the Amrican Microscopical Society, vol. L, No. 4. p. 281–296.

King, G. O., 1978. The late Quaternary history of Adrian Valley, Lyon County, Nevada: unpublished M.S. Thesis, Department of Geography, The University of Utah, 88 p.

Koch, D. L., J. J. Cooper, E. L. Lider, R. L. Jacobson & R. J. Spencer, 1979. Investigations of Walker Lake, Nevada: dynamic ecological relationships: Desert Research Institute, Publication No. 50010, University of Nevada, Reno, 191 p.

LaMarche, V. C., 1973. Holocene climatic variations inferred from treeline fluctuations in the White Mountains, California: Quat. Res. 3: 632–660.

Lao, Y. & L. V. Benson, 1988. Uranium-series age estimates and paleoclimatic significance of Pleistocene tufas from the Lahontan Basin, California and Nevada: Quat. Res. 30: 165–176.

Lockheed Ocean Science Labratories, 1982. Investigations of the effects of total dissolved solids on the principal components of the Pyramid Lake food chain: Final Report, Contract K51C14201130, U.S.D.I. Bureau of Indian Affairs, Lockheed Ocean Sci. Lab., San Diego, California, 545 p.

Mahood, A. D., R. D. Thompson & C. R. Goldman, 1984. Centric diatoms of Lake Tahoe: Great Basin Naturalist 44: 83–98.

Meyers, P. A. & L. V. Benson, 1988. Composition of organic matter in sediments of Walker Lake, Nevada: (abs) Abstracts of Papers for the 1988 Annual Meeting,

American Society of Limnology and Oceanography, Boulder, Colorado, p. 54.

Morrison, R. B. & J. O. Davis, 1984. Supplementary guidebook for field trip 13, Quaternary stratigraphy and archaeology of the Lake Lahontan area: a reassessment: Desert Research Institute, University of Nevada, Social Sciences Center Technical Report No. 41, 50 p.

Newton, M. S. & E. L. Grossman, 1986. Significance of some late Quaternary tufa deposits, Walker Lake, Nevada (abs): Geological Society of America, Abstracts with Programs, 99th Annual Meeting, p. 706.

Rush, F. E., 1970. Hydrologic regimen of Walker Lake, Mineral County, Nevada: U.S. Geological Survey Hydrological Investigations Atlas HA-415 (map with text).

Stine, S., S. Wood, K. Sieh & C. D. Miller, 1984. Holocene paleoclimatology and tephrochronology east and west of the central Sierran crest: Field Trip Guidebook for Friends of the Pleistocene, Pacific Cell, 12–14 October, 1984, 107 p.

Thompson, R. S., L. V. Benson & E. M. Hattori, 1986. A revised chronology for the last Pleistocene lake cycle in the central Lahontan basin: Quat. Res. 25: 1–9.

Oscillations of levels and cool phases of the Laurentian Great Lakes caused by inflows from glacial Lakes Agassiz and Barlow-Ojibway

C. F. M. Lewis[1] & T. W. Anderson[2]
[1] *Geological Survey of Canada, Bedford Institute of Oceanography, Box 1006, Dartmouth N.S., Canada, B2Y 4A2;* [2] *Geological Survey of Canada, 601 Booth Street, Ottawa, Ontario, Canada, K1A OE8*

Key words: Great Lakes history, meltwater effects, palynology, paleoclimate change, paleohydrology, paleogeography, late Wisconsinan, early Holocene, lacustrine stratigraphy, radiocarbon dates

Abstract

Two distinct episodes of increased water flux imposed on the Great Lakes system by discharge from upstream proglacial lakes during the period from about 11.5 to 8 ka resulted in expanded outflows, raised lake levels and associated climate changes. The interpretation of these major hydrological and climatic effects, previously unrecognized, is mainly based on the evidence of former shorelines, radiocarbon-dated shallow-water sediment sequences, paleohydraulic estimates of discharge, and pollen diagrams of vegetation change within the basins of the present Lakes Superior, Michigan, Huron, Erie and Nipissing. The concept of inflow from glacial Lake Agassiz adjacent to the retreating Laurentide Ice Sheet about 11–10 and 9.5–8.5 ka is generally supported, with inflow possibly augmented during the second period by backflooding of discharge from glacial Lake Barlow-Ojibway.

Although greater dating control is needed, six distinct phases can be recognized which characterize the hydrological history of the Upper Great Lakes from about 12 to 5 ka; 1) an early ice-dammed Kirkfield phase until 11.0 ka which drained directly to Ontario basin; 2) an ice-dammed Main Algonquin phase (11.0–10.5 ka) of relatively colder surface temperature with an associated climate reversal caused by greater water flux from glacial Lake Agassiz; 3) a short Post Algonquin phase (about 10.5–10.1 ka) encompassing ice retreat and drawdown of Lake Algonquin; 4) an Ottawa-Marquette low phase (about 10.1-9.6 ka) characterized by drainage via the then isostatically depressed Mattawa-Ottawa Valley and by reduction in Agassiz inflow by the Marquette glacial advance in Superior basin; 5) a Mattawa phase of high and variable levels (about 9.6–8.3 ka) which induced a second climatic cooling in the Upper Great Lakes area. Lakes of the Mattawa phase were supported by large inflows from both Lakes Agassiz and Barlow-Ojibway and were controlled by hydraulic resistance at a common outlet – the Rankin Constriction in Ottawa Valley – with an estimated base-flow discharge in the order of $200\,000 \; m^3 s^{-1}$. 6) Lakes of the Nipissing phase (about 8.3–4.7 ka) existed below the base elevation of the previous Lake Mattawa, were nourished by local precipitation and runoff only, and drained by the classic North Bay outlet to Ottawa Valley.

Originally published in
Journal of Paleolimnology **2**: 99–146.

Introduction

The basins and connecting lowlands of the Laurentian Great Lakes are centrally located in North America between a southward-draining watershed to Gulf of Mexico and a northward-draining watershed to Hudson Bay (Fig. 1a). During the last deglaciation, the northward retreating ice margin impounded proglacial lakes, first in the Laurentian basins, then in the northern watershed with overflow through the Great Lakes system after some basins had had no direct contact with a glacier for hundreds to thousands of years (Dyke & Prest, 1987). This system of northward migrating proglacial lake drainage from about 14 to 8 ka is a well-recognized pathway for the post-glacial dispersal of aquatic fauna, such as, the crustaceans *Mysis relicta*, *Pontoporeia 'affinis'*, *Limnocalanus macrurus*, *Senecella calanoides* and the fish *Myoxocephalus quadricornis* (Ricker, 1959; Martin & Chapman, 1965; Dadswell, 1974). Not demonstrated, however, are the potential physical effects of large glacial meltwater discharges on downstream non-glacial lakes, far from the glacial margin; a phenomenon which occurred during at least two periods in the later part of the glacial retreat between 11 and 8 ka. Events in these periods afford an opportunity to infer large-scale changes in water flux, level, sedimentation, temperature regime and local climate in downstream lakes as a response to input of glacial waters, and as a result, to better understand the evolution of lake systems during a deglacial cycle.

Recent studies of the glacial and proglacial lake history in central North America suggest that retreat of the Laurentide Ice Sheet in northwestern Ontario from Superior and Nipigon basins permitted glacial Lake Agassiz to discharge to the Great Lakes system (Fig. 1a & b) from about 9.5 to 8.5 ka (Clayton, 1983; Drexler *et al.*, 1983; Farrand & Drexler, 1985). From the dimensions and sediments of bedrock channels bearing this flow, Teller & Thorleifson (1983, 1987) and Teller (1985) estimate the equilibrium discharge of runoff and glacial melt to have been in the order of 10 000 to 100 000 m^3s^{-1}, on aver-

age about 30 000 m^3s^{-1}. (For comparison, the modern mean discharge of the Great Lakes is 6600 m^3s^{-1}. (International Great Lakes Levels Board, 1973)). During ice retreat, about 17 different outlets at successively lower elevations were opened, giving rise to rapid drawdowns of Lake Agassiz with catastrophic outflows possibly exceeding 200 000 m^3s^{-1} for limited periods. The same studies postulate an earlier period of similar discharge from Lake Agassiz to the Great Lakes from about 10.9 to 10 ka. Teller (1985) and Farrand & Drexler (1985) have speculated that surges may have occurred in Superior basin, but no evidence or theory has yet been advanced to document or account for the anticipated increases and fluctuations in levels of the Great Lakes due to eastward overflow from glacial Lake Agassiz.

The existing models for level changes in the Upper Great Lakes, through which the Agassiz discharge would have passed, generally describe a high Main Lake Algonquin phase (about 11–10.5 ka), preceded by a slightly lower Kirkfield phase (both lakes were ice-dammed), and generally followed by very low phases when ice receded from the Upper Ottawa Valley. Differential uplift of this outlet caused the low phases to transgress to the Nipissing high phase (about 4.7 ka). Then discharge was transferred to southern outlets and related lake levels generally descended to their present elevation as a consequence of continued differential uplift and outlet erosion (Leverett & Taylor, 1915; Hough, 1958, 1963; Lewis, 1969; Prest, 1970; Fullerton, 1980; Chapman & Putnam, 1984; Hansel *et al.*, 1985; Eschman & Karrow, 1985; Larsen, 1987; Dyke & Prest, 1987). The high Main Algonquin phase is an obvious correlative of the earlier eastward discharge from Lake Agassiz as described above. However, no model, at present, describes high levels or other effects that might be a response to the later Agassiz inflow, though, as will be shown later, evidence for higher early Holocene lake levels does exist in the Great Lakes basins.

Our thesis is that large oscillations in inflow, interacting with hydraulic resistance of outlet channels, caused rises and declines in levels of the Great Lakes or changes in the capacity of their

61

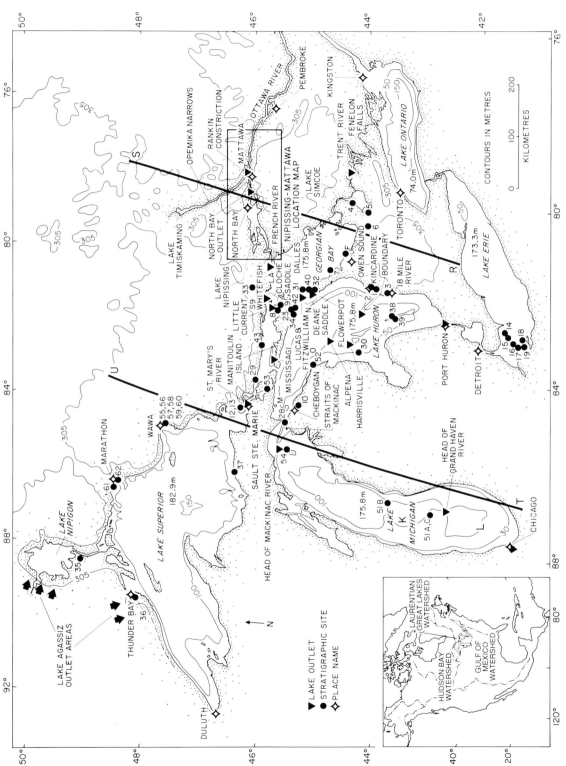

Fig. 1a. Location map of Laurentian Great Lakes showing positions of radiocarbon-dated sites (numbered dots keyed to Table 1) and key lake outlets (triangles). Also shown are sites mentioned in text (letters), present isobaths of basins with respect to indicated lake surface elevations above sea level and lines of sections (RS, TU) illustrated in Figures 7a and 7b. The topographic contour, 305 m asl, is shown onshore. The rectangle outlines the area of Fig. 1b.

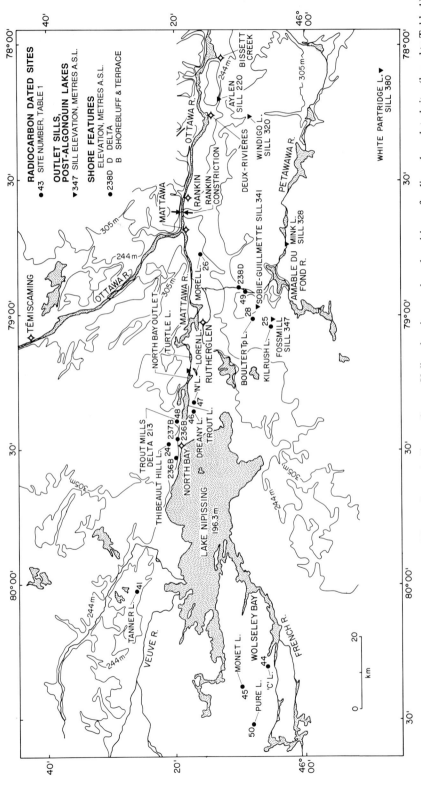

Fig. 1b. Location map of Nipissing basin, Mattawa River valley and adjacent Ottawa Valley showing numbered positions of radiocarbon-dated sites (keyed to Table 1), lake outlets identified as sills (Table 2), and selected shore features interpreted to be related to levels of the Mattawa Great Lakes phase.

outlets. In addition, we hypothesize that the inflows of glacial meltwater and proglacial lake water suppressed seasonal warming of the Upper Great Lakes, and induced local climate reversals compared with earlier and later periods without major inflow. In this paper, we examine the foregoing postulates by reconstructing former water planes and reinterpreting the lake-level history of the Upper Great Lakes from radiocarbon-dated evidence. Climate reversals are recognized from pollen diagrams and are correlated with two periods of cold water inflow. Summaries of early stages of this work have been published previously (Anderson & Lewis, 1987; Lewis & Anderson, 1987).

Methods

The data are mainly stratigraphic sections of deposits high in content of organic matter signifying former conditions of shallow wave- and current-protected waters, obtained offshore in the Great Lakes by gravity and piston coring, in small lake basins by piston coring, and by direct sampling of exposed onshore sections. The coring procedures are described elsewhere in more detail (Lewis, 1969; Thomas *et al.*, 1973; Mott & Farley-Gill, 1981).

The lithologies of sediment sequences were described using distinctions of color, grain size, sedimentary structure and composition. Samples were removed at selected levels for pollen analysis and conventional radiocarbon dating. The standard methods of sediment treatment for pollen extraction were followed (Anderson & Lewis, 1985) leading to percentage diagrams and occasionally to absolute frequency diagrams. For most pollen percentages, the pollen sums consist of 200 tree, shrub and herb pollen and exclude Cyperaceae and pollen of aquatic plants.

Radiocarbon dating was performed using standard methods (e.g. Lowdon & Blake, 1979) on organic materials such as wood, peat, plant detritus and gyttja which are assumed to be derived from organisms that directly or indirectly metabolized atmospheric carbon dioxide while living. However, as much of the Great Lakes region is underlain by carbonate bedrock or carbonate-bearing glacial sediments, radiocarbon datings are potentially susceptible to uptake of 'old' carbon (Karrow *et al.*, 1984; Karrow & Geddes, 1987); hence, wherever possible, the dates have been evaluated for consistency with the regional pollen stratigraphy.

Elevations of shorelines and other indicators of lake levels were determined by spirit levelling from benchmarks of the Geodetic Survey of Canada or by aneroid barometer measurements corrected for scale and diurnal changes by reference to two benchmarks. Elevations of underwater features were determined by echo sounding in relation to the altitude of the lake surface. The positions and elevations of key geomorphic features and radiocarbon-dated intervals were projected along uplift isobases of equivalent age onto sections generally oriented perpendicular to the isobases. Former water planes were identified on the sections by the alignment of shallow-water indicators of common radiocarbon age. Projections were made using trends interpolated from known isobases of the Algonquin, Washburn and Minong (about 10.7, 9.8 and 9.5 ka, respectively) and Nipissing (about 4.7 ka) water planes as shown in Fig. 2. The ages of the Algonquin and Nipissing water planes are discussed in a later section.

The sequence of lake phases is inferred from the radiocarbon-dated sequence of interpreted water planes with assistance from computed uplift histories of key outlets (Lewis & Anderson, 1985). Paleogeographic shoreline maps of key lake phases were constructed by 'back-projecting' the water plane elevations using appropriate isobase trends and noting where the water planes intersected the present topography or lake basin bathymetry.

Former river discharges were estimated indirectly by the slope-area (Manning formula) and contracted-width methods. These involved using inferred high water marks and channel dimensions as outlined by Chow (1959) and Benson (1968) and as used by Baker (1973) and by Teller & Thorleifson (1983; 1987) in the analysis of catastrophic drainage of glacial Lakes Missoula

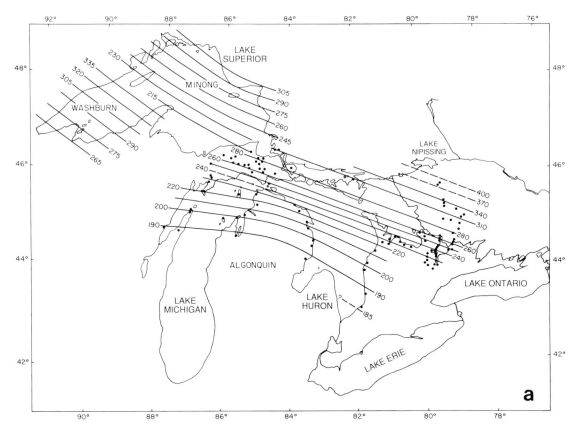

Fig. 2. Maps of isobases used to control projections of data from the lake basins to cross-sections and vice versa. a) Algonquin isobases which approximate the present elevation and configuration of the former water plane of Main Lake Algonquin at about 10.7 ka are contoured on shore feature elevations in the Huron, Georgian Bay and Michigan basins provided by Goldthwait (1907; 1908; 1910), Leverett & Taylor (1915), Johnston (1916), Stanley (1936; 1937a), Deane (1950), Chapman (1954; 1975), Harrison (1972), Futyma (1981), Warner *et al.* (1984), Kaszycki (1985), Finamore (1985), Cowan (1985) and Karrow (1987). Washburn (ca. 9.8 ka) and Minong (ca. 9.6 ka) isobases in the Superior basin are from Farrand & Drexler (1985).

and Agassiz, respectively. The Manning formula was also utilized by Hansel & Mickelson (1988) to demonstrate that changes up to 10 m in lake level of the Michigan basin could have been caused by changes in meltwater and precipitation supply from sources solely within the Great Lakes region.

Lake responses to Agassiz inflows

Evidence for increased inflow about 11–10 ka

The probable response of this first eastward discharge from Lake Agassiz was the relatively high Main Lake Algonquin phase in the basins of the

Upper Great Lakes because Lake Algonquin is of the same age as this Agassiz outflow. The period of discharge is inferred from the age of a drawdown in the Agassiz basin (Moorhead phase), estimated to have begun just after 11 ka on the basis of 2 wood dates (10.9 $\pm$ 0.3 and 10.8 $\pm$ 0.2 ka; Clayton, 1983) and to have ended about 1000 years later by readvance of ice across the Superior basin based on 11 wood dates which cluster about 10 ka (Clayton, 1983). Though not yet well dated, Main Lake Algonquin shoreline represents a large lake which existed about 10.7 $\pm$ 0.2 ka, as suggested by the available radiocarbon evidence.

Basal gyttja from small lake basins about 8 m and 14 m below the Main Algonquin shoreline

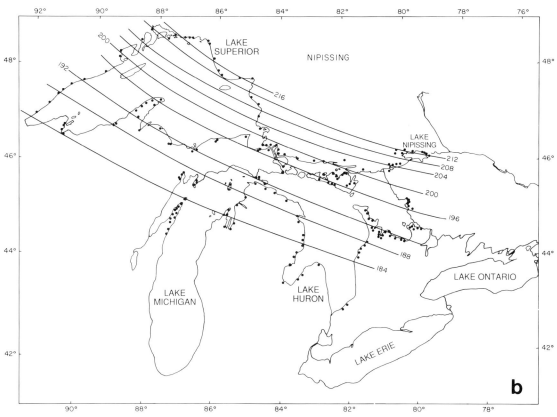

Fig. 2 continued. b) Nipissing isobases which represent the present elevation and configuration of the former water plane of the Nipissing Great Lakes about 4.7 ka, are contoured on shore feature elevations provided by Goldthwait (1907; 1908; 1910), Leverett & Taylor (1915), Stanley (1936; 1937a), Deane (1950), Farrand (1960) & Lewis (1969; 1970; this study).

near Sault Ste. Marie (as identified by Cowan, 1985; Saarnisto, 1974) and near Cheboygan (Futyma & Miller, 1986; R. P. Futyma, pers. commun., 1988) was dated 10.7 ka (Table 1: 12, 10). Though interpreted to be slightly too old on the basis of local pollen stratigraphy, a date of 10.8 ka was obtained on basal organic sediments from a second lake near Sault Ste. Marie about 20 m below the Algonquin shoreline (Table 1: 13) (Saarnisto, 1974). The Algonquin shore in this northwestern Huron-basin area is interpreted to be slightly older than these ages, as the dated small-lake organic sediments could only form after the larger Main Algonquin lake had regressed.

Because differential glacioisostatic uplift of the basin affected the Fenelon Falls outlet, the relative lake levels of Lake Algonquin would have regressed (fell) with time north of the isobase

(Fig. 2a) through the outlet, and transgressed south of the same isobase. Six radiocarbon dates, including four on wood, were obtained from plant layers embedded in estuarine silts east of southern Lake Huron and comprise a younging upward sequence from 11.3 ka at the base to 10.5 ka that is believed to reflect about 10 m of transgression (Table 1: 1, 2, 3) (Eschman & Karrow, 1985). However, some evidence closer to the outlet area where differential uplift effects are minimal could be interpreted to imply an absolute rise of water to the Main Algonquin level. A series of weakly developed shorelines near Cape Rich (Fig. 1a: F) on Georgian Bay span a vertical interval of at least 9 m below the Main Algonquin water plane (Stanley, 1937b). A relatively rapid rise of lake level is implied as these shorelines were subsequently impounded by a wave-built barrier of the Main Algonquin lake. Similar occurrences of

Table 1. Selected radiocarbon dates with significance for changes in former lake levels.

No.	Site, basin area[1], lake elevation m asl	Lat./Long.	Age in years B.P.	Laboratory No.	Material dated // Elevation m asl	Stratigraphy // Comments	References
1	Kincardine Bog seH	44° 09.0' 81° 39.0'	10,300 ± 200 10,600 ± 150 11,200 ± 170	GSC-1644 GSC-1366 GSC-1374	mollusc shells // 195 plant roots // 195 plant detritus // 194	Two shelly plant detritus beds and an intervening clay unit, all over sand and clay and under shelly marl. // Clay above lower plant detritus indicates abrupt transgression of a lagoonal deposit to Main Lake Algonquin.	Karrow et al., 1975
2	North Penetangore River seH	44° 10.0' 81° 38.0'	11,300 ± 140	GSC-1842	Picea or Larix wood // 191	Plant detritus and shell layer over stony silty clay (till) under sand and interbedded sand, silt and clay. // A lagoonal deposit (organics) overlain by inorganic sediments of Main Lake Algonquin.	Anderson, 1979
3	Eighteen Mile River seH	44° 01.3' 81° 43.6'	10,600 ± 160 10,500 ± 150	GSC-1127 GSC-1126	Picea wood // 188.2 Picea wood // 190	Wood and woody peat beds in stratified clay and silt over fine sand, silt and gravel under sand and silt. // Alluvial deposition graded to Main Lake Algonquin level.	Karrow et al., 1975; Karrow, 1986
4	Orillia sG	44° 34.6' 79° 26.3'	11,700 ± 250	n.r.	bone collagen // <251	Long bone in coarse and fine cross-bedded beach gravel interbedded with sand and silt. // Find is more than 4 m below level of Main Algonquin: here interpreted to predate it.	Peterson, 1965; Tovell & Deane, 1966
5	Cookstown Bog sG	44° 13.3' 79° 37.3'	10,200 ± 150	GSC-1111	basal peat // ca. 229	Peat over clayey gyttja over sandy clayey silt. // Indicates lake levels have fallen from Main Lake Algonquin.	Karrow et al., 1975
6	Wales Site, Everett sG	44° 12.2' 80° 57.0'	10,280 ± 100	WAT-493	wood // ca. 225	Woody peat in sandy gravel under woody peat. // Indicates lake levels have fallen from Main Lake Algonquin.	Fitzgerald, 1985
7	Hope Bay wG 175.8	44° 55.0' 81° 07.1'	8,785 ± 145 9,930 ± 250	I-7857 I-7858	top of peat // 147.3 basal peat // 147.0	Peat over laminated pinkish-grey clay under massive grey clay. // Indicates regression to Early L. Hough, marsh formation with interval of raised water level, then transgression in late Mattawa and Nipissing phases.	Figure 3c, This study
8	Greenbush Swamp neH	45° 56.0' 82° 00.5'	9,930 ± 90	WAT-579	basal gyttja // 307.5	Gyttja over silty gravel under peat. // Indicates lake levels have fallen from Main Algonquin and exposed this small isolated basin.	Warner et al., 1984

Table 1. (continued).

No.	Site, basin area[1], lake elevation m asl	Lat./Long.	Age in years B.P.	Laboratory No.	Material dated // Elevation m asl	Stratigraphy // Comments	References
9	Sheguiandah Bog *neH*	45° 53.7' 81° 55.4'	9,130 ± 350	W-345	basal peat // 219	Peat over clay. // Indicates Post Algonquin emergence of basin.	Lee, 1957; Terasmae & Hughes, 1960
10	Lake Sixteen *nwH* 216	45° 36.0' 84° 19.0'	10,690 ± 100	WIS-2000	basal gyttja // 211.9	Basal silty gyttja in 1.8 m water depth over clayey silt within sandy spit complex (225 m asl) 5 m below Algonquin water plane. // Indicates lake levels falling from Main Algonquin and isolation of small lake basin.	Futyma & Miller, 1986; R. P. Futyma, pers. commun., 1988
12	Upper Twin Lake *eS* 302	46° 32.5' 84° 35.0'	10,650 ± 265	HEL-400	basal gyttja // ca. 291.4	Gyttja over till. // Indicates lake levels falling from Main Algonquin and isolation of basins.	Saarnisto, 1974
13	Prince Lake *eS* 290	46° 33.5' 84° 33.0'	10,800 ± 360	GSC-1715	basal gyttja // ca. 280.8	Gyttja over clay. // Same as above.	Saarnisto, 1974
14	68–6, *Western Erie* 173.3	41° 55.0' 82° 45.4'	11,140 ± 160 12,650 ± 170	I-4041 I-4040	top plant det. // 159.5 base plant det. // 159.1	Plant detritus over clayey silt and fine sand under clayey silt. // Denotes a marsh environment transgressed because of Main Algonquin discharge.	Lewis, 1969; Coakley & Lewis, 1985; Figure 3b upper, This study
15	1244, *Western Erie* 173.3	41° 51.6' 82° 45.4'	11,140 ± 160 12,500 ± 310	GSC-1450 GSC-1438	top plant det. // 162.2 base plant det. // 162.0	Woody peat with sandy interbeds over stratified fine sand and silt under stratified clay. // Denotes a marsh environment transgressed because of Main Algonquin discharge.	Figure 3b lower, This study
16	1240, *Western Erie* 68–16 173.3	41° 45.9' 82° 57.3'	11,300 ± 160 11,430 ± 150	GSC-382 I-4035	plant detritus // 162.1 plant detritus // 162.6	Plant detritus over laminated silt with clay under laminated fine silty clay. // Same as above.	Lewis et al., 1966; Lowdon et al., 1967; Lewis, 1969; Coakley & Lewis, 1985; Figure 3b lower, This study
17	68–18 *Western Erie* 173.3	41° 42.4' 82° 53.9'	10,600 ± 160* 11,400 ± 160*	GSC-1180 GSC-1283	top plant det. // 163.5 base plant det. // 163.3	Plant detritus over clay till under silty clay. // Higher marsh environment not affected by discharge from Main Lake Algonquin.	This study
18	68–19 *Western Erie* 173.3	41° 40.0' 82° 50.8'	10,400 ± 190* 11,000 ± 160*	GSC-1125 GSC-1384	dissem. organics // 164.2 base pl. det. // 163.6	Marly plant detritus over silty, sandy clay under silty clay with disseminated organics. // Same as above.	This study
19	68–20 *Western Erie* 173.3	41° 36.6' 82° 53.5'	11,100 ± 180*	GSC-1136	base pl. det. // 167.0	Shelly marl and plant detritus over silt and clay under silty clay. // Same as above.	Lewis, 1969; Coakley & Lewis, 1985

Table 1. (continued).

No.	Site, basin area¹, lake elevation m asl	Lat./Long.	Age in years B.P.	Laboratory No.	Material dated // Elevation m asl	Stratigraphy // Comments	References
23	Tehkummah Lake *neH* 191.7	45°36.0' 81°59.9'	10,150 ± 190	GSC-1108	basal gyttja // 184.2	Gyttja over silty clay. // Indicates falling Post Algonquin Huron-basin lake levels to Early L. Stanley and isolation of small lake basin.	Lewis, 1969; Lowdon et al., 1971
24	Thibeault Hill Lake *neH* 312.4	46°21.0' 79°28.0'	9,820 ± 200	GSC-638	basal gyttja // n.r.	Gyttja over silty clay. // Indicates Post Algonquin lake levels have fallen to proto-L. Nipissing	Lewis, 1969; Lowdon & Blake, Jr., 1968
25	Kilrush Lake *sMA* 347	46°05.0' 79°03.0'	9,860 ± 270	GSC-1246	basal gyttja // 330.5	Gyttja over clay which overlies sand. // Same as above.	Harrison, 1972; Lowdon & Blake, 1975
26	Morel Lake *eM.i* 194	46°16.3' 78°48.0'	10,100 ± 240	GSC-1275	basal gyttja // 180.2	Gyttja over silty clay. // Indicates cessation of large Post Algonquin discharge via Mattawa valley, isolation of lake basin.	Harrison, 1972; Lowdon & Blake, 1975
27	Boulter Tp., lake *sMA* 345	46°19.3' 79°02.0'	11,500 ± 180 11,800 ± 400	GSC-1429 GSC-1363	gyttja // 329.2 basal gyttja // 329.1	Gyttja over sand. Too old by ca. 2000 yr according to pollen stratigraphy.	Harrison, 1972; Lowdon & Blake, 1979
28	Straits of Mackinac *nwH* 175.8	45°49.1' 84°43.8'	8,150 ± 300 9,780 ± 330	M-2337 M-1996	*Tsuga* root // 138.5 tree stump // 139.2	Tree stump and roots in place in 36.6 and 37.2 m water depth. // Evidence for low-level Early Lake Stanley and Late Lake Stanley-Hough phases.	Crane & Griffin, 1970; 1972
29	Bruce Mines Bog *nwH* 177.4	46°17.8' 83°42.8'	410 ± 160 8,160 ± 220 9,560 ± 160*	GSC-1361 GSC-1359 GSC-1360	basal plant det. // 176.5 plant det. // 180.3 plant det. // 178.6	Grey clay with sand zone beneath plant material (bog) (GSC-1361) over 3 beds (<8 cm) of plant detritus (upper GSC-1359; lowest GSC-1361) with interbeds of grey and blue-grey clay. // Low-level lakes are indicated by plant detritus; dates apply to Early Lake Stanley (9.6 ka) and Lake Chippewa-Hough (8.2 ka).	This study
30	SW. Lake Huron 175.8	44°30.3' 83°08.0'	8,460 ± 180 8,830 ± 410 9,370 ± 180 9,370 ± 220 9,170 ± 140 9,680 ± 110	GSC-1966 GSC-1943 GSC-1935 GSC-1982 GSC-1965 GSC-1983	gyttja // 125.3 top woody peat // 124.5 base woody peat // 124.3 top woody peat // 124.4 base woody peat // 124.2 base woody peat // 124.5	Plant detritus or gyttja over silty clay or sand. // Indicates a long period of emergence during Ottawa-Marquette and Mattawa phases.	This study

Table 1. (continued).

No.	Site, basin area[1], lake elevation m asl	Lat./Long.	Age in years B.P.	Laboratory No.	Material dated // Elevation m asl	Stratigraphy // Comments	References
31	Flummerfelt Basin wG 175.8	45° 22.0' 81° 31.7'	7,740 ± 360 9,770 ± 220	GSC-1847 GSC-1830	Populus wood // 122.7 Salix wood // 122.4	Driftwood enclosed in lacustrine clay with silt in local offshore basin, now deeply submerged. // Suggests low levels (Early L. Hough and L. Chippewa-Nipissing) while shoals adjacent to basin were emergent.	Sly & Sandilands, 1988
32	Middle Islands wG 175.8	45° 16.5' 81° 38.2'	10,305 ± 78	n.r.	wood // ca. 130	Tree parts embedded in lake-bottom sediment. // Not confirmed whether dated samples were in situ; not clear whether emergence is indicated.	Sly & Lewis, 1972
33	Wood Lake nH 218	46° 12.9' 81° 44.1'	9,620 ± 250	GSC-606	basal gyttja // 205	Gyttja over grey clay unusually free of grit or shells at contact. // Suggests isolation of basin by rapidly falling lake levels after Early Mattawa Flood.	Lewis, 1969; Lowdon et al., 1967
34	Smoky Hollow Lake nH 192.7	45° 38.1' 81° 04.3'	9,130 ± 140	I-4036	lowest basal gyttja // 187.7	Gyttja over silty clay. // Indicates possible scouring by Early Mattawa Flood and isolation of basin during Middle Lake Stanley.	Lewis, 1969
35	Nipigon nS	48° 56.5' 88° 23.0'	8,640 ± 130	BETA-9114	plant detritus? // 244	Organics in laminated lacustrine sand and silt deposited on Agassiz inflow sediment fan. // Probably deposited in post Minong lake.	Teller & Mahnic, 1988
36	Rosslyn nwS	48° 21.8' 89° 27.3'	9,380 ± 150	GSC-287	wood // 227	Wood from probable late Minong sediment. // Indicates age and level of probable post Minong nearshore zone.	Drexler & Farrand, 1983; Dyck et al., 1966
37	Grand Marais sS	46° 39.5' 85° 50.8'	9,345 ± 240	GX-4883	wood // n.r.	Probable driftwood in late Minong sediment. // Same as above.	Drexler & Farrand, 1983
38	S. Lake Huron 175.8	43° 53.3' 82° 14.7'	9,350 ± 90	GSC-3656	plant detritus // 107.3	Plant detritus under clayey silt. // Indicates emergence during Mattawa phase.	This study
39	S. Lake Huron 175.8	43° 53.7' 82° 17.2'	8,890 ± 100	GSC-3577	peat // 103.2	Plant detritus and shells in silty clay under silty clay. // Indicates emergence during Mattawa phase.	Woodend, 1983

Table 1. (continued).

No.	Site, basin area[1], lake elevation m asl	Lat./ Long.	Age in years B.P.	Laboratory No.	Material dated // Elevation m asl	Stratigraphy // Comments	References
40	Flummerfelt Patch wG 175.8	45° 21.0' 81° 32.9'	9,440 ± 160	GSC-1397	peat // 144.4	Subaerial peat (J. H. McAndrews, pers. commun., 1970) under laminated clay. // Indicates a lake rise following low-level Middle L. Hough.	Sly & Lewis, 1972; Sly & Sandilands, 1988
41	Tanner Lake nwN 250.5	46° 26.9' 80° 01.0°	9,420 ± 120*	BETA-19151	basal gyttja // 236.4	Gyttja over laminated clay (> 2 m). // Indicates basin isolation after Early Mattawa Flood.	This study
42	South Bay neH 175.8	45° 34.9' 81° 59.5'	8,310 ± 130 9,260 ± 290	GSC-1979 GSC-1971	plant detritus // 160.3 plant detritus // 159.5	Two shelly plant detritus layers and intervening clay, all beneath surficial silty clay, and all overlying firm then laminated basal clay. // Plant-rich layers date two low-level lakes (Middle L. Stanley and L. Stanley-Hough).	Anderson, 1978; Figure 4c, This study
43	Blind River Bog nH 218	46° 12.8' 82° 56.3'	8,760 ± 250	GSC-514	basal gyttja // 214	Gyttja over silty clay over fine sand in basin behind wave-built barrier at 221 m asl. // Minimum age for barrier and lagoon formation, in a high Mattawa phase lake.	Lowdon et al., 1967
44	Wolseley Bay, lake C swN 206.4	45° 10.0' 80° 16.0'	8,110 ± 170*	GSC-1178	basal gyttja // 191.6	Gyttja over clay. // Indicates regression of Lake Mattawa.	This study
45	Monet Lake swN 201.5	46° 09.8' 80° 21.0'	8,250 ± 180*	GSC-1389	basal gyttja // 190.2	Gyttja over silt. // Same as above.	This study
46	Dreany Lake eN 213.5	46° 17.4' 79° 21.8'	8,200 ± 160	GSC-815	basal gyttja // 204	Gyttja over clay. // Same as above.	Lewis, 1969; Lowdon et al.,1971
47	North Bay, lake N eN 211.8	46° 17.4' 79° 20.1'	8,320 ± 170	GSC-821	basal gyttja // 207	Gyttja over varved clay. // Same as above.	Lewis, 1969; Lowdon et al., 1971
48	Trout Mills, delta eN	46° 19.8' 79° 24.2'	8,050 ± 190	GSC-1263	wood // 212	Wood under clay between units of cross-bedded deltaic sands. // Sequence implies 2 low lake phases at delta level (Proto- and early Nipissing) and intervening high level Mattawa phase. Wood dates Early Lake Nipissing.	Harrison, 1972; Lowdon & Blake, 1975

Table 1. (continued).

No.	Site, basin area [1], lake elevation m asl	Lat./Long.	Age in years B.P.	Laboratory No.	Material dated // Elevation m asl	Stratigraphy // Comments	References
49	Amable du Fond R. sMA	46° 11.0' 78° 57.0'	8,750 ± 140	GSC-1097	wood // 242	Woody layer in sand and gravel over varved clay. // Implies alluvial aggradation graded to delta of Mattawa phase, 1.5 km downstream.	Harrison, 1972; Lowdon & Blake, 1975
50	Pure Lake swN 216.4	46° 08.7' 80° 32.7'	8,750 ± 140*	BETA-19153	basal gyttja // 206.0	Gyttja over massive clay over laminated clay. // Indicates falling lake level and basin isolation after a Mattawa-phase flood.	This study
51	Central Lake Michigan 175.8 (a) (b) (c)	43° 10.0' 86° 50.0' / 44° 00.0' 87° 14.0' / 43° 08.4' 86° 48.7'	7,400 ± 500 / 7,570 ± 250 / 7,580 ± 350	M-1571 / M-1972 / M-1736	shell // 71.5 / shell // 69 / shell // 80	Shelly sand above eroded clay and conformably below silty clay. // Shells and overlying silty clay denote deepening of Lake Chippewa-Nipissing following the Late Chippewa low-water phase.	Crane & Griffin, 1965, 1968, 1970
52	Thompson's Harbour wH 175.8	44° 23.0' 83° 36.0'	7,250 ± 300	M-1012	tree stump // 171.3	Tree stump in growth position on lake bottom. // Tree drowned by Nipissing transgression.	Crane & Griffin, 1961
53	Rains Lake nwH 192.7	46° 06.1' 83° 54.1'	7,020 ± 200* / 7,090 ± 150*	GSC-1365 / GSC-1368	basal gyttja // 183.7 / top plant det. // 183.4	Sandy silt over coarse plant detritus and under stiff basal gyttja. // Silt marks Nipissing transgression prior to barrier and lagoon (now Rains Lake) formation.	This study
54	Beaver Island nM 175.8	45° 42.1' 85° 25.2'	6,788 ± 250	M-1888	tree stump (*Pinus resinosa*) // 166.1	Tree stump in growth position on lake bottom, partially sediment buried. // Tree drowned by Nipissing transgression.	Crane & Griffin, 1968
55	Last Lake neS 265.2	47° 52.4' 84° 52.4'	9,220 ± 100	GSC-1851	basal gyttja // 252.2	Gyttja over clay. // Basin isolation by regressing Post Minong lake level.	Saarnisto, 1975
56	Alfies Lake neS 288.3	47° 53.0' 84° 51.5'	9,210 ± 100	GSC-1851	basal gyttja // 272	Gyttja over clay. // Basin isolation by regressing Post Minong lake level.	Saarnisto, 1975
57	Antoine Lake neS 271	47° 54.0' 84° 54.5'	8,830 ± 200	HEL-398	basal gyttja // 259	Gyttja over clay. Same as above.	Saarnisto, 1975
58	Blackington Lake neS 260.9	47° 53.8' 84° 50.4'	8,640 ± 280	HEL-399	basal gyttja // 248.3	Gyttja over clay. // Same as above.	Saarnisto, 1975

72

Table 1. (continued).

No.	Site, *basin area* ¹, lake elevation m asl	Lat./ Long.	Age in years B.P.	Laboratory No.	Material dated // Elevation m asl	Stratigraphy // Comments	References
59	Fenton Lake *neS* 253.3	47° 51.3' 84° 52.8'	8,100 ± 180	HEL-397	basal gyttja // 232.4	Gyttja over clay. // Same as above.	Saarnisto, 1975
60	Crozier Lake *neS* 223.1	47° 54.2' 84° 49.3'	7,590 ± 180	HEL-396	basal gyttja // 210.9	Gyttja over clay. // Basin isolation by regressing post Houghton lake.	Saarnisto, 1975
61	Jock Lake *neS* 290	48° 41.0' 86° 27.0'	9,060 ± 200	HEL-402	basal gyttja // 275	Gyttja over clay. // Basin isolation by regressing post Minong lake.	Saarnisto, 1975
62	Rous Lake pit *neS*	48° 42.0' 86° 02.7'	8,310 ± 100	WAT-1508	wood // ca. 256	Wood in silt of delta bottomset beds. // Deltaic deposition in Post Minong lake, 274 m.	Bajc, 1987

¹ E = Erie, G = Georgian Bay, H = Huron, M = Michigan, MA = Mattawa River Valley, N = Nipissing, O = Ottawa Valley, S = Superior, T = Timiskaming; n = north, e = east, s = south, w = west, ne = northeast, se = southeast, sw = southwest, nw = northwest.
* Date corrected for carbon isotopic composition.
n.r. Not reported.

impounded pre-Main Algonquin shorelines are known from Lake Simcoe area (Deane, 1950) and Straits of Mackinac area (Spurr & Zumberge, 1956). Early low-water conditions with a subsequent rise are also possibly suggested by the recovery of grizzly bear (*Ursus arctos horribilis*) remains, dated 11.7 ka, from beach gravels more than 4 m below the Main Algonquin shore northwest of Lake Simcoe (Fig. 1a, Table 1: 4) (Peterson, 1965; Tovell & Deane, 1966). Though originally thought to be associated with the Post Algonquin Ardtrea strand, it is possible that the dated fossil was exhumed from an earlier pre-Main Algonquin beach deposit which was later occupied, coincidentally, by the Ardtrea water plane. Such evidence is potentially significant but needs verification.

A core from southwestern Lake Huron indicates the presence of Lake Algonquin until about 10.5 ka (Fig. 3a). In this core, a non-erosional transition from glaciolacustrine laminated clay to massive silty clay is coincident with the well-known pollen change from a spruce (*Picea*) – dominated pollen assemblage to one characterized by pine (*Pinus*) dated at about 10.5 ka (Anderson & Lewis, 1974; Karrow *et al.*, 1975).

Further evidence of high Algonquin-age lake levels in the southern Lake Huron basin and their overflow via Port Huron is implied by the stratigraphy of western Lake Erie sediments. An extensive layer of plant material beneath laminated silty clay (Fig. 3b) (Table 1: 14, 15, 16) below about 163 m asl dates between about 12.5 and 11 ka. Abundant pollen of shallow-water plants extracted from this deposit indicate that it accumulated under marsh-like conditions. The plant-rich layer therefore represents a low-water phase; the overlying laminated silty clay was deposited by a subsequent deeper lake phase which flooded and eroded the previous marsh-like habitats after 11 ka. At about 11 ka, too, the marsh environment apparently migrated upslope and became established 1 to 4 m higher (Table 1: 17, 18, 19) above about 163 ml asl during the period of higher water. This higher water is interpreted as a response to new inflow from Huron basin when water levels were raised in that basin; the hydraulic resistance to the resulting increased flow over sills in the island region of western Lake Erie would have caused the increase of water level in western Erie basin.

The foregoing evidence, taken together, suggests that the Main Algonquin phase of the Upper Great Lakes existed about 11–10.5 ka and was contemporaneous with the early part of the Moorhead phase of glacial Lake Agassiz and its eastward discharge. This discharge began about 11 ka and possibly became catastrophic about 10.7 ka when the Laurentide Ice might have retreated sufficiently to expose lower outlets west of Lake Nipigon (Farrand & Drexler, 1985: Fig. 2). The high Main Algonquin shoreline in the northern Huron basin is about the same age, and this correspondence is consistent with the idea that the relatively high Main Algonquin lake and its additional outlet at Port Huron are responses to increased inflow from Lake Agassiz, though much more dating is needed.

The switching of drainage from Agassiz basin to Lake Algonquin at about 11 ka represents roughly a two- to three-fold increase in catchment area for precipitation and glacial meltwater, implying a substantial increase in supply for the Great Lakes water budget. Hansel & Mickelson (1988) have shown that an approximate doubling of drainage area (with appropriate allowances for the relative contributions of precipitation and ice melt) is sufficient to account for a 6 m level change in Michigan basin between the earlier Glenwood and Calumet lake phases. Thus, for an unchanged outlet channel an increase in the level of Lake Algonquin as a result of the addition of Agassiz discharge is to be expected. However it is not possible, at present, to estimate this increase on hydraulic and hydrologic grounds as the division of flow between the Algonquin Fenelon Falls and Port Huron outlet channels is not known, and the channel morphology itself, particularly for the Port Huron outlet, has probably been altered by subsequent erosion. The major response of Lake Algonquin at 11 ka to Agassiz inflow appears to be inauguration of discharge via Port Huron and spillover through Lake Erie, although an increase

LAKE HURON M-17

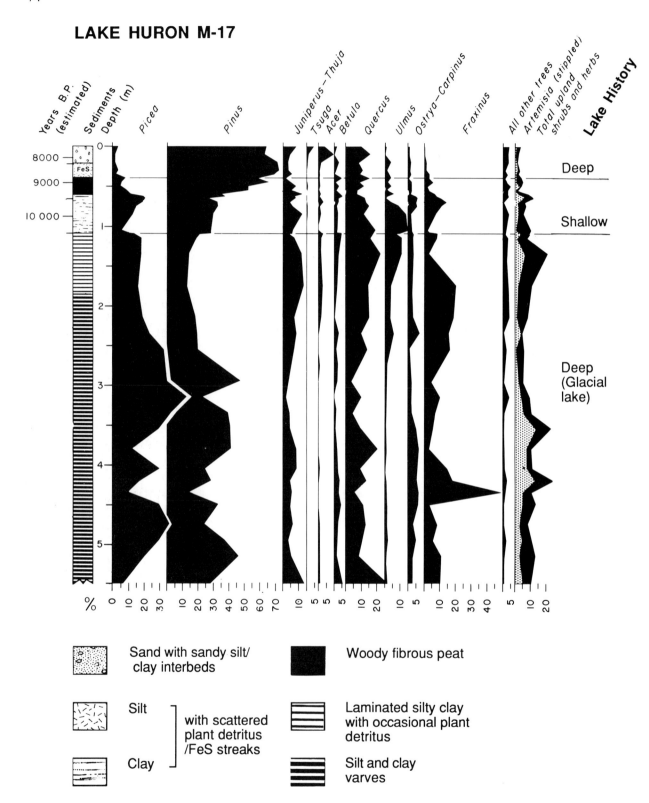

Fig. 3a. Summary lithology and palynological diagrams of a piston core from southwestern Lake Huron (Site 30 on Fig. 1a).

PELEE BASIN CORE 68-6   LAKE ERIE

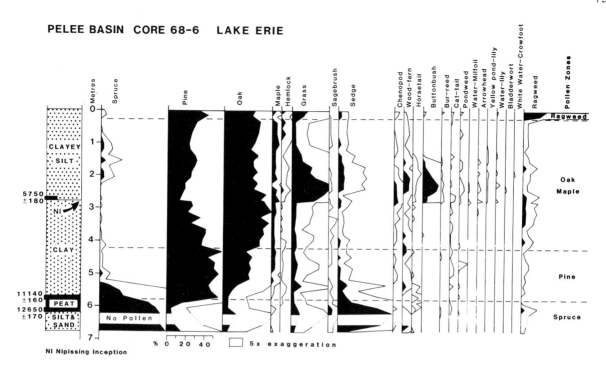

SISTER BASIN CORE 1240
PELEE BASIN CORE 1244 } LAKE ERIE

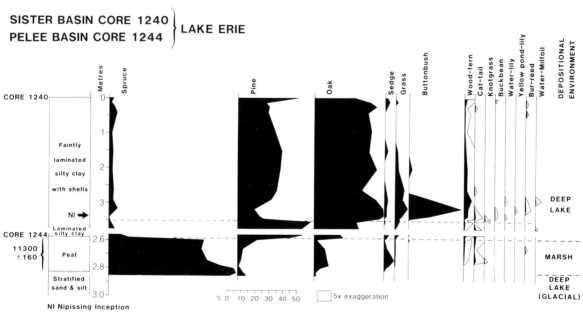

Fig. 3b. Summary lithology and palynological diagrams of piston cores from western Lake Erie showing evidence of marsh (peat) deposition: upper – core 68–6 at site 14 on Fig. 1a (provided courtesy of J. H. McAndrews), and lower – cores 1240 and 1244 located at sites 16 and 15 respectively on Fig. 1a (adapted from Lewis *et al.*, 1966). For peat section, radiocarbon age applies to core 1240 and pollen data to core 1244. In core 1244, the basal part of the peat (plant detritus) dates 12500 ± 310 and the upper part of the peat dates 11140 ± 160 (Table 1: 15). In both diagrams, NI refers to the inception of Nipissing phase overflow via Port Huron into Erie basin.

76

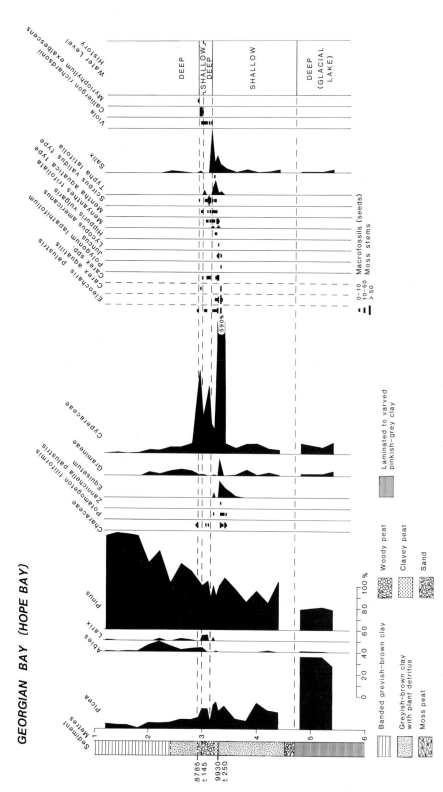

Fig. 3c. Summary lithology, palynological and plant macrofossil diagram of a piston core from Hope Bay in southwestern Georgian Bay (Site 7 on Fig. 1a).

in the order of 5–10 m for Lake Algonquin is compatible with the available geomorphic and stratigraphic evidence of water level change about 11 ka.

As the Agassiz eastward discharge continued to about 10 ka (Teller & Thorleifson, 1983; 1987), this flow must have also supplied the Post Algonquin lakes which drained to Ottawa Valley after 10.5 ka during the retreat or breakup of the Algonquin icedam (Harrison, 1972; Chapman & Putnam, 1984; Ford & Geddes, 1986; Veillette, 1988). After the Post Algonquin phases, low level lakes were established in the Upper Great Lakes basins (Hough, 1955; 1962), illustrated, for example, by peat accumulation in Georgian Bay (Fig. 3c) dated 9.9 ka, by erosional truncation of glaciolacustrine sediments in Huron basin (Fig. 4a), and by river(?) erosion of bedrock surfaces at sills north of Manitoulin Island (Fig. 4b).

Evidence for increased inflows from about 9.6 to 8.3 ka

A Great Lakes response is most clearly suggested for the second period of Agassiz inflow by radiocarbon-dated sediment sequences in the northern Lake Huron basin, at South Bay, Wood Lake and Bruce Mines bog, at Hope Bay in Georgian Bay, and at several places in the Lake Nipissing-Mattawa River basins.

South Bay of Manitoulin Island (Figs 1a: 42, 4c), opening to northeastern Lake Huron by a narrow but deep channel, consists of two separate basins, an inner deep basin and an outer shallower basin separated by a sill 10 m below present water surface. The present water level in the outer basin of South Bay lies about 100 m below the Algonquin water plane, about 45 m above the inferred low-level Stanley water plane (Hough, 1962), and about 19 m below the later Nipissing water plane. Piston coring in the shallower outer basin revealed laminated glaciolacustrine clay overlain by firm silty clay; these units were succeeded by two couplets of inorganic silty clay under plant detritus, all overlain by a surficial unit of silty clay. The plant detritus layers date 9.3 and 8.3 ka (Table 1:

42) and contain abundant pollen of pondweeds, waterlilies, grasses, sedges and buttonbush (*Potamogeton*, *Nuphar*, *Nymphaea*, Gramineae, Cyperaceae, *Cephalanthus*), suggesting that shallow pond conditions existed in South Bay (with correlative low lake levels in Huron basin) at these times. The inorganic silty clay units which alternate with the plant detritus layers imply high lake levels and deposition of suspended sediments instead of shallow pond conditions. For example, under conditions of moderately high levels, such as at present, local waves and shore erosion result in deposition of inorganic silty clay in the outer basin (Kemp & Harper, 1977).

A similar sediment sequence exists at Bruce Mines bog (Fig. 1a: 29) beside the northwestern shore of Lake Huron. The surface of this bog (177.4 m asl) lies about 115 m below the Algonquin water plane, about 35 m above the inferred Stanley low level waterplane, and about 25 m below the Nipissing water plane. The site has recently emerged from Lake Huron owing to continuing glacioisostatic uplift. The surface bog (basal age of 400 years) which is an accumulation of plant material over grey lacustrine clay is evidence of recent terrestrial exposure and of descending (low) relative lake levels. Similarly, two periods of former low water levels at about 9.6 and 8.2 ka (Table 1: 29) are indicated at this site by two beds of radiocarbon-dated plant detritus separated by inorganic lacustrine clay, all of which underlie the grey clay beneath the surface bog. The same low lake levels are suggested by evidence in the Straits of Mackinac where *in situ* tree stumps dated 9.8 and 8.2 ka occur in about 37 m water depth (Table 1: 28). The earlier episode of low levels also correlates with deposition of the lower peat (9.9 ka) at Hope Bay (Fig. 3b) and with onset of gyttja deposition (10.1 ka) in Morel Lake in Mattawa River Valley (Table 1: 26). The inorganic clay units at Bruce Mines bog signify periods of higher lake levels following the low lakes at 10.1–9.6 and 8.2 ka.

A lower limit for the altitude of the higher lake levels following the low stage at 10.1–9.6 ka is indicated by sediments of Wood Lake (218 m asl) north of Manitoulin Island (Fig. 1a: 33). This

78

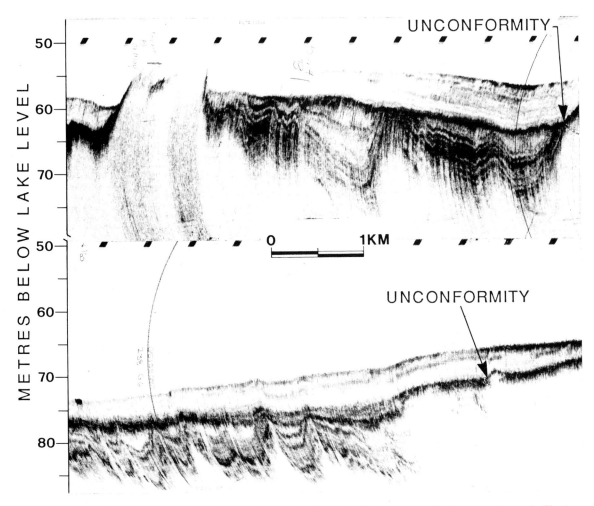

Fig. 4a. Selected evidence of lake level fluctuations in the Huron basin. Location of evidence is shown in Fig. 1a. Lakebed acoustic reflection profiles (14.25 kHz) from the eastern (site N on Fig. 1a) and western (site O) sides of the northern basin of Lake Huron showing pronounced subsurface unconformities at 64 and 73 m below present lake level respectively. This evidence of erosion is interpreted to have arisen by wave abrasion in a former low-level lake.

basin, between hills of quartzite bedrock which rise locally over 270 m asl, contains basal gyttja dated 9.6 ka (Table 1: 33) over a thick section (>4 m) of grey and red clay. The clay and its contact with the gyttja are entirely free of sand and grit, suggesting an episode of sedimentation from suspensates in a large lake which inundated the region. The clay sedimentation and high waters were of relatively short duration (interpreted to postdate the deposition of plant detritus (9.6 ka) at Bruce Mines bog and to predate the accumulation of gyttja (9.6 ka) in Wood Lake and organics (9.3 ka) at South Bay). The high levels

which inundated the South Bay area after 9.3 ka may not have risen to the level of the Wood Lake basin as no further clay beds were observed in the Wood Lake gyttja section.

Hope Bay is one of several embayments along the southwestern Georgian Bay coast (Fig. 1a: 7). The basal 7 m of sediment in a 12 m piston core from the centre of the bay (at water depth 25.6 m) consists of laminated to varved clay of glacial Lake Algonquin. An abbreviated pollen and plant macrofossil diagram (Fig. 3c) shows *Picea* percentages are highest in these glacial sediments and they decline abruptly across the overlying sand

Fig. 4b. Aerial photograph showing a bare bedrock surface (light tone) just north of the town of Little Current, Ontario, on the northern shore of Manitoulin Island. This surface is interpreted to have been scoured by a former river draining the Huron basin into a low phase of the Georgian Bay basin (Little Current Sill, Table 2). Photograph (No. 73–45 41–6–87) courtesy of the Ontario Ministry of Natural Resources.

80

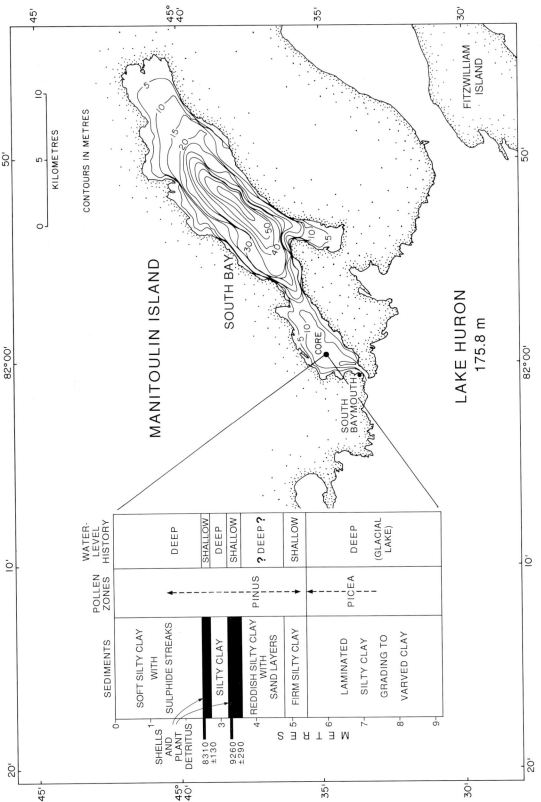

Fig. 4c. Stratigraphy of sediments from South Bay, Manitoulin Island, showing two shelly plant detritus horizons within a mud sequence. The plant-rich zones represent episodes of marsh ponding in South Bay while low-lake levels existed in Huron basin. The firm clay unit at 5 m core depth probably represents an earlier low-water phase following deposition of the underlying laminated clay in a high-level proglacial lake. The intervening mud units are interpreted to reflect major rises in lake level which flooded South Bay and imported suspended sediment or allowed local wave generation, shore erosion and inorganic sedimentation in the Bay.

a) A wave-cut shorebluff in the Gateway district of North Bay, Ontario. The base of this bluff at 236 m asl defines the present elevation of a former lake surface which correlates with other features dated about 8.5–9 ka (see Fig. 7a) (GSC 204626-C). This well-developed shore feature has been traced over 3 km in the North Bay area by Harrison (1972: Map 3-1971).

b) Stratified sand in a delta whose surface at 238 m asl marks the elevation of a former lake in the Amable du Fond River valley, about 21 km southwest of Mattawa, Ontario. A log from alluvium, just 1.5 km upstream and dated 8.7 ka (site 49), suggests this delta and probably the Gateway shore (as above) at a similar elevation were formed in a widespread early Holocene lake which must have inundated the entire Nipissing-Mattawa lowland (GSC 204626).

d) Upper limit at 233 m asl of a ubiquitous boulder lag, suggesting an elevation of former rapid river flow, on the south bank of the Rankin Constriction in Ottawa Valley, about 4.5 km downstream (east) from Mattawa, Ontario. Boulders also occur sporadically up to an elevation of 248 m asl (GSC 204626-A).

◄c) Laminated red clay in a borrow pit 6 m above the classic Nipissing shore level about 10 km east of North Bay, Ontario. The clay, containing an early Holocene pollen assemblage, was possibly transported as suspended sediment during a period of higher lake levels between the pre-Holocene Algonquin and mid-Holocene Nipissing lake phases (GSC 204626-B).

Fig. 5. Selected evidence of an early Holocene high lake (Mattawa) phase in the North Bay-Mattawa area. See Figure 1b for location of evidence.

layer. The sand and overlying clay (with concentrations of plant detritus) accumulated during the post-Algonquin period of reduced water levels. Much of the clay is probably reworked glacial clay that was washed into the basin by slope wash from nearby high elevations. Once the slopes around the site had become vegetated, clay deposition ceased in the basin and ponding was initiated as indicated by plant macrofossil occurrences of the Characeae (*Chara*) and the pondweeds, *Potamogeton filiformis* and *Zannichella palustris* in the lower half of the peat. Substantial peaks in spores of *Equisetum* (horsetail), and pollen of Gramineae (grass) and Cyperaceae (sedge) and seeds of *Eleocharis palustris* (spikerush), *Carex aquatilis* (sedge), *Polygonum lapathifolium* (smartweed), *Juncus* (rush), *Lycopus americana* and *Mentha aquatica* type (water mints), *Hippuris vulgaris* (mare's tail) and *Menyanthes trifoliata* (buckbean) show that the shallow nearshore areas of the pond were well vegetated with emergent aquatic plants. The upward appearance of seeds of *Scirpus validus* type (bullrush), *Typha latifolia* (cat-tail) and peak percentages of *Salix* (willow) pollen indicate increased shallowing and gradual transformation to a reed/shrub-dominated swamp.

The abrupt decline of *Salix* mid-way in the peat (Fig. 3c) reflects an interval of higher water levels which drowned out the inferred reed/shrub swamp at about 9.6 to 9.5 ka (based on the assumption of a constant sedimentation rate between the dated levels at top and bottom of the peat). The lack of fossils of shallow-water plants except redeposited seeds and seed fragments of *Carex*, *Menyanthes trifoliata* and *Scirpus validus* type) in the upper part of the peat supports such an interpretation.

Water levels fell again at Hope Bay according to fossil evidence from near the top of the peat (Fig. 3c). Moss remains of *Calliergon richardsonii* suggest the site had become either a wet but emergent peaty habitat with temporary pools or less likely a shallow pond (Unpublished GSC Peat Report LO 25 by L. Ovenden). This second episode of reduced water levels (prior to the late Mattawa and Nipissing transgressions which deposited the overlying clay sequence) is also implied by a resurgence in Cyperaceae and *Typha latifolia* pollen and in seeds of *Carex aquatilis*. The clayey peat layer at the top of the peat layer suggests that the peat record was eroded and thus does not resolve the possible re-emergence of the site following Lake Mattawa and prior to the Nipissing transgression, as described later.

Wave-cut shore bluffs which probably relate to the post −9.3 ka high water levels inferred from the South Bay stratigraphy were found at three separate sites in the North Bay area. A well-developed shore terrace and bluff is located in the Gateway district of North Bay at 236 m asl, with its type exposure in St. Mary's Cemetery (Fig. 1b: near North Bay; Fig. 5a). Correlative shorebluffs were found 5 km to the west at Duchesnay River (236 m) and 5 km to the east at Trout Mills (237 m) (Fig. 1b). The lake implied by these shorebluff occurrences appears to correlate with a delta (Fig. 5b), a further 41 km southeast, in the Amable du Fond River valley at 238 m asl where a log in gravel alluvium above varved clay, just 1.5 km upstream, was dated 8.7 ka (Fig. 1b, Table 1: 49) (Harrison, 1972). In addition, laminated red clay (Fig. 5c) containing an early Holocene pollen assemblage was found in a borrow pit well above (6 m) the classic Nipissing shore 10 km east of North Bay, Ontario. The occurrence of the foregoing features suggests the presence of an early Holocene, relatively high-level lake in the Nipissing-Mattawa lowland.

The termination of this moderately high, early Holocene lake phase is indicated by the onset of gyttja sedimentation in small lake basins of the Nipissing lowland. Basal gyttja samples overlying silt and laminated clay in four lakes at relatively low elevations in the Nipissing lowland (Wolseley Bay lake (Table 1: 44), Monet Lake (Table 1: 45), Dreany Lake (Table 1: 46) and lake 'N' at North Bay (Table 1: 47)) are dated between 8.1 and 8.3 ka. As these sites are at or above the local elevation of the Nipissing waterplane, the basins should have been exposed immediately after the Post Algonquin phase almost 2000 years earlier. The delay in onset of gyttja deposition in these lakes is attributed to inundation by a moderately high, early Holocene lake phase, characterized by dominantly inorganic sedimentation implied by the silt and laminated clay found below the gyttja. This inorganic sedimentation in the Nipissing lowland correlates with the inorganic silty clay between beds of plant detritus in the northern Lake Huron area at South Bay and Bruce Mines bog. And in both regions the correlated onset of the subsequent organic sedimentation is interpreted to result from regression of the early Holocene moderately high lake.

Overall, the evidence from South Bay, Bruce Mines bog, Wood Lake and Hope Bay suggests that the northern Huron and Nipissing region was inundated by a high lake at about 9.6 ka. The data further suggest that the high lake was short-lived and subsided sufficiently to permit organic sedimentation in Wood Lake after 9.6 ka and in South Bay by 9.3 ka. Lake levels transgressed South Bay again and remained moderately high throughout the Huron-Nipissing region until after 8.4 ka when a substantial regression induced widespread organic sedimentation in small lake basins. This period of generally high but variable lake phases in the Upper Great Lakes correlates with, and is attributed to, the catastrophic drainages of glacial Lake Agassiz into Nipigon and Superior basins from about 9.5 to 8 ka as described by Teller & Thorleifson (1983, 1987) and by Clayton (1983).

Evidence for lake control at the Gateway level

With the Nipissing-Mattawa lowland long deglaciated, the question arises: what held the lake waters up to the level of the Gateway shore at North Bay? Possibly, the earliest high-level flows were held up by residual morainic dams in the Ottawa Valley, for example at Bissett Creek (Veillette, 1988), but these unlithified deposits would have likely been quickly breached and channelized by the high flows. The classic North Bay outlet for the Nipissing phase (213 m asl) at the head of Mattawa River (Taylor, 1897; Harrison, 1972) is an inadequate control for the higher Gateway level. However, other constrictions in the Mattawa and Ottawa River

valleys are evident on topographic maps of the North Bay-Mattawa region (Fig. 1b). For example, the valley contours at 244 m asl narrow at Rutherglen and become even more constrictive in the Ottawa Valley 5 km downstream of Mattawa, towards Rankin. The constriction near Rankin is unique and especially severe; the valley width is only 990 m at 250 m asl, 100 m above the present Ottawa River (Fig. 6). There, too, the valley is extraordinarily deep, down to 97 m below the present river level, suggesting scour by a rapid river flow. Above 250 m elevation, however, the valley width increases rapidly, and at 300 m, the valley is 9.5 km wide. Although the area is now heavily forested, large boulders appear to be common on the south slope of the bedrock constriction up to about 233 m asl (Fig. 5d); the limit of sporadic boulder occurrence is about 248 m asl.

A possible discharge capacity for this section at the assumed equilibrium or base flow elevation of 244 m asl, which is equivalent to the Gateway shore and Amable du Fond delta levels, has been tentatively estimated in the range of 210 000 to 600 000 $m^3 s^{-1}$. This discharge is based on the contracted-width method of calculation in which the Rankin Constriction is modelled as a fully eccentric contraction of Type III with rough valley side slopes of 2 to 1 and a Manning resistance coefficient of 0.04 (Chow, 1959; Benson, 1968). The higher discharge value is probably excessive as it assumes a maximum slope of paleoflow through the constriction equal to the full 11-m difference in altitude between inferred lake level (244 m) and highest common boulder occurrences (233 m) observed in the constriction. The lesser estimate was obtained using an average river gradient (1.1 m/km) between Rankin and the Chalk River area (Catto *et al.*, 1982; see below) and may be more realistic.

A second estimate of discharge amounting to about 250 000 $m^3 s^{-1}$ was obtained by applying the slope-area method to a section of channel below 159-m elevation about 75 km downstream from the Rankin Constriction in the Chalk River area. There, Catto *et al.* (1982) had mapped river terraces and determined that the 159-m terrace

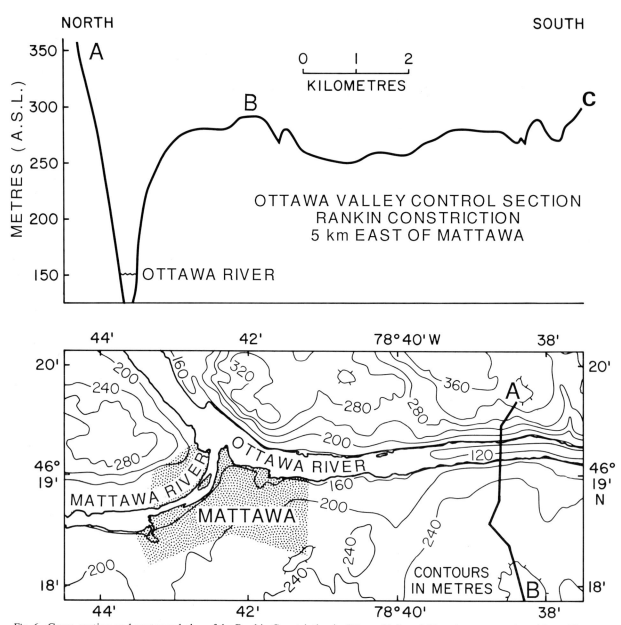

Fig. 6. Cross-section and contoured plan of the Rankin Constriction in Ottawa Valley, 4.5 km downstream (east) from Mattawa, Ontario. Mattawa is located at the junction of Mattawa River valley (from left) and Ottawa Valley (top left to right). Note that the plan illustrates only the deeper portion AB of the valley profile ABC.

probably related to river flows between 9.5 and 8.0 ka. This level was also correlated by Catto *et al.* (1982) to the 137-m terrace a further 60 km downstream which had been earlier mapped and related to a 9.5-ka bog by French & Hanley (1975). Catto *et al.* (1982), however, calculated a much lower discharge for this terrace level, but the reasons for the discrepancy with our estimates are

not clear. Because we have obtained comparable estimates at different sections of the outlet valley using different methods and because one of our methods is clearly similar to those employed by Teller & Thorleifson (1983) who estimated Agassiz inflow to the Great Lakes, we have adopted the preliminary estimate of 200 000 m³s⁻¹ for discharge at the Rankin Constriction as a basis for comparison, pending a more detailed evaluation.

Though large, this preliminary value is reasonable as it represents the base discharge from Lake Agassiz with additions contributed by runoff from the Upper Great Lakes and meltwater from the rapidly wasting Laurentide Ice Sheet bordering Lake Ojibway in northern Ontario and Quebec. The actual discharge would likely have originated as a combination of net ablation, annual precipitation and storage reduction in the upstream basins. The base flow estimated above may be a seasonal maximum resulting from summer ice melt; the winter flows and winter lake levels could have been much less.

Interpretation of water planes and lake history

The radiocarbon-dated sequences implying former lake levels within the Upper Great Lakes and Lake Nipissing system are listed in Table 1. These data are projected along their isobases and plotted as cross-sections in Figs 7a and 7b. The surfaces of the modern lakes and the well-known inferred water planes for the Lakes Main Algonquin, Minong (Superior basin only) and Nipissing Great Lakes (Fig. 2) are also shown for reference. The interpretation and placement of other water planes are controlled by 1) the alignment of water level indicators of common age, 2) the position and elevation of potential controlling outlets, and 3) an assumption that successively younger water planes should decrease in slope. The estimated accuracy of the water-plane reconstructions is about ± 2–5 m depending on the density of control data. Some water planes are controlled by sills on the low parts of divides (saddles) between local basins. The sills which could have exerted control upon levels within the Upper Great Lakes system are listed in Table 2. These sills were selected following a comparison of the computed uplift histories of many possible controls identified in an examination of detailed topographic maps and bathymetric charts (Lewis & Anderson, 1985). The uplift histories of the key outlets are plotted in Fig. 7c relative to Port Huron, con-

sequently the Port Huron curve appears static (level). These uplift histories assume exponential uplift, a commonly applied function for describing glacioisostatic rebound of a formerly depressed zone inside its peripheral area of a potential forebulge (Andrews, 1968a, 1968b; 1970; Walcott, 1970, 1972; Anderson & Lewis, 1985: 252–253; Lewis & Anderson, 1985; Larsen, 1985b, 1987).

The relationship among ice margin, lake outline and discharge routes, which have evolved since about 11.5 ka, are depicted in a series of paleogeographical reconstructions at selected times (Fig. 8) and in a table of correlations of lake phases (Fig. 9). The ice margin positions have been largely interpolated from maps of Laurentide Ice Sheet retreat by Dyke & Prest (1987) except in the Nipissing-Mattawa area where our interpretations are based on reconstructions by Ford & Geddes (1986) and Veillette (1988). Anomalously old radiocarbon dates (11.8 and 11.4 ka) on basal gyttja from Boulter Township lake are rejected (Table 1: 27). The criteria for lake phases as correlated in Fig. 9 and shown in Fig. 8 are essentially based on interpreted changes in outlet control, namely 1) a significant change in lake level at an outlet, as occurred, for example, in the transition from Early Mattawa Flood to Early Lake Mattawa, or 2) a change of outlet or use of additional outlets, as occurred in the change from Kirkfield to Main Lake Algonquin. The latter criteria was commonly met by changes in outlet dominance, for example, when an uplifting outlet backfloods upstream lake outlets and brings water levels into confluence in contiguous basins, as occurred during the Mattawa and Nipissing transgressions. The changes in outlet control were deduced from the regional stratigraphic evidence of water-plane succession (Figs 7a, 7b) and from the evidence of outlet dominance based on a comparison of the uplift histories of outlets (Fig. 7c). The detail of some lake phase transitions shown in Fig. 9 are theoretical, based on histories of outlet uplift, and thus are sensitive to the exponential parameters of uplift.

86

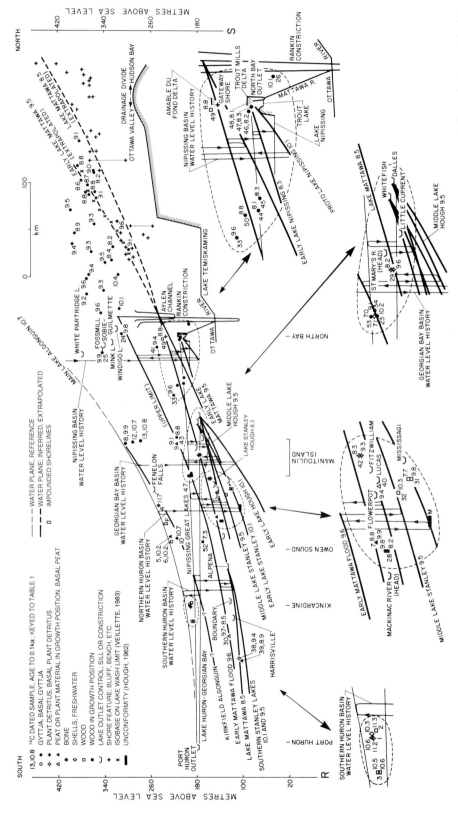

Fig. 7a. Former water planes interpreted for the Huron, Georgian Bay and Nipissing basins. Line of section (RS) for this figure, perpendicular to isobases (Fig. 2a) of glacioisostatic uplift, is shown on Fig. 1a. North of North Bay, the basal lake gyttja dates are from Veillette (1983, 1988). Numerals beside the dated symbols refer to entries in Table 1.

87

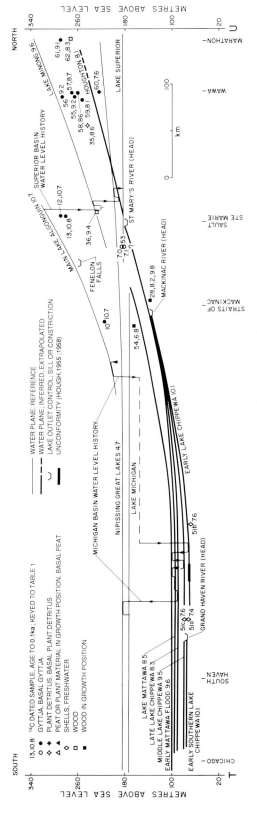

Fig. 7b. Former water planes inferred for the Michigan and Superior basins. Line of section (TU) for this figure is plotted on the location map, Figure 1a, perpendicular to the Algonquin and Minong isobases (Fig. 2a). Numerals beside the dated symbols refer to entries in Table 1.

88

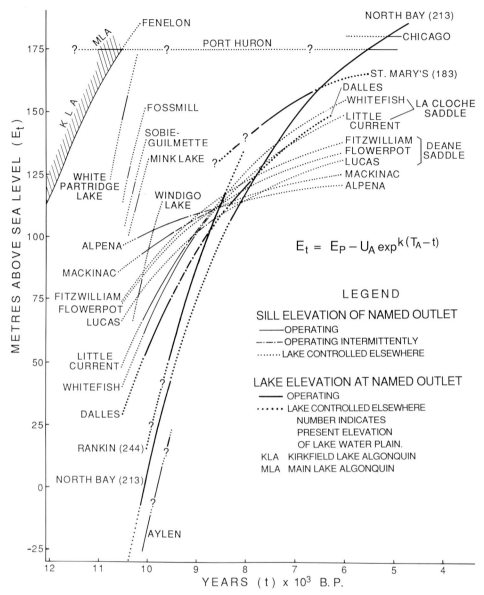

Fig. 7c. Uplift histories of lake outlet sills (bases of outlet channels) (Table 2). For some outlets, the water-plane elevation was better known or more meaningful than the sill elevation; for these, the relevant water-plane uplift history is illustrated instead, and the outlets are identified by a numeral representing the present elevation of the water plane at the outlet. The period during which an outlet may have operated as a significant control of a lake is indicated. The expression for elevation, E(t), of a sill at time, t, is from Lewis & Anderson (1985). E(p) is the present elevation of the sill and U(A) is the excess elevation of the Algonquin water plane at the sill relative to that at Port Huron (see Table 2). The relaxation coefficient, k (-0.38 ± 0.01), was computed from the differences in elevations of the Algonquin and Nipissing water planes between North Bay and St. Mary's River, Mackinac River, Whitefish, Lucas and Fenelon Falls outlets by the procedures of Anderson & Lewis (1985: 252–253).

Kirkfield water plane 11.2 ka (Figs 7a, 8a). The earliest water plane discussed in this paper, the Kirkfield, is interpreted in Fig. 7a to lie below the Port Huron outlet at about 140–150 m asl. This is based on the uplift history of its outlet, Fenelon Falls (Fig. 7c). It then passes just below the 11.2–11.3-ka bog deposits (Table 1: 1, 2) in southern Huron basin and the pre-Main Algon-

Table 2. Key sills for control of drainage through Upper Great Lakes region.

Name and *basin* area	Present sill elevation m asl	Interpreted water plane elevations	
		Algonquin m asl	Nipissing m asl
Fenelon Falls *seG*[1]	253	268	194
Port Huron *sH*	175 i	164	183
White Partridge Lake *wO*	380	435	n.d.
Fossmill *sMA*	347	435	210
Sobie-Guillmette *sMA*	341	445	211
Mink Lake *sMA*	329	440	211
Windigo Lake *wO*	320	480	n.d.
Aylen *O*	220	495	n.d.
North Bay (east end of Trout Lake) *eN*	213 w	450	213
Rankin *O*	244 w	484	215
Dalles *neG*	178 w	345	202
Whitefish *neH*	180	335	201
Little Current *neH*	171	315	199
Fitzwilliam *neH*	152	268	194.5
Lucas *neH*	145	268	194.5
Flowerpot *neH*	147	263	194
Mississagi *nwH*	125	276	195.5
Alpena *wH*	126	217	188
Boundary *cH*	122	204	186
Harrisville *wH*	116	198	185
Mackinac River (head) *nM*	134	235	190
Grand Haven River (head) *eM*	74	n.d.	n.d.
Chicago *swM*	180	n.d.	n.d.
St. Mary's River (head) *seS*	183 w	304	198

[1] E = Erie, G = Georgian Bay, H = Huron, M = Michigan, MA = Mattawa River valley, N = Nipissing, O = Ottawa Valley, S = Superior, T = Timiskaming; n = north, e = east, s = south, w = west, ne = northeast, se = southeast, sw = southwest, nw = northwest.
i Inferred.
w Mean or inferred lake elevation related to outlet, not outlet threshold elevation.
n.d. Not determined.

90

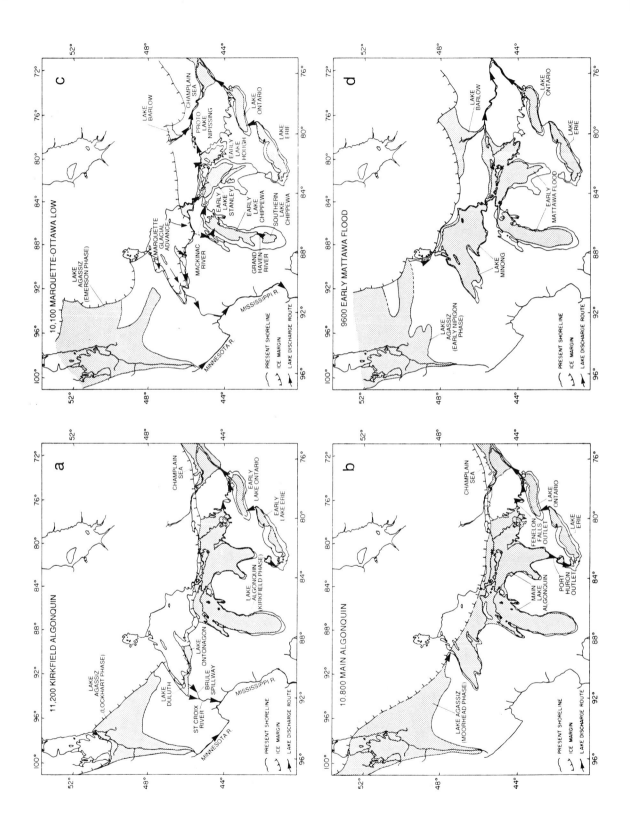

Fig. 8. Paleogeographic maps showing interpreted shorelines of former lake phases. The elevations of the former lakes are given in Figs 7a and 7b. Figure 9 illustrates the interpreted ages and durations of the former lake phases.

a) Kirkfield Algonquin lake at a relatively late stage at about 11.2 ka when ice had retreated north of the Straits of Mackinac area, allowing waters to become confluent in the Michigan, Huron and Georgian Bay basins. Drainage was entirely southeastward via the Fenelon Falls outlet to Ontario basin and thence to Champlain Sea in the St. Lawrence Valley.

b) Main Lake Algonquin high phase at about 10.7–10.8 ka showing inflow from the Moorhead phase of Lake Agassiz and outflow via Fenelon Falls and via Port Huron to Erie basin. Note the broad open connections allowing relatively unrestricted circulation between Superior basin and Michigan and Huron basins.

c) Ottawa-Marquette low phase at about 10.1 ka caused by withdrawal of an ice dam in the Mattawa-Ottawa area and drawdown of the previous Algonquin phases, and by advance of ice in the Superior basin, preventing inflow from Lake Agassiz.

d) Early Mattawa Flood phase about 9.6 ka caused by retreat of ice in the Superior basin to its northeastern sector allowing a rapid drawdown of glacial Lake Agassiz (Nipigon phase) into the Upper Great Lakes. This large flow augmented by inflow from glacial Lake Barlow in Temiskaming basin was resisted hydraulically at the Rankin Constriction in Ottawa Valley, causing a major, but presumedly short-term, increase in lake level.

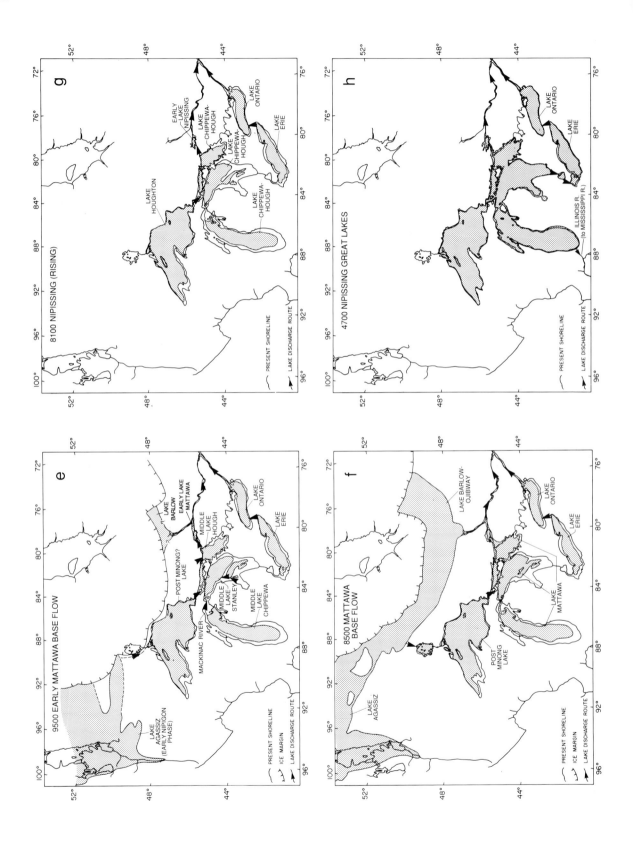

Fig. 8. e) Early Mattawa phase shown for base flow conditions about 9.5 ka after the effects of the initial drawdown of glacial Lake Agassiz (Nipigon phase) has ceased.

f) Late Mattawa phase shown for base flow conditions about 8.5 ka with inflow from glacial Lake Agassiz and backflow from glacial Lake Barlow-Ojibway controlled by hydraulic resistance at the Rankin Constriction, Ottawa Valley.

g) Early Nipissing low phase about 8.1 ka caused by the cessation of large meltwater inflows to the Great Lakes system when breakup of the Laurentide Ice Sheet induced drawdown of Lakes Agassiz and Ojibway to Hudson Bay.

h) Nipissing Great Lakes culmination about 4.7 ka caused by differential uplift of the classic North Bay outlet and the transfer of drainage to the southern outlets at Chicago and Port Huron.

94

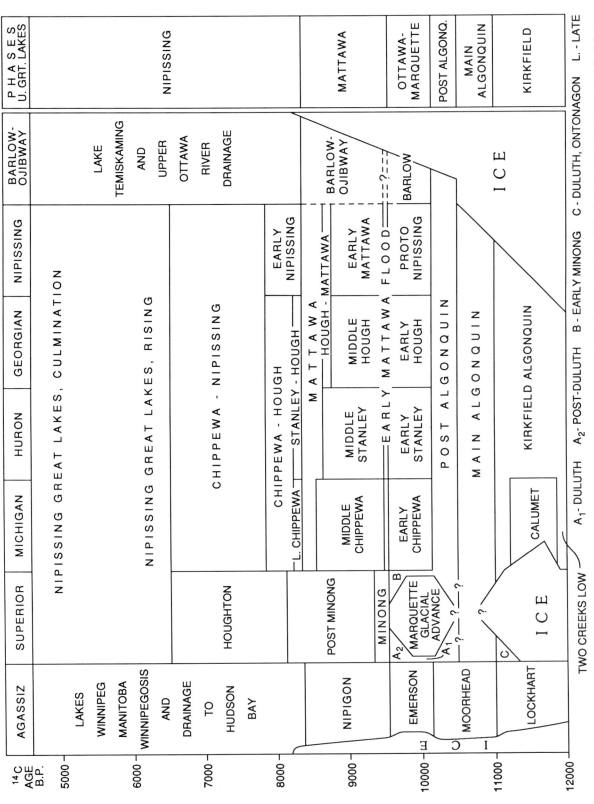

Fig. 9. A correlation table of interpreted lake phases in the Upper Great Lakes basins and the adjacent Agassiz and Barlow basins. Lake histories of the Mattawa phase were likely punctuated by catastrophic floods as successively lower outlets from glacial Lake Agassiz opened. Except for the first, the Early Mattawa Flood, the floods have not been interpreted. Many lake phase transitions are theoretical, based on outlet-uplift histories, and thus, are sensitive to the exponential parameters of uplift, especially in the period 9 to 7.5 ka when outlet dominance changed relatively rapidly.

A₁ - DULUTH   A₂ - POST-DULUTH   B - EARLY MINONG   C - DULUTH, ONTONAGON   L - LATE

quin beach strands in Georgian Bay (Fig. 1a: F). The Kirkfield water plane passes well above (>30 m) the Lake Huron site (Fig. 3a) of Kirkfield-age glaciolacustrine sediments. Finally, it projects between the elevations of the 11.7-ka grizzly bear remains (Table 1: 4) and the lake outlet at Fenelon Falls.

Main Algonquin water plane 10.7–10.8 ka (Figs 7a, 7b, 8b). This water plane is a smoothed surface through shore bluffs, terraces and beaches attributed to Main Algonquin which is mapped throughout much of the Georgian Bay, Huron, Michigan and extreme southeastern Superior basins by those persons listed in the caption to Fig. 2a. Our reconstruction of Main Lake Algonquin which is based on isobase trends of Main Algonquin-age shore features and the new evidence for discharge at both Fenelon Falls and Port Huron differs from the interpretations of Raphael & Jaworski (1982), Kaszycki (1985) and Larsen (1987) which deny the Port Huron outlet. The 'highest' Algonquin shore features are of slightly different age, for example, slightly older in the north and slightly younger in the south, consistent with the generally held view that relative levels of Lake Algonquin fell (regressed) north, and rose (transgressed) south of the outlet at Fenelon Falls owing to differential crustal rebound (Finamore, 1985; Kaszycki, 1985).

Low-level water planes about 10.1 ka (Figs 7a, 7b, 8c). Following the Main Algonquin phase, lake levels fell rapidly through the Post Algonquin phase as the ice dam retreated and morainic dams were eroded in the Mattawa-Upper Ottawa Valley (Prest, 1970; Harrison, 1970; 1972; Chapman, 1975; Chapman & Putnam, 1984; Ford & Geddes, 1986; Veillette, 1988). As a result of the new outlet to Ottawa Valley and the reduction of Agassiz inflow caused by the Marquette glacial advance in Superior basin, low phases were established by about 10.1 ka in each sub-basin of the Upper Great Lakes region. The water planes of these phases are shown in Figures 7a and 7b with their controlling outlets. Early Lake Stanley is defined by Hough's unconformity

(1962) (Fig. 1a: M) and the lowest sill (Whitefish, Table 2) allowing drainage at the time. The water plane just clears the Mississagi Sill (Table 2) between Lake Huron and North Channel west of Manitoulin Island. The water plane for the proto-Lake Nipissing is drawn with a similar slope to intersect the level of the lower cross-bedded unit of the Trout Mills delta (Table 1: 48) (Harrison, 1972) in Nipissing basin. In Michigan basin, the low-level southern Lake Chippewa and Early Lake Chippewa water planes are tied to their respective controlling outlets, the heads of Grand Haven and Mackinac rivers (Stanley, 1938; Hough, 1955, 1958). The water plane for Early Lake Hough is drawn more steeply from its controlling sill, inferred to be the Dalles Sill (Table 2) northeast of Georgian Bay, to account for indications of low-lake levels (Tovell *et al.*, 1972), including the occurrence of subaerially exposed glaciolacustrine clay about 100 m below present lake level in western Georgian Bay near site 40 (Fig. 1a) (Sly & Sandilands, 1988).

Early Mattawa Flood water plane 9.6 ka (Figs 7a, 7b, 8d). This surface is drawn well above (>20 m) the grit-free grey clay of Wood Lake and Tanner Lake (Table 1: 33, 41), allowing suspended sediment into these northern basins. The presence of continuous organic sedimentation from 9.7 and 9.8 ka in small basins in southern Lake Huron (Fig. 1a, Table 1: 30) and near North Bay (Fig. 1b, Table 1: 24) respectively, provide an upward limit for this water plane (Fig. 7a). The water plane also passes well above the Hope Bay site in southern Georgian Bay where an abrupt decline of *Salix* suggests an episode of inundation (Fig. 3c). This high level correlates with the initial drawdown of Lake Agassiz's Nipigon phase. Early Mattawa Flood, like Early Lake Mattawa and Lake Mattawa to be described next, is named after the town of Mattawa located just upstream of its presumed outlet control, the Rankin Constriction in Ottawa Valley.

As noted previously, the water plane of the Early Mattawa Flood appears to be higher than that of subsequent floods. While this might be caused by a larger drawdown and greater dis-

glacial meltwater, a long period of relatively stable lake levels was inaugurated.

A water plane of similar slope to that of the earlier Lake Mattawa water plane and slightly younger age (about 8.3 ka) is clearly indicated in the Nipissing basin by re-establishment of deltaic sedimentation over wood at 213 m asl at Trout Mills (Table 1: 48) and the inauguration of gyttja sedimentation in 'C', Monet, Dreany and 'N' lakes (Table 1: 44, 45, 46, 47). This water plane coincides with the well-known North Bay outlet and 'Nipissing beach' (Taylor, 1897; Leverett & Taylor, 1915; Lewis, 1969; Harrison, 1972). It marks the earliest phase of the traditional Nipissing transgression, hence we apply the name Early Lake Nipissing to it. A water plane for the contemporaneous low phase in the Georgian Bay and northern Huron basins (Lake Stanley-Hough) is inferred at the Dalles outlet. This lake backflooded the Huron basin via a narrow connection at the Lucas Sill (Fig. 7a). The latter sill is located on the broad 'low' between Huron and Georgian Bay basins southeast of Manitoulin Island, herein termed the Deane Saddle (Fig. 1a). The Lucas Sill may have co-operated with other low points on the Deane Saddle, for example, the Fitzwilliam and Flowerpot Sills, in controlling water levels in the Huron basin immediately following the regression of Lake Mattawa, and in providing a connection with Georgian Bay waters during the ensuing Nipissing transgression. The 8.3-ka water plane passes below the upper plant detritus bed at South Bay (8.3 ka) and below the stump dating 8.2 ka in the Straits of Mackinac. Water planes for separate water bodies, which would have existed in southern Huron basin, impounded by the Alpena Sill, and in Michigan basin, impounded by the head of Mackinac River (Late Lake Chippewa), were nearly similar to the previous Lake Mattawa surface. Hence the stratigraphic data in southern Lake Huron and Lake Michigan show a long period of slow transgression without a distinctive regression between the 'Mattawa' and 'Nipissing' transgressions. The contemporaneous 'Houghton' water plane (Farrand & Drexler, 1985) is illustrated in the Superior basin (Fig. 7b).

Differential rebound caused lakes with outlets in northern sectors of their basins to backflood and slowly transgress upstream lakes (Fig. 7c, 9). By 8.1 ka, the Dalles Sill had uplifted sufficiently to backflood its upstream basins, Georgian Bay, Huron and Michigan. The shoreline of the resulting water body, Lake Chippewa-Hough, is illustrated in Fig. 8g.

Nipissing transgression after 8.1 ka and Nipissing Great Lakes about 4.7 ka (Figs 7a, 7b, 8h). Continued rebound of northern outlets caused lakes to transgress their basins, so that by 7.7 ka all the Upper Great Lakes, except Superior, had coalesced with waters of the Nipissing basin. Then water levels rose more rapidly throughout the Michigan and Huron basins under the influence of the controlling North Bay outlet, an event marked in the Michigan basin by the burial of shallow-water molluscs (Table 1: 51) under deep-water mud. The Nipissing transgression back-flooded St. Mary's River about 6.5 ka and finally brought all the Upper Great Lakes to a common level (Fig. 7c).

The Nipissing Great Lakes reached their maximum elevation and began to regress about 4.7 ka when differential uplift finally completed the transfer of discharge from North Bay to southern outlets at Chicago and Port Huron. The age of the Nipissing culmination is well established by radiocarbon-dated detrital organics in alluvium and shore sediment, transgressed *in situ* plant remains and basal gyttja from Nipissing- and post-Nipissing-age deposits (Dreimanis, 1958; Lewis, 1969, 1970; Cowan, 1978; Terasmae, 1979; Karrow, 1980, 1987; Larsen, 1985a, b and Monaghan *et al.*, 1986). Some dates which had been interpreted previously to suggest an earlier age of Nipissing culmination were re-evaluated and found to be of less significance than previously thought. A 5.5-ka stump on Manitoulin Island (Lewis, 1969) was found to be rooted on compressed peat under dune sand and thus was probably living at a higher relative elevation than previously thought; this level is correlative with a late, but not final, phase of the Nipissing trans-

charge from Lake Agassiz, we postulate that exceptionally high initial flow levels were also supported by greater hydraulic resistance in the Ottawa Valley until glacial deposits in the Rankin Constriction and morainic dams, for example, at Bissett Creek (Veillette, 1988) were scoured away. The Early Mattawa Flood water plane cleared the Mackinac River channel and penetrated the Michigan basin (Fig. 7b). It also cleared the lowland at Sault Ste. Marie and probably existed in the extreme eastern Superior basin. The Minong water plane, recognized by a shoreline throughout the remainder of the Superior basin, is of similar age and elevation (Fig. 7b).

Early Mattawa and correlative low-level water planes 9.5 ka (Figs 7a, 7b, 8e). Several floods ensued at intervals as new lower outlets for Lake Agassiz were opened during the following 1000 years (Teller & Thorleifson, 1983). We interpret an equilibrium or base flow between floods which was controlled at about the 240–250-m level at the Rankin Constriction in Upper Ottawa Valley. Water planes for an early (e.g., 9.5 ka) and a late (8.5 ka) base flow condition are illustrated in Figs 7a and 7b. Because of its similar age the slope of the 9.5-ka water plane is similar to that of the 9.6-ka water plane. Thus the Early Lake Mattawa plane is defined by its slope and its interpreted altitude at the Rankin Constriction. It is located entirely within the Nipissing-Mattawa basin. Similarly in the Georgian Bay basin, the correlative Middle Lake Hough plane is defined and tied to its outlet sill at Dalles. This position is supported by the presence of terrestrial peat on Flummerfelt Patch (Table 1: 40, 9.4 ka) and driftwood (Table 1: 31, 9.4 ka) in Flummerfelt basin. In the Huron basin, the correlative water planes of Middle Lake Stanley and the southern Stanley-age lakes are inferred to be controlled at Whitefish and at Alpena, respectively. Small water bodies (water planes not shown) were also impounded by Harrisville and Boundary Sills in southern Huron basin. The correlative Michigan basin level, Middle Lake Chippewa, is controlled at the head of the Mackinac River (Fig. 7b).

Lake Mattawa water plane 8.5 ka (Figs 7a, 7b, 8f). The base-flow water plane at about 8.5 ka is defined by dated stratigraphic sites. This surface must pass above a sill between internal basins of South Bay (Fig. 4c, Table 1: 42) on Manitoulin Island to allow wave development and mud sedimentation there. To the north it is tightly constrained. The water plane must pass beneath the site (Table 1: 50) of Pure Lake where gyttja sedimentation was in progress, and above the sites of Bruce Mines bog (Table 1: 29), Wolseley Bay lake 'C' (Table 1: 44) and Monet Lake (Table 1: 45) where peat and gyttja sedimentation were suppressed until 8.1–8.3 ka. In Georgian Bay, the water plane passes over the Hope Bay site (Fig. 3c) where inorganic sedimentation had succeeded peat deposition about 8.8 ka (Table 1: 7). This surface projects further north directly to the Gateway shore at North Bay and to the 240–250-m level in the Rankin Constriction of Ottawa Valley. Thus, this surface defines an extensive waterbody at a common level throughout the Huron, Georgian Bay and Nipissing-Mattawa River basins which is here called Lake Mattawa after the town near its outlet. To the south and west, Lake Mattawa flooded the stump sites (Table 1: 28) in the Straits of Mackinac and, just barely, the basins of Lake Michigan and southern Lake Huron (Figs. 7a and 7b). Lake Mattawa nearly penetrated the Superior basin; its water plane passes just 2–3 m below the present level of water impounded by the St. Mary's River Sill (Fig. 7b).

Early Nipissing & correlative water planes 8.3 and 8.1 ka (Figs 7a, 7b, 8g). Large inflows to the Great Lakes system ceased when the Laurentide Ice Sheet disintegrated and glacial Lakes Agassiz and Barlow-Ojibway drained into Hudson Bay shortly after 8.4 ka (Dyke & Prest, 1987). The reduction in flow at the Rankin Constriction lowered Lake Mattawa and reinstated the traditional North Bay outlet (channels out of Trout Lake through Turtle and Loren Lakes at head of Mattawa River (Fig. 1b)) (Taylor, 1897; Harrison, 1972) as control for Early Lake Nipissing. Nourished only by precipitation in the absence of

gression. An analysis of uplift for Little Pic River, northeastern Superior basin, showed that detrital organics of age 5.9–6.1 ka from a Nipissing-level terrace at that location (Prest, 1970) do not constrain the age of the Nipissing culmination, as this location is close to the isobase through the North Bay outlet and little change in relative lake level occurred at Little Pic River throughout the Nipissing phase.

The principal events in the post Nipissing history of lake levels comprise the re-emergence of the St. Mary's River Sill as control for Lake Superior (Farrand & Drexler, 1985), erosion of the Port Huron outlet and diversion of all Michigan-Huron discharge to Lake Erie (Eschman & Karrow, 1985).

Comparison of interpreted Great Lakes water planes with shore features of glacial Lake Barlow-Ojibway

The recognition of the Rankin Constriction as a lake control in the Ottawa Valley implies an hydraulic connection and similar water levels in both the Upper Great Lakes basins and the Upper Ottawa-Lake Timiskaming (Barlow-Ojibway) basins. This would be the case during periods of high flow once the connecting channel through Opemika Narrow (Fig. 1a) had been opened.

Figure 7a contains northward projections of the Great Lakes water planes as well as most Barlow-Ojibway shore features, compiled by Vincent & Hardy (1979), with a selection of dated lakes and lacustrine wash limits in the northern Lake Timiskaming area from Veillette (1983; 1988). The wash limit may have been initiated by wave erosion in Lake Barlow at its Lake McConnell Moraine dam phase while glacial deposits blocked the Ottawa Valley at Opemika Narrows south of Lake Timiskaming as interpreted by Veillette (1988), for example. However, the correspondence between the Early Lake Mattawa (9.5 ka) water plane and the isobases on the upper lacustrine wash limit is striking (Fig. 7a). Thus, the wash limit could also reflect

lake levels controlled at Rankin Constriction provided the morainic dam at Opemika Narrows had been eroded and channelized prior to Early Lake Mattawa. As the Narrows were likely channelized early by high glacial lake outflows, the much later Lake Mattawa water plane (8.5 ka), which demarcates the zone of upper shore features more than 70 km north of Lake Timiskaming, was almost certainly present in the Barlow-Ojibway basin. The coincidence of the Mattawa water planes and the altitudinal distribution of high lacustrine shore features in the Barlow-Ojibway basin is consistent with our interpretations both of the water planes themselves and of the significance of hydraulic resistance in an outlet channel as a control of lake level.

Lake-induced climate effects

From the history interpreted earlier it is clear that the Great Lakes basins experienced periods of massive inflow from proglacial lakes bordering the retreating Laurentide Ice Sheet. A question arises: what effects did this flow impart to the local environment and climate of the Great Lakes? Although it is beyond the scope of this paper to examine this question in detail, it is interesting to note a correlation between the inflows of northern waters and occurrences of suppression or reversal in vegetation development.

First, Shane (1987) interpreted a climatic reversal on the Till Plains of Ohio, Indiana and Illinois near the southern margins of the Erie and Michigan basins. This reversal occurred between about 11.0 and 10.4 ka and is recorded in pollen diagrams as a recurrence in spruce (*Picea*) with some loss of deciduous trees and an expansion of fir (*Abies*) populations. The implied cooling of this vegetation change correlates exactly with the Main Algonquin phase of the Great Lakes when cold proglacial waters were circulating directly into Michigan basin and overflowing through Erie basin. A re-examination of pollen diagrams from Lake Huron and Lake Erie (Figs 3a and 3b) and from western Ontario between the Huron and Georgian Bay basins, for example, at Kincardine

(Fig. 1a: 1) and Cookstown (Fig. 1a: 5) bogs (Karrow *et al.*, 1975) and at Edward Lake 60 km southeast of Owen Sound (McAndrews, 1981), disclosed a correlative cooling effect also associated with Main Lake Algonquin. The onset of cooling is subtly expressed as a suppression of spruce and a concurrent lingering persistence of herbs and shrubs, such as grass (Gramineae), sedge (Cyperaceae) and sagebrush (*Artemisia*), after about 11 ka; the termination is marked by the well-known spruce-pine transition at 10.5–10.6 ka and the disappearance of significant percentages of herbs and shrubs. This climatic change and vegetation suppression had previously been proposed to have resulted from a poorly known ice advance (Karrow *et al.*, 1975), which is now thought to be an unlikely cause. Bjorck (1985) also postulates a climate reversal in northwestern Ontario from 11.1 to 10.2 ka on the basis of the record from one lake; confirmation of this event in other records is desirable though it is consistent with the concept of regional temperature suppression near proglacial lakes, in this case Lake Agassiz.

Existing lacustrine faunal records are broadly consistent with the implication for cold Algonquin waters. For example, in Michigan basin the ostracod assemblage was limited to the cold-tolerant species *Candona subtriangulata* (Buckley, 1975) and in Huron basin the presence of cold-species molluscs *Pisidium conventus* from shallow-water estuarine sediments suggests that the surface Algonquin water was no warmer than about 3.6 degrees C, even in summer (Miller *et al.*, 1985). The termination of cooling is coincident with the diversion of discharge of northern waters to the Ottawa Valley and the lowering of Lake Algonquin.

A second climate reversal is documented by Warner *et al.* (1984) at Greenbush Swamp (Fig. 1a: 8) on Manitoulin Island from about 10 to 8 ka when spruce returned as the dominant pollen type, more abundant than red and jack pine (*Pinus banksiana/resinosa*) following a period of pine-dominated woodland from about 10.5 to 10 ka. The spruce was almost always accompanied by an increase in fir (*Abies*). Spruce was

then succeeded at about 8 ka by increased frequency of fir, tamarack and birch (*Abies*, *Larix*, and *Betula*). On the basis of the changed interpretation of lake history presented here, we correlate the cooling implied by their record with the drainage of cold waters past Manitoulin Island under the influence of Agassiz inflow during the Mattawa Great Lakes phase from about 9.6 to 8.3 ka. The onset of cooling may have been influenced by the earlier passage of meltwater from the Marquette surge of ice at about 10 ka in the Superior basin about 220 km to the west. Similar evidence for this cooling is shown by a second peak in *Picea* at 0.55–0.75 m depth at the Lake Huron M-17 site (Fig. 3a) and at 3.0–3.1 m depth at the Hope Bay site (Fig. 3c). We interpret additional evidence of this cooling in pollen diagrams of other offshore cores from northeastern Georgian Bay (McAtee, 1977), western Georgian Bay 5.6 km east of Flowerpot Sill (J. H. McAndrews, pers. commun., 1969) and northwestern Lake Huron (Zilans, 1985) in which slightly elevated but highly fluctuating spruce pollen percentages are indicated within sediments of equivalent stratigraphic position. An increase in spruce was radiocarbon-dated at 8.7 ka from Charles Lake, 20 km north of Owen Sound on the peninsula between Lake Huron and Georgian Bay (R. Bailey, pers. commun., 1973), an event that is correlative with the transgressive expansion of Lake Mattawa (Figs 8f, 9). The termination of this cooling event correlates with the extinction of glacial Lakes Agassiz and Barlow-Ojibway by drawdown into Hudson Bay and the establishment of the Nipissing phase in the Great Lakes, nourished by inflows from local precipitation and runoff only.

Although more information is desirable to establish the geographic distribution of these observed pollen anomalies and to ensure the cooling was not caused by an unknown perturbation in atmospheric circulation, the evidence is sufficient to postulate an association with the Great Lakes history, namely that large inflows of cold glacial meltwater or proglacial lake waters have induced climatic cooling events in the Great Lakes region. The cooling was probably accom-

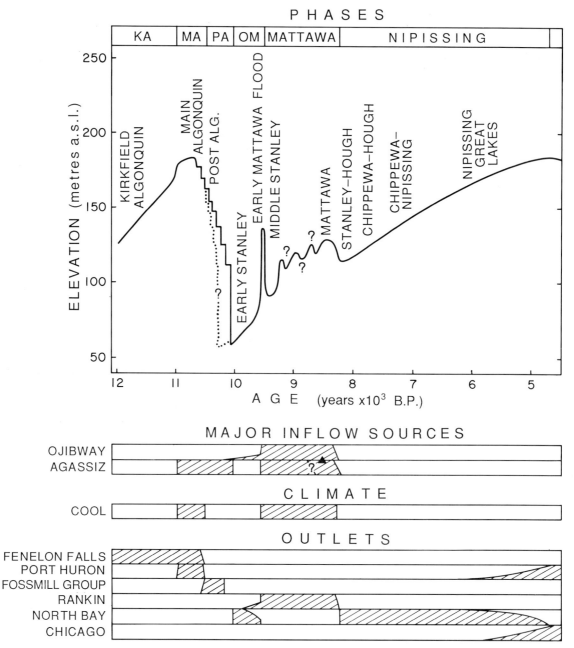

Fig. 10. Summary diagram showing reconstructed and interpreted water levels with lake names (vertical) for the northern Huron basin from 12 to 4.5 ka. The lake elevations have been restored by removing the estimated differential uplift relative to Port Huron. Also shown are timespans for phases of the Upper Great Lakes, major inflows, cool climate episodes, and operational periods of outlets. The Fossmill group of outlets refers to the northern Post Algonquin outlets and comprises the White Partridge Lake Sill to Aylen Sill inclusive as listed in Table 2. Although the sequence of the Post Algonquin lakes is known, their ages are not; consequently they are shown above as equally spaced on the age axis between 10.5 and 10.1 kyr. An alternative, more rapid drawdown sequence is also shown. The level of Early Lake Stanley is speculative. Elevations of the Mattawa phase lakes are also speculative except for the Early Mattawa Flood and Lake Mattawa. The initialed phases of the Upper Great Lakes are as follows: KA = Kirkfield Algonquin, MA = Main Algonquin, PA = Post Algonquin, and OM = Ottawa-Marquette. A ?-mark and upward-directed arrowhead in the rectangle of 'major inflow sources' indicate a possible diversion of late Agassiz drainage via Lake Ojibway basin; if this happened, the levels of the Huron-basin lakes would have been unchanged as the same outlet in Ottawa Valley (Rankin) was in effect.

plished by suppression of seasonal warming influences, for example, by the high albedo of seasonally persisting lake ice and by rapid advection of insolation-absorbing surface waters due to high inflow and rapid water flux. Though the first reversal has been correlated with the European Younger Dryas cooling (Shane, 1987), there is no European equivalent for the second reversal. Moreover, the first event terminated a few hundred years earlier than the Younger Dryas, which existed from 11 to about 10.2 ka (J. Mangerud, pers. commun., 1987), but closed exactly contemporaneously with the demise of the Main Algonquin phase of the Great Lakes. This difference is small, close to the resolution of the radiocarbon dating technique, but it is probably significant. Thus it is probable that the Great Lakes climate reversals were generated locally, mainly by the year-round presence of cold lake waters. This conclusion is further supported by a numerical atmospheric circulation experiment which showed that a reduction of temperature over the northeastern Atlantic Ocean, equivalent to the Younger Dryas cooling, did not induce cooling in North America west of its eastern seaboard (Rind *et al.*, 1986). Local cooling by meltwater is further supported by a preliminary investigation of pollen records which show similar vegetation responses along the down stream route from the Great Lakes during periods of meltwater drainage through the St. Lawrence estuary and Gulf of St. Lawrence (Lewis *et al.*, 1988).

Summary and conclusions

1. The evidence of pollen, other fossils, and offshore and shallow-water sediment sequences clearly support the hypothesis that inflows from upstream proglacial lakes induced two protracted periods of limnological change within the Great Lakes and probably beyond at distances of several hundred kilometres from the retreating ice margin. The first, Main Lake Algonquin (11–10.5 ka), received discharge from glacial Lake Agassiz (Moorhead phase) which supported Algonquin overflow to Erie basin and possibly

raised lake levels, cooled surface waters and induced a local climate reversal in the remainder of the Upper Great Lakes basins as well as in Erie basin. The lake waters also discharged through Ontario basin but possible effects here have not been evaluated. The second episode, the Mattawa phase (9.6–8.3 ka) of the Upper Great Lakes, received massive inflow from both glacial Lakes Agassiz and Ojibway. This combined inflow raised lake levels substantially and induced a second climate reversal by keeping the surface waters of the Upper Great Lakes cool. A summary of changes in the northern Huron basin is illustrated in Fig. 10. The events in the Great Lakes generally support the interpretation of two periods of eastward discharge of glacial Lake Agassiz as interpreted by Clayton (1983), Teller & Thorleifson (1983), Teller (1985) and Farrand & Drexler (1985). Further work to obtain better dating control, a greater understanding of crustal movement, and more precise environmental interpretations is recommended to verify and refine the foregoing postulates.

2. The climate reversal at 11 to 10.5 ka is recognized from fossil pollen evidence of a resurgence in more cold-tolerant vegetation. It is postulated to result from local atmospheric cooling caused by the cold waters (heat sink) of Main Lake Algonquin. The termination of cooling is disclosed by the well-known spruce-pine transition at about 10.5 ka and is coincident with the drawdown of Main Lake Algonquin. Though the correspondence in timing of the onset of this reversal with that of the European Younger Dryas cooling has been previously noted, it is more likely that an increased flux of glacial meltwater from Lake Agassiz through the Great Lakes system caused this deterioration of central North American climate.

3. Following the drawdown of Lake Algonquin during the Post Algonquin phase (10.5 to about 10.1 ka), a period of low lake levels, here termed the Ottawa-Marquette low phase, was initiated about 10.1 to 9.6 ka by the opening of a lower outlet via the Mattawa and Ottawa Valleys and by the diversion of Agassiz discharge by the Marquette glacial advance in Superior basin.

4. The clearest episode of upstream influence was the previously unrecognized Mattawa phase of increased lake level from about 9.6 to 8.3 ka when the combined flows of glacial Lake Agassiz, entering Superior basin, glacial Lake Barlow-Ojibway, draining via Timiskaming basin, and the Upper Great Lakes were controlled hydraulically at the Rankin Constriction in the Ottawa Valley near Mattawa, Ontario, with an estimated base-flow discharge in the order of $200\,000$ m^3s^{-1}. This base flow supported Lake Mattawa about 20 m above the ensuing nonglacial Nipissing phase whose control was at the classic North Bay outlet. Differential glacioisostatic rebound of the control area caused Lake Mattawa to transgress and backflood all basins of the Upper Great Lakes except possibly Superior. The Mattawa phase ended with the extinction of glacial Lakes Agassiz and Barlow-Ojibway caused by breakup of the Laurentide Ice Sheet in Hudson Bay.

5. Floods resulting from high flows and catastrophic drawdowns of Lake Agassiz caused short-term rises of lake level of several 10 s of metres above the base-flow elevation of Lake Mattawa in Huron-Georgian Bay basin. The Early Mattawa Flood, the first and highest of these floods was caused by hydraulic resistance in the Ottawa Valley to the initial large drawdown and discharge of the Nipigon phase of Lake Agassiz; the resistance was probably enhanced by the presence of incompletely scoured valley glacial deposits. The Early Mattawa Flood may have created high water marks that overlapped shore features of the earlier Post Algonquin glacial lakes and may also have influenced the formation of the Minong shore in Superior basin.

6. The climate of the Upper Great Lakes during the Mattawa phase was probably cooled by the year-round (especially summer) presence of cold lake waters, as evidenced in pollen diagrams by sporadic increases and lingering presence of spruce pollen percentages. Further study is needed to confirm and to establish the geographic extent of this climatic cooling.

7. Levels of even the large lakes of the Great Lakes system were affected significantly by large inflows of meltwater owing to the hydraulic resist-ance of their outlet channels. This was particularly evident for lakes controlled by flow through deep narrow valleys, less so for those draining by broad outlet channels or by the opening of additional channels. The joint influence of water flow and morphology of an outlet channel is an important consideration in the reconstruction of former lake levels.

8. Discovery of downstream (Gulf of St. Lawrence and North Atlantic Ocean) climate and oceanic effects caused by high discharge melt-water flows from Lake Agassiz and the Great Lakes is expected.

Acknowledgements

This work has been supported by the Geological Survey of Canada. The Canada Centre for Inland Waters (National Water Research Institute) provided office and laboratory facilities during our visits and supplied excellent research vessels, M. V. Martin Karlsen and CSS Limnos; we thank their staff for assistance in the collection of offshore cores and acoustic profiles. Our early work, some of which has been used in this paper, was made possible by the then Great Lakes Institute of the University of Toronto and its research vessel CCGS Porte Dauphine; we thank their staff for enthusiastic support. Acquisition of radiocarbon dates for the Hope Bay core was made possible by the Royal Ontario Museum. Many persons have assisted us in the past, some over the lengthy period, by locating information or by sharing perspectives on interpretation; in particular, we thank P. J. Barnett, W. S. Broecker, L. J. Chapman, J. P. Coakley, W. R. Farrand, W. A. Gorman, A. K. Hansel, P. F. Karrow, C. E. Larsen, R. N. McNeely, D. M. Peteet, V. K. Prest, C. N. Raphael, P. L. Storck, J. Terasmae & W. M. Tovell. We have benefited from unpublished notes on SCUBA dive observations by the late R. E. Deane, who recognized long ago the paleolimnological significance of the submerged sill area between Huron and Georgian Bay basins as a former lake-level control, a region we refer to as the Deane Saddle. C. DeCuypere assisted us ably with field studies in the North Bay-Mattawa area. L. Ovenden kindly identified the fossil mosses in the Hope Bay peat. We are grateful also to G. D. Cameron, D. C. Mosher & S. E. Young who helped with extensive compilations for this study, and to G. V. Sonnichsen who helped us prepare revisions of this paper. Many of the figures were prepared under the direction of the Illustrations Group at the Bedford Institute of Oceanography. R. E. Bailey, R. P. Futyma, J. H. McAndrews & R. J. Mott kindly provided unpublished data. In addition to the journal's reviewers, we thank R. J. Mott, L. H. Thorleifson, B. H. Luckman, D. L. Forbes, D. R. Sharpe, P. G. Sly, J. T. Teller & J. J. Veillette for reading earlier drafts and for providing comments which have helped to improve this paper.

References

Anderson, T. W., 1978. South Bay, Manitoulin Island, a preliminary report. In Guidebook for the Michigan Basin Geological Society Field Trip on Manitoulin Island, September 29–October 1, 1978.

Anderson, T. W., 1979. Stratigraphy, age, and environment of a Lake Algonquin embayment site at Kincardine, Ontario. Geological Survey of Canada Paper 79–1B: 147–152.

Anderson, T. W. & C. F. M. Lewis, 1974. Chronology and paleoecology of an Holocene buried plant detritus bed in Lake Huron (Abst.). 17th Conference on Great Lakes Research, Hamilton, Ontario, Abstracts, p. 4.

Anderson, T. W. & C. F. M. Lewis, 1985. Postglacial water-level history of the Lake Ontario basin. In P. F. Karrow & P. E. Calkin (eds), Quaternary Evolution of the Great Lakes. Geological Association of Canada, Special Paper 30: 231–253.

Anderson, T. W. & C. F. M. Lewis, 1987. Late Quaternary oscillations of Laurentian Great Lakes levels (Abst.). 30th Conference on Great Lakes Research, Ann Arbor, Michigan, Program and Abstracts, p. A-1.

Andrews, J. T., 1968a. Postglacial rebound: similarity and prediction of uplift curves. Can. J. Earth Sci. 5: 39–47.

Andrews, J. T., 1968b. Pattern and cause of variability of postglacial uplift and rate of uplift in arctic Canada. J. Geology 76: 404–425.

Andrews, J. T., 1970. A geomorphological study of postglacial uplift with particular reference to arctic Canada. Institute of British Geographers Special Publication 2, 156 pp.

Baker, V. R., 1973. Paleohydrology and sedimentology of Lake Missoula flooding in eastern Washington. The Geological Society of America, Special Paper 144, 79 pp.

Bajc, A. F., 1987. Rous Lake pit. In R. S. Geddes, F. J. Kristjansson, J. T. Teller, H. M. French & P. Richard (eds), Quaternary features and scenery along the north shore of Lake Superior. Twelfth Int. Union for Quaternary Research Congress Field Excursion Book C-12: 22–25.

Benson, M. A., 1968. Measurement of peak discharge by indirect methods. World Meteorological Organization, Technical Note 90, 161 pp.

Bjorck, F., 1985. Deglaciation and revegetation in northwest Ontario. Can. J. Earth Sci. 22: 850–871.

Buckley, S. B., 1975. Study of post-Pleistocene ostracod distribution in the soft sediments of southern Lake Michigan. Ph.D. dissertation, University of Illinois, Urbana, Illinois, 293 pp.

Catto, N. R., R. J. Patterson & W. A. Gorman, 1982. The Late Quaternary geology of the Chalk River region, Ontario and Quebec. Can. J. Earth Sci. 19: 1218–1231.

Chapman, L. J., 1954. An outlet of Lake Algonquin at Fossmill, Ontario. Geological Association of Canada Proceedings 6: 61–68.

Chapman, L. J., 1975. The Physiography of the Georgian Bay-Ottawa Valley Area of Southern Ontario. Ontario Division of Mines, Geoscience Report 128, 33 pp.

Chapman, L. J. & D. F. Putnam, 1984. The Physiography of Southern Ontario. Ontario Geological Survey Special Volume 2, Ontario Ministry of Natural Resources, Toronto, Canada, 270 pp.

Chow, V. T., 1959. Open-Channel Hydraulics. McGraw–Hill Book Co., N.Y., 680 pp.

Clayton, L., 1983. Chronology of Lake Agassiz drainage to Lake Superior. In J. T. Teller & L. Clayton (eds), Glacial Lake Agassiz. Geological Association of Canada, Special Paper 26: 291–307.

Coakley, J. P. & C. F. M. Lewis, 1985. Postglacial lake levels in the Erie basin. In P. F. Karrow & P. E. Calkin (eds), Quaternary Evolution of the Great Lakes. Geological Association of Canada, Special Paper 30: 195–212.

Cowan, W. R., 1978. Radiocarbon dating of Nipissing Great Lakes events near Sault Ste. Marie, Ontario. Can. J. Earth Sci. 15: 2026–2030.

Cowan, W. R., 1985. Deglacial shorelines at Sault Ste. Marie, Ontario. In P. F. Karrow & P. E. Calkin (eds), Quaternary Evolution of the Great Lakes. Geological Association of Canada, Special Paper 30: 33–37.

Crane, H. R. & J. B. Griffin, 1961. University of Michigan Radiocarbon Dates VI. Radiocarbon 3: 105–125.

Crane, H. R. & J. B. Griffin, 1965. University of Michigan Radiocarbon Dates X. Radiocarbon 7: 123–152.

Crane, H. R. & J. B. Griffin, 1968. University of Michigan Radiocarbon Dates XII. Radiocarbon 10: 61–114.

Crane, H. R. & J. B. Griffin, 1970. University of Michigan Radiocarbon Dates XIII. Radiocarbon 12: 161–180.

Crane, H. R. & J. B. Griffin, 1972. University of Michigan Radiocarbon Dates XV. Radiocarbon 14: 195–222.

104

Dadswell, M. J., 1974. Distribution, ecology, and postglacial dispersal of certain crustaceans and fishes in eastern North America. Canada Nat. Mus. of Nat. Sci., Pub. Zool. 11, 110 pp.

Deane, R. E., 1950. Pleistocene geology of the Lake Simcoe district, Ontario. Geological Survey of Canada Memoir 256, 108 pp.

Dreimanis, A., 1958. Beginning of the Nipissing phase of Lake Huron. J. Geology 66: 591–594.

Drexler, C. W., W. R. Farrand & J. D. Hughes, 1983. Correlation of glacial lakes in the Superior basin with eastward discharge events from Lake Agassiz. In J. T. Teller & L. Clayton (eds), Glacial Lake Agassiz. Geological Association of Canada, Special Paper 26: 309–329.

Dyck, W. J., J. A. Lowdon, J. G. Fyles & W. Blake, Jr., 1966. Geological Survey of Canada Radiocarbon Dates V. Geological Survey of Canada Paper 66–48, 32 pp.

Dyke, A. S. & V. K. Prest, 1987. Late Wisconsinan and Holocene history of the Laurentide ice sheet. Géographie physique et Quaternaire 41: 237–263.

Eschman, D. F. & P. F. Karrow, 1985. Huron basin glacial lakes: a review. In P. F. Karrow & P. E. Calkin (eds), Quaternary Evolution of the Great Lakes. Geological Association of Canada, Special Paper 30: 79–93.

Farrand, W. R., 1960. Former shorelines in western and northern Lake Superior basin. Ph.D. dissertation, University of Michigan, Ann Arbor, Michigan, 266 pp.

Farrand, W. R. & C. W. Drexler, 1985. Late Wisconsinan and Holocene history of the Lake Superior basin. In P. F. Karrow & P. E. Calkin (eds), Quaternary Evolution of the Great Lakes. Geological Association of Canada, Special Paper 30: 17–32.

Finamore, P. F., 1985. Glacial Lake Algonquin and the Fenelon Falls outlet. In P. F. Karrow & P. E. Calkin (eds), Quaternary Evolution of the Great Lakes. Geological Association of Canada, Special Paper 30: 125–132.

Fitzgerald, W. D., 1985. Postglacial history of the Minesing basin, Ontario. In P. F. Karrow & P. E. Calkin (eds), Quaternary Evolution of the Great Lakes. Geological Association of Canada, Special Paper 30: 133–146.

Ford, M. J. & R. S. Geddes, 1986. Quaternary Geology of the Algonquin Park Area. Ontario Geological Survey, Open File Report 5600, 87 pp.

French, H. M. & P. T. Hanley, 1975. Post Champlain Sea drainage evolution near Pembroke, Upper Ottawa Valley. Canadian Geographer 19: 149–158.

Fullerton, D. S., 1980. Preliminary correlation of post-Erie interstadial events (16000–10000 radiocarbon years before present), central and eastern Great Lakes region, and Hudson, Champlain, and St. Lawrence Lowlands, United States and Canada. United States Geological Survey Professional Paper 1089, 52 pp.

Futyma, R. P., 1981. The northern limits of Lake Algonquin in Upper Michigan. Quat. Res. 15: 291–310.

Futyma, R. P. & N. G. Miller, 1986. Stratigraphy and genesis of the Lake Sixteen peatland, northern Michigan. Can. J. Bot. 64: 3008–3019.

Goldthwait, J. W., 1907. Abandoned shorelines of eastern Wisconsin. Wisconsin Geological and Natural History Survey Bulletin 17, 134 pp.

Goldthwait, J. W., 1908. A reconstruction of water planes of the extinct glacial lakes in the Lake Michigan basin. J. Geology 16: 459–476.

Goldthwait, J. W., 1910. An instrumental survey of the shorelines of extinct lakes in southwestern Ontario. Geological Survey of Canada Memoir 10, 57 pp.

Hansel, A. K., D. M. Mickelson, A. F. Schneider & C. E. Larsen, 1985. Late Wisconsinan and Holocene history of the Lake Michigan basin. In P. F. Karrow & P. E. Calkin (eds), Quaternary Evolution of the Great Lakes. Geological Association of Canada, Special Paper 30: 40–53.

Hansel, A. K. & D. M. Mickelson, 1988. A reevaluation of the timing and causes of high lake phases in the Lake Michigan basin. Quat. Res. 29: 113–128.

Harrison, J. E., 1970. Deglaciation and proglacial drainage: North Bay-Mattawa region, Ontario. 13th Conference on Great Lakes Research, Ann Arbor, Michigan, Proceedings, Int. Assoc. Great Lakes Res., pp. 756–767.

Harrison, J. E., 1972. Quaternary geology of the North Bay-Mattawa region. Geological Survey of Canada Paper 71–26, 37 pp.

Hough, J. L., 1955. Lake Chippewa, a low stage of Lake Michigan indicated by bottom sediments. Geol. Soc. America Bull. 66: 957–968.

Hough, J. L., 1958. Geology of the Great Lakes. University of Illinois Press, Urbana, Illinois, 313 pp.

Hough, J. L., 1962. Lake Stanley, a low stage of Lake Huron indicated by bottom sediments. Geol. Soc. America Bull. 73: 613–620.

Hough, J. L., 1963. The prehistoric Great Lakes of North America. Am. Scientist 51: 84–109.

International Great Lakes Levels Board, 1973. Regulation of Great Lakes Levels, Appendix A – Hydrology and Hydraulics. Report to the International Joint Commission, December 7, 1973. 58 pp.

Johnston, W. A., 1916. The Trent valley outlet of Lake Algonquin and the deformation of the Algonquin waterplane in Lake Simcoe district, Ontario. Geological Survey of Canada Museum Bulletin 23, 27 pp.

Karrow, P. F., 1980. The Nipissing transgression around southern Lake Huron. Can. J. Earth Sci. 17: 1271–1279.

Karrow, P. F., 1986. Valley terraces and Huron basin water levels, southwestern Ontario. Geol. Soc. America Bull. 97: 1089–1097.

Karrow, P. F., 1987. Glacial and glaciolacustrine events in northwestern Lake Huron, Michigan and Ontario. Geol. Soc. America Bull. 98: 113–120.

Karrow, P. F., T. W. Anderson, A. H. Clarke, L. D. Delorme & M. R. Sreenivasa, 1975. Stratigraphy, paleontology and age of Lake Algonquin sediments in southwestern Ontario, Canada. Quat. Res. 5: 49–87.

Karrow, P. F. & R. S. Geddes, 1987. Drift carbonate on the Canadian Shield. Can. J. Earth Sci. 24: 365–369.

Karrow, P. F., B. G. Warner & P. Fritz, 1984. Corry Bog,

Pennsylvania: a case study of the radiocarbon dating of marl. Quat. Res. 21: 326–336.

Kaszycki, C. A., 1985. History of Glacial Lake Algonquin in the Haliburton region, south central Ontario. In P. F. Karrow & P. E. Calkin (eds), Quaternary Evolution of the Great Lakes. Geological Association of Canada, Special Paper 30: 110–123.

Kemp, A. L. W. & N. S. Harper, 1977. Sedimentation rates in Lake Huron and Georgian Bay. J. Great Lakes Res. 3: 215–220.

Larsen, C. E., 1985a. A stratigraphic study of beach features on the southeastern shore of Lake Michigan: new evidence for Holocene lake level fluctuations. Illinois State Geological Survey, Environmental Geology Notes 112, 31 pp.

Larsen, C. E., 1985b. Lake level, uplift, and outlet incision, the Nipissing and Algoma Great Lakes. In P. F. Karrow & P. E. Calkin (eds), Quaternary Evolution of the Great Lakes. Geological Association of Canada, Special Paper 30: 64–77.

Larsen, C. E., 1987. Geological history of Glacial Lake Algonquin and the Upper Great Lakes. United States Geological Survey Bulletin 1801, 36 pp.

Lee, T. E., 1957. The antiquity of the Sheguiandah Site. The Canadian Field-Naturalist 71; 117–147.

Leverett, F. & F. B. Taylor, 1915. The Pleistocene of Indiana and Michigan and the history of the Great Lakes. United States Geological Survey Monograph 53, 529 pp.

Lewis, C. F. M., 1969. Late Quaternary history of lake levels in the Huron and Erie basins. 12th Conference on Great Lakes Research, Ann Arbor, Michigan, Proceedings, Int. Assoc. Great Lakes Res., pp. 250–270.

Lewis, C. F. M., 1970. Recent uplift of Manitoulin Island, Ontario. Can. J. Earth Sci. 7: 665–675.

Lewis, C. F. M. & T. W. Anderson, 1985. Postglacial lake levels in the Huron basin: comparative uplift histories of basins and sills in a rebounding glacial marginal depression (Abst.). In P. F. Karrow & P. E. Calkin (eds), Quaternary Evolution of the Great Lakes. Geological Association of Canada, Special Paper 30: 147–148.

Lewis, C. F. M. & T. W. Anderson, 1987. Early Holocene oscillation of Laurentian Great Lakes levels (Abst.). Int. Union for Quaternary Research, XII Int. Congress, Ottawa, Ontario, Programme with Abstracts, p. 210.

Lewis, C. F. M., T. W. Anderson & A. A. Berti, 1966. Geological and palynological studies of Early Lake Erie deposits. 9th Conference on Great Lakes Research, Ann Arbor, Michigan, Proceedings, Pub. No. 15, Great Lakes Research Division, The University of Michigan, pp. 176–191.

Lewis, C. F. M., T. W. Anderson & A. A. L. Miller, 1988. Lake, ocean and climate response to meltwater discharge, Great Lakes and western Atlantic Ocean. Amer. Quat. Association, Program and Abstracts, 10th Biennial Meeting, 6–8 June, 1988, Amherst MA, U.S.A., p. 81.

Lowdon, J. A. & W. Blake, Jr., 1968. Geological Survey of Canada Radiocarbon Dates VII. Radiocarbon 10:

207–245, and Geological Survey of Canada Paper 68–2, Part B: 207–248.

Lowdon, J. A. & W. Blake, Jr., 1975. Geological Survey of Canada Radiocarbon Dates XV. Geological Survey of Canada Paper 75–7, 32 pp.

Lowdon, J. A. & W. Blake, Jr., 1979. Geological Survey of Canada Radiocarbon Dates XIX. Geological Survey of Canada Paper 79–7, 57 pp.

Lowdon, J. A., J. G. Fyles & W. Blake, Jr., 1967. Geological Survey of Canada Radiocarbon Dates VI. Geological Survey of Canada Paper 67–2, Part B, 42 pp.

Lowdon, J. A., I. M. Robertson & W. Blake, Jr., 1971. Geological Survey of Canada Radiocarbon Dates XI. Geological Survey of Canada Paper 71–7: 255–324.

Martin, N. V. & L. J. Chapman, 1965. Distribution of certain crustaceans and fishes in the region of Algonquin Park, Ontario. J. Fish. Res. Bd. Can. 22: 969–976.

McAndrews, J. H., 1981. Late Quaternary climate of Ontario: temperature trends from the fossil pollen record. In W. C. Mahaney (ed.), Quaternary Paleoclimate. Geo Abstracts Ltd., Norwich, England, pp. 319–333.

McAtee, C. L., 1977. Palynology of late-glacial and postglacial sediments in Georgian Bay, Ontario, Canada, as related to Great Lakes history. M.Sc. thesis, Brock University, St. Catharines, Ontario, 153 pp.

Miller, B. B., P. F. Karrow & G. L. Mackie, 1985. Late Quaternary molluscan faunal changes in the Huron basin. In P. F. Karrow & P. E. Calkin (eds), Quaternary Evolution of the Great Lakes. Geological Association of Canada, Special Paper 30: 95–107.

Monaghan, G. W., W. A. Lovis & L. Fay, 1986. The Lake Nipissing transgression in the Saginaw Bay region, Michigan. Can. J. Earth Sci. 23: 1851–1854.

Mott, R. J. & L. Farley-Gill, 1981. Two late Quaternary pollen profiles from Gatineau Park, Quebec. Geological Survey of Canada Paper 80–31, 10 pp.

Peterson, R. L., 1965. A well-preserved grizzly bear skull recovered from a late glacial deposit near Lake Simcoe, Ontario. Nature 208: 1233–1234.

Prest, V. K., 1970. Quaternary geology of Canada. In R. J. E. Douglas (ed.), Geology and Economic Minerals of Canada. Geological Survey of Canada Economic Geology Report 1, pp. 675–764.

Raphael, C. N. & E. Jaworski, 1982. The St. Clair Delta, a unique lake delta. The Geographical Bull. 21: 7–27.

Ricker, K. E., 1959. The origin of two glacial relict crustaceans in North America, as related to Pleistocene glaciation. Can. J. Zool. 37: 871–893.

Rind, D., D. Peteet, W. Broecker, A. McIntyre & W. Ruddiman, 1986. The impact of cold North Atlantic sea surface temperatures on climate: implications for the Younger Dryas cooling (11–10 k). Climate Dynamics 1: 3–33.

Saarnisto, M., 1974. The deglaciation history of the Lake Superior region and its climatic implications. Quat. Res. 4: 316–339.

Saarnisto, M., 1975. Stratigraphical studies on the shoreline displacement of Lake Superior. Can. J. Earth Sci. 12: 300–319.

Shane, L. C. K., 1987. Late glacial vegetational and climatic history of the Allegheny Plateau and the Till Plains of Ohio and Indiana, U.S.A. Boreas 16: 1–20.

Sly, P. G. & C. F. M. Lewis, 1972. The Great Lakes of Canada – Quaternary Geology and Limnology. 24th Int. Geol. Congress, Montreal, Quebec, Guidebook for Field Excursion A43, 92 pp.

Sly, P. G. & R. G. Sandilands, 1988. Geology and environmental significance of sediment distributions in an area of the submerged Niagara escarpment, Georgian Bay. In M. Munawar (ed.), Limnology and Fisheries of Georgian Bay/North Channel Ecosystem. Kluwer Academic Publishers. Hydrobiologica 163: 47–76.

Spurr, S. H. & J. H. Zumberge, 1956. Late Pleistocene features of Cheboygan and Emmet counties, Michigan. Am. J. Sci. 254: 96–109.

Stanley, G. M., 1936. Lower Algonquin beaches of Penetanguishene Peninsula. Geol. Soc. America Bull. 47: 1933–1959.

Stanley, G. M., 1937a. Lower Algonquin beaches of Cape Rich, Georgian Bay. Geol. Soc. America Bull. 48: 1665–1686.

Stanley, G. M., 1937b. Impounded Early Algonquin beaches at Sucker Creek, Grey County, Ontario. Papers of the Michigan Academy of Science, Arts & Letters XXIII: 477–495.

Stanley, G. M., 1938. The submerged valley through Mackinac Straits. J. Geology 46: 966–974.

Taylor, F. B., 1897. The Nipissing-Mattawa River, the outlet of the Nipissing Great Lakes. The American Geologist 20: 65–66.

Teller, J. T., 1985. Glacial Lake Agassiz and its influence on the Great Lakes. In P. F. Karrow & P. E. Calkin (eds), Quaternary Evolution of the Great Lakes. Geological Association of Canada, Special Paper 30: 1–16.

Teller, J. T. & P. Mahnic, 1988. History of sedimentation in the northwestern Lake Superior basin and its relation to Lake Agassiz overflow. Can. J. Earth Sci. 25: 1660–1673.

Teller, J. T. & L. H. Thorleifson, 1983. The Lake Agassiz – Lake Superior connection. In J. T. Teller & L. Clayton (eds), Glacial Lake Agassiz. Geological Association of Canada, Special Paper 26: pp. 261–290.

Teller, J. T. & L. H. Thorleifson, 1987. Catastrophic flooding into the Great Lakes from Lake Agassiz. In L. Mayer & D. Nash (eds), Catastrophic Flooding. Allen & Unwin London, pp. 121–138.

Terasmae, J., 1979. Radiocarbon dating and palynology of Glacial Lake Nipissing deposits at Wasaga Beach, Ontario. J. Great Lakes Res. 5: 292–330.

Terasmae, J. & O. L. Hughes, 1960. Glacial retreat in the North Bay area, Ontario. Science 131: 1444–1446.

Thomas, R. L., A. L. W. Kemp & C. F. M. Lewis, 1973. The surficial sediments of Lake Huron. Can. J. Earth Sci. 10: 226–271.

Tovell, W. M. & R. E. Deane, 1966. Grizzly bear skull: site of a find near Lake Simcoe, Science 154: 158.

Tovell, W. M., J. H. McAndrews, C. F. M. Lewis & T. W. Anderson, 1972. Geological reconnaissance of Georgian Bay – a preliminary statement (Abst.). 15th Conference on Great Lakes Research, Abstracts, Ann Arbor, Michigan, Int. Assoc. for Great Lakes Res., p. 15.

Veillette, J., 1983. Déglaciation de la vallée supérieure de l'Outanouais, la lac Barlow et le sud du Lac Ojibway, Québec. Géographie physique et Quaternaire 37: 67–84.

Veillette, J., 1988. Déglaciation et évolution des lacs proglaciaires Post-Algonquin et Barlow au Témiscaminque, Québec et Ontario. Géographie physique et Quaternaire 42: 7–31.

Vincent, J.-S. & L. Hardy, 1979. The evolution of Glacial Lakes Barlow and Ojibway, Quebec and Ontario. Geological Survey of Canada Bulletin 316, 18 pp.

Walcott, R. I., 1970. Isostatic response to loading of the crust in Canada. Can. J. Earth Sci. 7: 716–727.

Walcott, R. I., 1972. Late Quaternary vertical movements in eastern North America: quantitative evidence of glacio-isostatic rebound. Reviews of Geophysics and Space Physics 10: 849–884.

Warner, B. G., R. J. Hebda & B. J. Hann, 1984. Postglacial paleoecological history of a cedar swamp, Manitoulin Island, Ontario, Canada. Palaeogeography, Palaeoclimatology, Palaeoecology 45: 301–345.

Woodend, S. L., 1983. Glacial and post-glacial history of Lake Huron as defined by lithological and palynological analysis of a core in southern Lake Huron, with special regard to a low-lake level phase. B.Sc. thesis, Carleton University, Ottawa, Ontario, 74 pp.

Zilans, A., 1985. Quaternary geology of the Mackinac basin, Lake Huron. M.Sc. thesis, University of Waterloo, Waterloo, Ontario, 275 pp.

Water levels in Lake Ontario 4230–2000 years B.P.: evidence from Grenadier Pond, Toronto, Canada

Francine M. G. McCarthy[1] & John H. McAndrews[2]

[1] *Department of Geology, University of Toronto, Toronto, Canada, M5S 1A1;* * [2] *Department of Botany, Royal Ontario Museum, 100, Queen's Park Crescent, Toronto, Canada, M5S 2C6, and Departments of Geology and Botany, University of Toronto, Toronto, Canada, M5S 1A1; Present address: Department of Geology, Dalhousie University, Halifax, Canada, B3H 3J5*

Key words: Great Lakes water levels, Lake Ontario, pond sedimentation

Abstract

Lake Ontario water levels have been rising for the past 11 500 years due to differential isostatic rebound of the St. Lawrence outlet. Small scale fluctuations in water level superimposed on this general trend have received little study, with the exception of the 'Nipissing Flood'.

The transgression of a Grenadier Pond was studied from cores along a transect from the bar that separates the pond from Lake Ontario to the marsh on the north shore. Radiocarbon dates of the transition from swamp peat to pond marl in five cores provide estimates of the rate of water level rise since 4230 years B.P. These estimates are supported by changes in sediment type and in abundance of pollen and seeds of aquatic plants. There were three short intervals of accelerated water level rise in Grenadier Pond, around 4200, 3000, and 2000 years B.P., when water levels rose up to 2 m instantaneously, within the resolution of radiocarbon dating. Sedimentological and paleobotanical data suggest that Grenadier Pond was an open embayment of Lake Ontario until 1970–1850 years B.P., when it was isolated by the bar, and therefore sediments deposited prior to this time reflect water levels in Lake Ontario.

Short term departure of up to 2 m from the average rate of water level rise over the past 4000 years, as observed in the record at Grenadier Pond, is of the same range as historically observed departures from the mean lake stage of Lake Ontario. This implies that a threshold discharge exists above which broadening of the outflow channel occurs to accommodate further increase in discharge with little rise in lake level. The intervals of accelerated water level rise in Lake Ontario broadly coincide with periods of cool, wet climate, suggesting that increased moisture may have caused the short term fluctuations in water level.

Originally published in
Journal of Paleolimnology **1**: 99–113.

108

Introduction

The level of Lake Ontario was long assumed to have risen at an exponentially decreasing rate solely in response to differential isostatic rebound of the St. Lawrence outlet since the Admiralty Phase (or Early Lake Ontario) 11 500 years B.P. (Muller & Prest, 1985). Recent work indicates that the Holocene water level history of Lake Ontario is more complex than the simple rebound model suggests. Sutton *et al.* (1972) and Anderson & Lewis (1982, 1985) indicate that periods of accelerated water level rise followed by temporary stabilization occurred around 5000 to 4000 B.P. The accelerated water level rise, called the 'Nipissing Flood', was attributed to the capture of Upper Great Lakes drainage.

Larsen (1985) in his stratigraphic study of shorelines of Lakes Michigan and Huron showed that water level fluctuations occurred in the late Holocene. The high water levels lasted 200–300 years and were 1–2 m above the historic mean of the lakes. He suggests that the fall in water level from the Nipissing to the Algoma and later Holocene levels, as well as the fluctuations in the late Holocene, were climate-related, citing pollen, archeological, and neoglacial evidence. Our objective is to describe small-scale, post-Nipissing Flood water level fluctuations in Grenadier Pond and to show how these fluctuations reflect the late Holocene water-level history of Lake Ontario.

This study
Evidence for small-scale fluctuations in water level in Lake Ontario can be found and dated in embayments containing organic authigenic sediments (Fig. 1), as was done by Otto & Dalrymple (1983) in the St. Catherines area. Rates of transgression can be measured by ^{14}C dating of the inception of pond sediments at successively higher elevations in the embayment away from the lake.

The sediments in embayments contain pollen and plant macrofossils that provide additional data for estimating water depths, record progradation of fringing marshes during intervals of stable or falling water levels, and permit differentiation between gradual and sudden transgression of the shoreline. Vegetational succession reflects shoreline transgression and increasing water depth as upland species are replaced by emergent aquatic marsh species. If transgression continues, these are in turn replaced by floating and submerged aquatic species, commonly found in water to 4 m depth in Ontario lakes, below which there is a sharp decline in species richness and biomass (Crowder *et al.*, 1977). This depth varies with physical limnological conditions in each basin. Because aquatic pollen and plant macrofossils are locally deposited, an abundance of emergent aquatic fossils reflects sedimentation in the littoral zone, the part of the basin shallow enough to support rooted vegetation.

Agitated water and high energy conditions would be expected to accompany rapid water level rise in Lake Ontario. Increased clastic deposition resulting from accelerated longshore drift and shoreline erosion is easily noted in marl-dominated embayments.

Grenadier Pond (Fig. 2) was chosen to study the transgression of Lake Ontario embayments due to its long sedimentary record, its long, narrow morphology perpendicular to Lake Ontario, and its location in one of the five areas of sand concentration in Lake Ontario (Thomas *et al.*, 1972).

The pond is isolated by a bar and its level in May, 1985, was 1.3 m above the level of Lake Ontario. The bar has been widened in historic times by landfilling, but maps and records indicate that the pond was isolated before European exploration (Robinson, 1933). The level of the pond is maintained by a culvert today, and so does not necessarily reflect natural conditions. Prior to this artificial stabilization, the elevation of the pond relative to Lake Ontario depended on the elevation of the natural outlet stream.

Methods

Five cores were taken using a modified Livingstone piston sampler (Wright, 1967) along

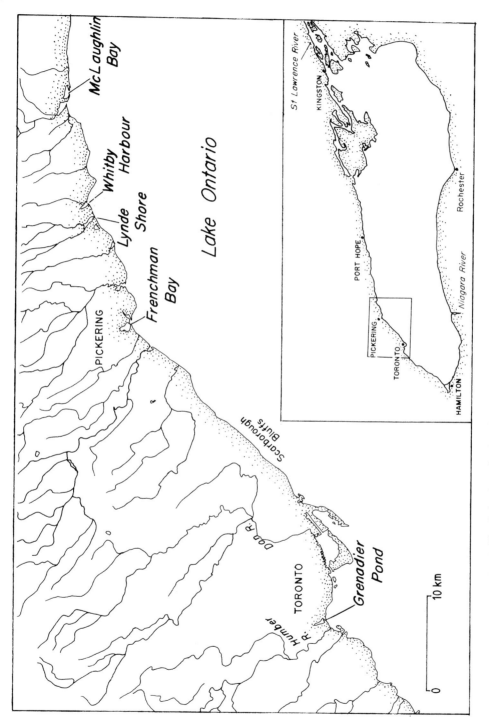

Fig. 1. The Lake Ontario shoreline in the Toronto area, showing several embayments.

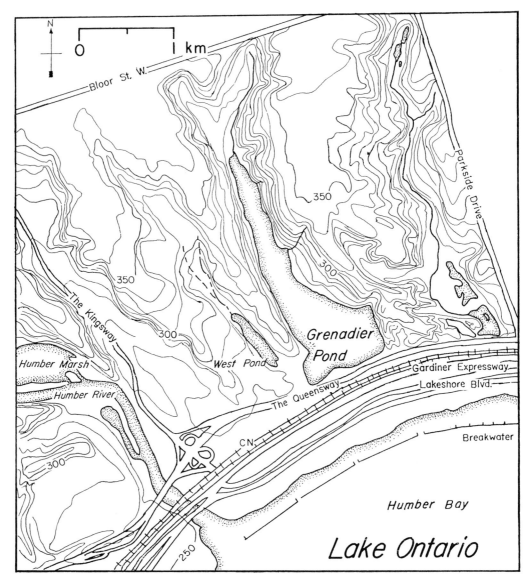

Fig. 2. Grenadier Pond lies in a bedrock valley which has been filled with Pleistocene lake sediments and subsequently planed and covered with the sandy, shallow water sediments of Glacial Lake Iroquois. Grenadier Pond is separated from Lake Ontario by a sandy baymouth bar. Contours are in feet.

a transect from the bar that separates the pond from Lake Ontario to the cat-tail marsh on the north shore (Fig. 3). The upper metre of sediment at each coring site was sampled with a Rowley-Dahl sampler (Rowley & Dahl, 1956) to retain the sediment-water interface, preserve stratigraphy of upper watery sediments, and allow water depths to be accurately measured.

Sediments were described visually and analysed using differential thermal analysis (Dean, 1974). Absolute chronology was provided by radiocarbon dating. Both accelerator dates on picked terrestrial plant macrofossils and beta emission dates on bulk sediment samples were obtained.

Sediment samples were prepared for pollen

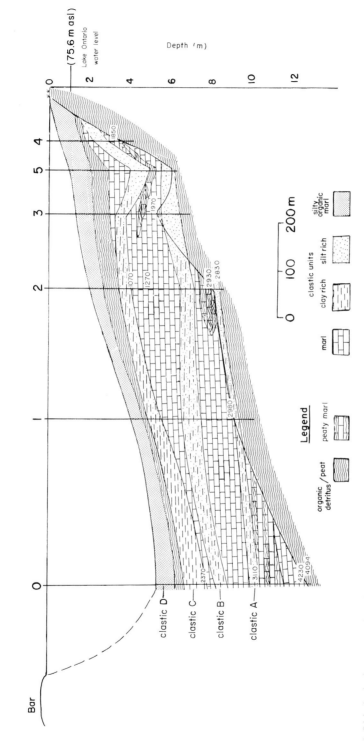

Fig. 3. Lithological cross-section along S-N transect in Grenadier Pond. North is to the right. Numbers 0 to 5 refer to coring sites. Note the peat/marl contact which represents transgression and the four clastic units in the marl-dominated embayment.

analysis using standard procedures (Faegri & Iversen, 1975). Coarse clastics and peats were sieved through 0.15 and 0.015 mm mesh in addition to the chemical treatment. Spores of *Lycopodium clavatum* (Stockmarr, 1971) were used as marker particles, permitting pollen density to the estimated. The pollen and spores were identitied using the key of McAndrews *et al.*, (1973). A sum of over 200 pollen grains per sample was used to generate pollen diagrams.

Selected 10 cm core intervals were sieved through 0.5 mm mesh to concentrate macrofossils. Seeds were identified using Montgomery (1977) and Martin & Barkley (1961) and by comparison with the reference seed collection of the Royal Ontario Museum.

Results

The stratigraphic succession typical of Lake Ontario lagoons is from basal glaciolacustrine sediments to peat, peaty marl, and marl (Anderson & Lewis, 1982). The marl is overlain by a dark silty organic marl which persists to the present. The change in sedimentation to foul-smelling silty organic marl is detectable in all cores taken from Lake Ontario bays and lagoons, and corresponds to the rise in *Ambrosia* pollen, about A.D. 1850 (McAndrews & Boyko-Diakanow, in press). The *Ambrosia* rise was ^{14}C dated in the Humber Marsh, about 1 km west of Grenadier Pond, at 150 ± 50 years B.P. (Weninger & McAndrews, submitted).

A similar stratigraphy exists in Grenadier Pond, where a humic, woody peat is overlain by peaty (organic) marl, marl, coarse organic detritus, and silty organic marl. Three intervals of clastic sedimentation, clastic units A, B, and C in order of decreasing age, occur below the *Ambrosia* zone, clastic unit D (Fig. 3). Two factors make the identification of these clastic units difficult without thermal analysis (Fig. 4): the decrease in grain size away from the shoreline, and the dilution by authigenic marl production, generally highest at intermediate water depths (ca. 1–3 m) where carbonate-secreting organisms are most abundant.

Radiocarbon dates are listed in Table 1. Note that accelerator mass spectrometry (AMS) dates on terrestrial plant samples date approximately 3% younger than bulk sediment dates from the same interval in Grenadier Pond.

The contact between woody peat and marl records transgression. Transgression is diachronous, dating at 4230 ± 60 years B.P. (AMS 4094 ± 60 years B.P.) in core 0, 2980 ± 60 years B.P. in core 1, 2930 ± 80 years B.P. (AMS 2830 ± 50 years B.P.) in core 2, 1970 ± 100 years B.P. in core 3 and 1850 ± 60 years B.P. (AMS) in core 4 (Figs. 3, 4). Dates at successively higher elevations along the transect provide quantitative estimates of the rates of water level rise between the transgression of each core site along the transect (Table 2). The accuracy of the rate estimates depends on the dating error (± 50 to 100 years) and on the accuracy of measurement of the peat/marl contact, which is ± 5cm.

The vegetational succession was determined in cores 0, 1 and 2 from pollen and plant macrofossils (Fig. 5). Pollen zones 4 and 3 of McAndrews (1981) were identified in each core (McCarthy, 1986). In addition to these regional pollen zones a local pollen zone, containing a high percentage of pollen of emergent aquatic plants, was identified in the lower part of each core. Above this 'emergent aquatic pollen zone', the sediment contains low background concentrations of emergent aquatic pollen grains.

The top of the emergent aquatic pollen zone parallels the transgressive contact between humic peat and marl (Fig. 3). In core 2, the top of this zone occurs just below a date of 1270 ± 50 years B.P., suggesting that the water level at this site became too deep for emergent aquatics about 1500 years after its initial transgression. Figure 5 shows that the pollen and macrofossils of emergent aquatic species generally decrease upcore, and further, that there is a succession from emergent aquatic species (e.g. Cyperaceae, *Typha, Sparganium*) to floating aquatic species (*Nuphar* and *Nymphaea*) and then to submerged aquatic species (e.g. *Potamogeton* and *Myriophyllum*). In general, this succession occurs steadily upward in the sequence, reflecting rising water level in the

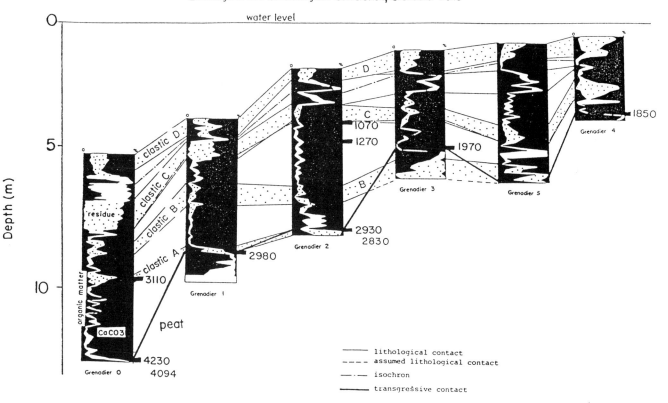

Fig. 4. Lithological and chronological correlation along S-N transect in Grenadier Pond. The clastic units contain more than 15% allocthonous residue after ignition. Heavy lone connecting cores near their bases indicates the top of the peat. Core positions are not shown to horizontal scale.

littoral zone due to the continuous isostatic uplift of the St. Lawrence outlet. Periods of accelerated water level rise are reflected by abrupt decreases in emergent aquatic pollen and macrofossils, whereas periods of stable water level are marked by an increase in emergent aquatic pollen and macrofossils as marsh plants grow outward into the pond.

Clastic unit A was dated in core 0 at 3110 ± 80 years B.P. using accelerator dating. It occurs as a fine sand in peat in cores 1 and 2. It was dated at 2980 ± 60 years B.P. in core 1 and lies just below a date of 2930 ± 80 years B.P. (AMS 2830 ± 50 years B.P.) in core 2. Given the errors involved, these dates are essentially of the same age, suggesting that clastic unit A was deposited synchronously at all three sites. The base of clastic unit C was accelerator dated in core 2 at

1070 ± 100 years B.P. A bulk C ^{14}date of 2370 ± 60 years B.P. at the base of clastic unit C in core 0 was rejected as being too old based on sedimentological and pollen density data (McCarthy, 1986). This is probably due to recycling of older sediments suggested by high percentages of *Picea* and Cyperaceae in the pollen record of clastic unit C (Fig. 5), since these pollen types dominate the late Pleistocene Glacial Lake Iroquois sediments. The clastic units parallel isochrons, i.e., the surface and base of the *Ambrosia* zone, and cut across diachronous facies boundaries reflecting transgression (i.e., the lithological boundary between woody peat and marl and the top of the aquatic pollen and plant macrofossil zone). The deposition of each clastic unit thus appears to have been synchronous throughout the basin, and these units represent relatively short periods of increased clastic input.

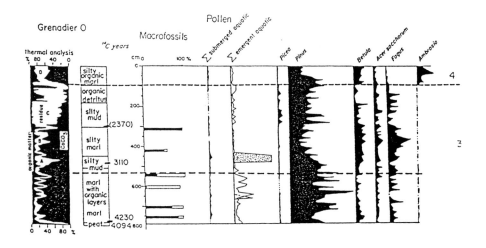

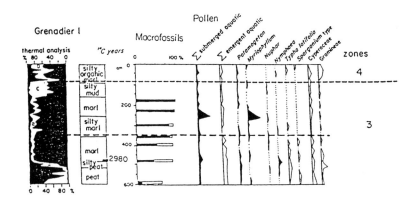

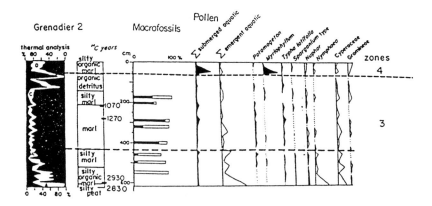

Fig. 5. Summary of water level indicators in cores 0, 1, and 2. An abbreviated upland pollen diagram is also shown for core 0, zoned following McAndrews (1981). Pollen percentages of individual aquatic taxa are shown for cores 1 and 2. On plant macrofossil and aquatic pollen diagrams, black represents submerged aquatic plant species. Dashed lines indicate sudden changes in water depth reflected by pollen and plant macrofossils; stipple represents percentages of grass (wild rice) pollen, excluded from the emergent aquatic pollen sum. Percentages of grass pollen in clastic unit A in core 0 range from 334% at 500 cm to 2010% at 448 cm.

Table 1. Radiocarbon dates from Grenadier Pond cores. Dates with TO- lab numbers are accelerator dates on terrestrial plant macrofossils, while dates with WAT- lab numbers are beta emission dates on bulk sediment samples.

Core	Lab No.	Depth		^{14}C date (years B.P.)	Material dates	Significance
		In core (cm)	Below datum* (cm)			
0	WAT-1332	771–781	1171–1181	4230 ± 60	marl	transgression
0	TO-160	780	1180	4094 ± 60	twig (unidentified)	transgression
1	WAT-1330	500–510	760–770	2980 ± 60	marl	transgression
2	WAT-1328	608–618	668–678	2930 ± 80	marl	transgression
2	TO-164	618	678	2830 ± 50	Angiosperm wood	transgression
3	WAT-1334	380–390	375–385	1970 ± 100	marl	transgression
4	TO-165	299	216	1850 ± 60	marl	transgression
0	TO-159	480	880	3110 ± 80	pine needles birch seeds, wood fragments	Clastic A
0	WAT-1591	290–300	690–700	2370 ± 60	silty mud	Clastic C
2	TO-162	200	260	1070 ± 100	pine needles leaf petiole	Clastic C
2	TO-163	275	335	1270 ± 50	pine needles cedar needles	% aquatics falls

* Modern level of Lake Ontario.

Table 2. Rates of water level rise based on ^{14}C dates of transgression. The rate of rise is considered instantaneous where three times the standard errors of the dates overlap.

Interval			rate (cm/y)		
(cm)	years B.P.	cores	rate	minimum rate	maximum rate
420	4230–2980	0 to 1	0.336	0.307	0.373
90	2980–2930*	1 to 2	instantaneous		
300	2930–1970**	2 to 3	0.313	0.288	0.441
170	1970–1850*	3 to 4	instantaneous		

* dates indistinguishable using ^{14}C dating
** sediment and vegetational succession indicates that most of this water level rise occurred in the last few hundred years of this interval. Water level appears to have been stable until ca. 3200 B.P.; the rate of water level rise from 2300 to 1970 B.P. was nearly instantaneous.

Discussion

The dates of the transgression in cores 0, 1, 2, 3, and 4 are plotted (Fig. 6), and are compared with the dates predicted by the water level curve of Sly & Prior (1984) (Fig. 7). There are two anomalies: 1) water levels rose 90 cm as the shoreline retreated about 250 m from site 1 to site 2 instantaneously, within the resolution of radiocarbon dating, and 2) the date for transgression in core 4 lies 1.7 m above the predicted value for Lake Ontario.

The second anomaly is attributed to isolation of Grenadier Pond. Rapid transgression from site 3 to site 4 was not accompanied by clastic sedimentation, suggesting that bar buildup and ponding

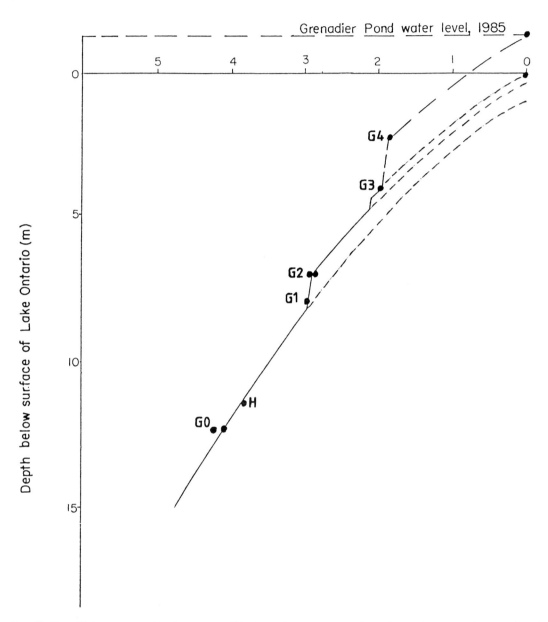

Fig. 6. Detailed late Holocene water levels based on ^{14}C dates of transgression along Grenadier Pond (G) and of the inception of aggradation in the Humber Marsh (H) (Weninger & McAndrews, submitted). The dashed lines represent rising water levels in Grenadier Pond due to isolation from Lake Ontario.

may have been responsible for the sudden quiet rise in water level. The 1.7 m water level rise is consistent with the present elevation of Grenadier Pond of 1.3 m above Lake Ontario.

The first anomaly cannot be attributed to pond isolation because high energy conditions are indicated by the contemporaneous deposition of clastic unit A during the rapid transgression. Nor

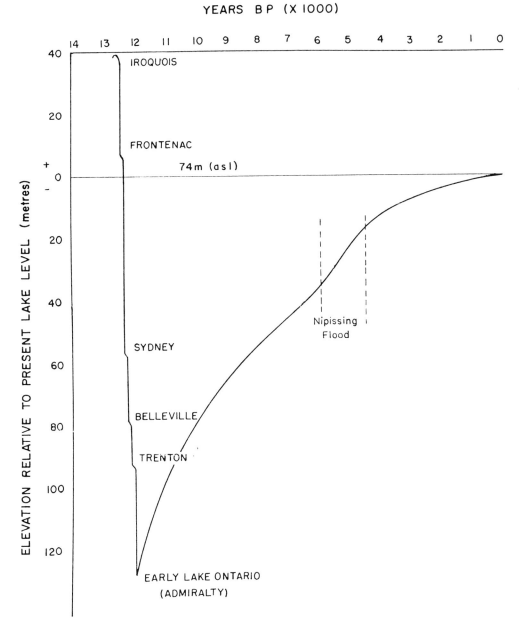

YEARS B P (X 1000)

Fig. 7. Postglacial water levels in the western Lake Ontario basin, redrafted from Sly & Prior (1984).

can it be attributed to dating error because of independent paleobotanical evidence of rapid transgression at this time.

In core 0, the organic content of the marl decreased around 3300 years B.P. and emergent aquatic pollen fell abruptly (Fig. 5), implying rapid water level rise. Clastic unit A was deposited simultaneously, or shortly afterward, 3110 ± 80 years B.P. This unit contains very high percentages (up to 2000%) of grass pollen. Based on pollen morphology and size, this is probably the pollen of wild rice *(Zizania aquatica).* This aquatic grass grows in moving or agitated water, usually in streams and in the bays of large lakes, such as Long Point Bay of Lake Erie, suggesting agitated water in Grenadier Pond during the deposition of clastic unit A. Around 2950 years B.P. clastic deposition ('residue' in Fig. 5) in core 0 was succeeded by marl, and grass pollen fell sharply to background levels of less than 1%.

The site of core 1 was transgressed 2980 ± 60 years B.P. at an elevation of 7.6 m below datum (the modern elevation of Lake Ontario), and the transgression is contemporaneous with the deposition of clastic unit A (Fig. 5). The silty peat is immediately succeeded by marl with relatively low percentages of emergent aquatic macrofossils and pollen, which remain relatively stable until just prior to the deposition of clastic unit B.

Clastic unit A also intersects the transgression in core 2, which dates at 2930 ± 80 years B.P., 6.7 m below datum. In this core, however, the humic peat is succeeded by silty, peaty (organic) marl, indicating the persistence of very shallow water, shoreline conditions at this site for eight hundred years, until about 2100 years B.P., when water levels again rose abruptly, depositing clastic unit B. The percentage of emergent aquatic and pine pollen in the silty peaty (organic) marl is very high, relative to that of all arboreal taxa excluding pine (Mc Carthy, 1986). This suggests that this site was at the edge of the shoreline, because marshes differentially trap *Pinus* pollen. The fluctuation in percentages of organic matter and $CaCO_3$ and the high percentage of clastics supports this interpretation.

The evidence for anomalously high energy conditions contemporaneous with paleobotanical and lithological evidence for rapid transgression between 3300 and 2900 years B.P. indicates that Grenadier Pond was an open embayment of Lake Ontario at the time. Water levels therefore rose 90 cm nearly instantaneously in Lake Ontario around 3000 years B.P. as the shoreline retreated about 250 m from site 1 to site 2 in Grenadier Pond.

Because clastic unit A indicates high energy conditions and transgression in the embayment, we suggest that other clastic units in the pond reflect intervals of rapid transgression, at least prior to pond isolation around 1900 years B.P.

Clastic sediments in lagoons, however, could also result from increased erosion and runoff from the upland due to deforestation. To distinguish between clastics of shoreline erosion and slopewash origin on sedimentological grounds McCarthy (1986) found that the grainsize distribution of clastic unit C, deposited after pond isolation, was similar to that of clastic unit D which resulted from slopewash accompanying European land clearing. The grainsize distribution of clastic unit B, however, resembled that of clastic unit A, which we have shown above was the result of accelerated shoreline erosion accompanying rapid transgression.

Pollen and plant macrofossil data also support accelerated shoreline erosion due to rapid transgression as the origin of clastic unit B, deposited about 2100 years B.P. Marl rapidly succeeded silty organic marl just prior to the deposition of clastic unit B at site 2, and an abrupt decrease in the pollen and macrofossils of emergent aquatic plants accompanied the increase in clastics at sites 1 and 2 (Fig. 5).

High clastic sedimentation contemporaneous with paleobotanical evidence for rapid transgression is also found at the base of core 0. Site 0 was transgressed 4230 years B.P., and the woody peat was succeeded by silty marl rich in pollen and macrofossils of submerged aquatic species. This indicates rapid water level rise from terrestrial to deep water conditions, without the intermediate stage of emergent aquatic plants. Around 3700 B.P., the organic content of the marl increased and peaty layers rich in pollen and seeds of emergent aquatic species alternated with marl. This succession probably records the end of the 'Nipissing Flood' followed by stabilization of the water level, permitting the progradation of a shoreline-fringing, sedge-dominated marsh to site 0 (McCarthy, 1986).

The periods of accelerated water level rise in Lake Ontario 3000 and 2100 years B.P. broadly coincide with periods of cool, wet climate in northeastern North America. Figure 8 compares regional climatic interpretations for the late Holocene with water level reconstructions for Lake Ontario. The climatic interpretations were

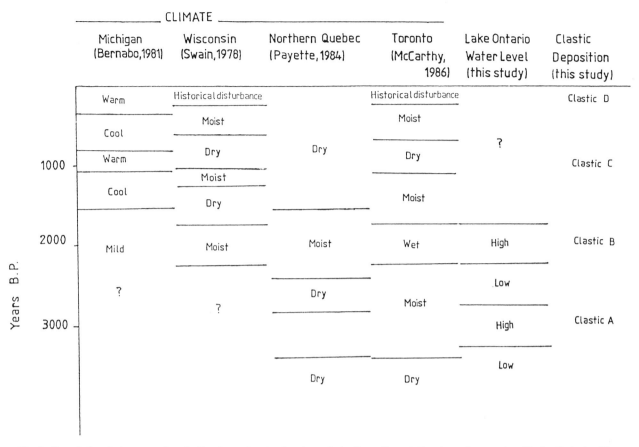

Fig. 8. Correlation between regional climate and water levels and clastic sedimentation in embayments of Lake Ontario. The two intervals of rapid transgression, accompanied by the deposition of clastic units A and B in Grenadier Pond, correlate with moist climates around 2000 and 3000 years B.P., as reconstructed from pollen, plant macrofossil, charcoal, and peat inception data.

based on pollen, plant macrofossil, charcoal, and peatland inception data. In addition, the pollen assemblage in Grenadier Pond changed from pine dominance to mesic deciduous tree dominance around 3400 years B.P., implying increased moisture (Fig. 5). *Pinus* comprised an average of 45% of the pollen assemblage between 4230 and 3400 years B.P., then fell to about 20% as percentages of mesic taxa such as *Fagus, Acer saccharum* and *Betula* approximately doubled, just prior to the deposition of clastic unit A in Grenadier Pond. Maximum percentages of *Fagus, Acer saccharum,* and *Betula* coincide with the deposition of clastic unit B, suggesting highest moisture during this interval (McCarthy, 1986).

Increased inflow resulting from wetter climate could account for the fluctuations documented for Lake Ontario around 3000 and 2000 years B.P. Short term departure of up to 2 m from the average rate of water level rise over the past 4000 years, as observed in the record at Grenadier Pond, is of the same range as historically observed departures from the mean lake stage of Lake Ontario. Water levels in Lake Ontario have fluctuated about a mean elevation of 246.5 ft. (75 m) with a range of ±3 ft. (1 m) from 1860 to 1958 (US Army Corps of Engineers, 1956). Apparently, water levels cannot rise much beyond this limit, for any additional increase in input beyond this threshold discharge will be accommodated by broadening of the outflow channel with little rise in lake level. Because of continued differential isostatic rebound of the outlet, this outflow channel-widening process must be repeated for discrete increases in inflow, resulting in the step-like pattern of lake level rise observed in the record at Grenadier Pond.

Summary and conclusions

Radiocarbon dates of transgression along a transect in Grenadier Pond suggest that water levels rose quickly for short periods (up to 300 years) around 3000 and 2100 years B.P. This resulted in accelerated shoreline erosion and turbid water depositing clastic sediments, and rapid transgression displacing the shoreline flora. These intervals were followed by longer periods of stable or slowly rising water levels and low energy conditions, when marshes grew out into the pond, and marls and peaty marls low in clastics were deposited.

The deposition of clastic sediments in marly embayments, and moderate to very high percentages of wild rice pollen, a grass that grows abundantly only in flowing or agitated water, suggest that high energy conditions accompanied the intervals of rapid transgression prior to ca. 1900 years B.P. This argues against bar construction and pond isolation as a viable cause for the rise in water level. Grenadier Pond was probably isolated by barrier bar construction between 1970 and 1850 years B.P. when rapid transgression occurred, unaccompanied by evidence for high energy conditions, suggesting a quiet rise in water level. The 1.7 m water level rise is consistent with the present elevation of Grenadier Pond of 1.3 m above Lake Ontario.

The periods of accelerated water level rise in Lake Ontario correlate with periods of wetter, cooler climate, inferred from pollen and peat inception data. Short term departure of up to 2 m from the average rate of water level rise over the past 4000 years, as observed in the record at Grenadier Pond, is of the same range as historically observed departures from the mean lake

stage of Lake Ontario. This implies that a threshold discharge exists above which broadening of the outflow channel occurs to accommodate further increase in discharge with little rise in lake level. Because of continued differential isostatic rebound of the outlet, this outflow channel-widening process must be repeated for discrete increases in inflow, resulting in the step-like pattern of lake level rise observed in the record at Grenadier Pond.

Acknowledgements

This work formed part of a thesis supported by an NSERC scholarship to F. McCarthy. Part of the work was also supported by NSERC grant A5699 to J. McAndrews. Discussions with N. Eyles, G. Norris, and A. Zimmerman of the University of Toronto provided important insights. The helpful suggestions of R. B. Davis and two anonymous reviewers are gratefully acknowledged. We also thank the staff of the Department of Botany of the Royal Ontario Museum for assistance in the field and in the laboratory.

References

Anderson, T. W. & C. F. M. Lewis, 1982. The mid-Holocene Nipissing flood into Lake Ontario. Abstract. In: Seventh Biennial Conference, American Quaternary Association: 60.

Anderson, T. W. & C. F. M. Lewis, 1985. Post-glacial water level history of the Ontario basin. Geol. Assoc. Canada Spec. Pap. 30: 231–253.

Bernabo, J. C., 1980. Quantitative estimates of temperature changes over the last 2700 years in Michigan based on pollen data. Quat. Res. 15: 143–159.

Corps of Engineers, US Army, 1956. Great Lakes Pilot US Lake Survey, Detroit.

Crowder, A. A., J. M. Bristow & M. R. King, 1977. Distribution, seasonality, and biomass of aqautic macrophytes in Lake Opinicon (eastern Ontario). Nat. Canadien 104: 441–456.

Dean, W. E., 1974. Determination of carbonate and organic matter in calcareous sediments and sedimentary rocks by loss on ignition: comparison with other methods. J. Sed. Pet. 44: 242–248.

Faegri, K. & J. Iversen, 1975. Textbook of Pollen Analysis (3rd edition). Munksgard, Copenhagen, 237 pp.

Kite, G. W., 1972. An engineering study of crustal movement around the Great Lakes. Dept. Of Environment, Inland Waters Branch, Technical Bulletin No. 63, 57 pp.

Larsen, C. E., 1985. Lake level, uplift, and outlet incision, the Nipissing and Algoma Great Lakes. Geol. Assoc. Canada Spec. Pap. 30: 63–75.

Martin, A. C. & W. D. Barkley, 1961. Seed identification manual. Univ. California Press, Berkeley. 221 p.

McAndrews, J. H., 1981. Late Quaternary climate of Ontario: temperature trends from the fossil pollen record. In: Quaternary Paleoclimate, W. Mahaney, Ed., Geo. Abstracts, Norwich: 319–333.

McAndrews, J. H., A. A. Berti & G. Norris, 1973. Key to the Quaternary pollen and spores of the Great Lakes region. Royal Ontario Museum Life Sciences Miscellaneous Publication, 61 pp.

McAndrews, J. H. & M. Boyko-Diakonow, in press. Pollen analysis of varved sediment at Crawford Lake, Ontario: evidence of Indian and European farming. In: R. J. Fulton & J. A. Higinbottom Eds. Quaternary Geology of Canada and Greenland. Geological Survey of Canada, Ottawa.

McCarthy, F., 1986. Late Holocene water levels in Lake Ontario: evidence from Grenadier Pond. M.Sc. thesis, University of Toronto, Toronto, Canada. 114 pp.

Montgomery, F. H., 1977. Seeds and Fruits of Plants of Eastern Canada and Northeastern United States. University of Toronto Press, Toronto, 232 pp.

Muller, E. H. & V. K. Prest, 1985. Glacial lakes in the Ontario basin. Geol. Assoc. Canada Spec. Pap. 30: 213–229.

Otto, J. E. & Dalrymple, R. W., 1983. Terrain characteristics and physical processes in small lagoon complexes. Ontario Geoscience Research Grant No. 78, Ontario Geological Survey Open File Report 5463, 89 pp.

Payette, S., 1984. Peat inception and climatic change in northern Quebec. In: Climatic changes on a Yearly to Millenial Basis. N. A. Morner and W. Karlen, Eds. D. Reidel Publishing Co., Boston: 173–180.

Robinson, P. J., 1933. Toronto during the French regime (from 1615–1793). Ryerson Press, Toronto, 254 pp.

Rowley, J. R. & A. D. Dahl, 1956. Modifications in design and use of the Livingstone piston sampler. Ecology 37: 849–852.

Sly, P. G. & J. W. Prior, 1984. Late glacial and postglacial geology in the Lake Ontario basin. Can. J. Earth Sci. 21: 802–821.

Stockmarr, J., 1971. Tablets with spores used in absolute pollen analysis. Pollen et Spores 13: 615–621.

Sutton, R. G., T. L. Lewis & D. L. Woodrow, 1972. Post-Iroquois lake stages and shoreline sedimentation in Eastern Ontario Basin. J. Geol. 80: 346–356.

Swain, A. M., 1978. Environmental changes during the past 2000 years in north central Wisconsin: analysis of pollen, charcoal and seeds from varved lake sediments. Quat. Res. 10: 55–68.

Thomas, R. L., A. L. W. Kemp & C. F. M. Lewis, 1972. Distribution, composition and characteristics of the surficial sediments of Lake Ontario. J. Sed. Pet. 42: 66–84.

Weninger, J. M. & J. H. McAndrews, submitted. Late Holocene evolution of the Humber River Marshes. Can. J. Earth Sci.

Wright, H. E. Jr., 1967. A square rod piston sampler for lake sediments. J. Sed. Pet. 37: 975–976.

Periods of rapid environmental change around 12500 and 10000 years B.P., as recorded in Swiss lake deposits

Brigitta Ammann
Systematisch-Geobotanisches Institut Universität Bern, Switzerland

Key words: biostratigraphy, late-glacial, early Holocene, radiocarbon stratigraphy, rapid climatic changes

Abstract

In the sediment of three Swiss lakes at a range of altitude from 514 to 2017 m, the Bölling and the Preboreal are recognized as two periods of rapid biotic changes. The main reason is rapid climatic change that triggered shifts in different groups of aquatic and terrestrial organisms. Although these groups would be expected to have very different response times, e.g., to increasing summer temperature, their assemblages responded with surprizing synchroneity. For extracting climatic signals from stratigraphies, the quickly responsive indicators like oxygen isotopes or beetles are useful. For understanding ecological dynamics under a changing climate, the comparison of biota with various response time are important.

Introduction

Over 50 years ago, in describing and interpreting vegetation history, Oberdorfer (1937: 526–527) had already expressed what is still striking in late-glacial pollen diagrams from central Europe, i.e., two steps of very rapid changes, first the reforestation and then the transition to the Holocene: 'Die Klimaäanderung, die von der arktischen Zeit in die subarktische überleitete, muss sehr plötzlich und ruckartig vor sich gegangen sein und war vielleicht von ähnlicher Bedeutung und ähnlichem relativem Ausmass, wie jene zweite Klimabesserung, die zur endgültigen borealen Wärmezeit führte'. And later, he said: 'Explosionsartige Ausbreitung der Birken-

und Kieferngehölze...' and 'Zweite unvermittelt einsetzende Klimamilderung, die zum vollen Einbruch der wärmeliebenden Gehölze führte'.

In reviewing Swiss material recently published in Lang (1985), we can now ask: how are these two steps recorded in biostratigraphies other than pollen? How synchronous or metachronous were changes in terrestrial and lalcustrine ecosystems? And, how good a time scale do we have for distinguishing between synchronism and metachronism?

Out of the 13 sites presented in Lang (1985) I will concentrate here on only three, all from the main transect, viz. Hobschensee, Amsoldingersee and Lobsingensee (Fig. 1).

Originally published in
Journal of Paleolimnology **1**: 269–277.

124

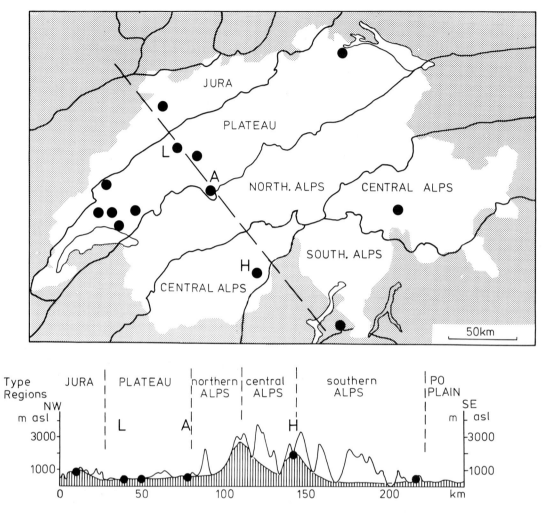

Fig. 1. Map of Switzerland (white area) indicating type regions and 13 sites of the Bernese research program (Lang, 1985). H = Hobschensee, A = Amsoldingersee, L = Lobsigensee.

Hobschensee at Simplon Pass – results and interpretations

Simplon Pass (2005 m asl) connects the central alpine (longitudinal) part of the Rhone Valley (Valais) with the southern alpine (transverse) valley of the Toce River (Italy). Crystalline bedrock types including gneisses and schists dominate. During the Würm Glaciation, confluence of local glaciers from mountains around the pass with the Rhone ice in the north and with the Ticino ice in the south, resulted in at least 200 m depth of ice over the pass (Müller, 1984). Hobschensee (Fig. 1), at 2017 m altitude is

at today's timber line, and therefore at a sensitive elevation. This sensitivity is one reason why several scientists have chosen it for investigation (Keller, 1935; Küttel, 1979; Welten, 1982; Lang & Tobolski, 1985). Two findings are striking: (1) the site became ice-free as early as during the Bölling (i.e. after 13 000 BP), in spite of the lake's situation in the Central Alps at 2000 m altitude (Welten, 1982); and (2) the reforestation occurred very rapidly and as early as at the transition from Younger Dryas to Preboreal (i.e., at the beginning of the Holocene at ca. 10 000 yr B.P.). This was demonstrated by Küttel (1979) and Welten (1982) who identified stomata of *Juniperus, Larix* and

Pinus, and by Tobolski in Lang & Tobolski (1985) who recorded a large variety of macrofossils of *Juniperus, Larix, Betula 'alba'* and *Pinus cembra* (Fig. 2). Time control is not yet good as the only two late-glacial radiocarbon dates (B-529 and B-608) in Welten's diagram considered of Alleröd age are 12 580 ± 200 yr B.P. and 10 430 ± 250 yr B.P., older and younger, respectively, than expected. Palaeolimnological studies are in progress (Cladocera by M. Boucherle; diatoms by B. Marciniak).

Amsoldingersee at the boundary Plateau/Prealps – results and interpretations

Where the rivers (the glaciers, during the Ice Ages) leave the limestone Prealps and enter the lowland of the Tertiary Molasse-Plateau, a fringe of large piedmont lakes (such as Zürichsee and Thunersee) and of small kettle hole lakes (e.g., Amsoldingersee, Fig. 1) occur. Littoral cores from Amsoldingersee (641 m asl) were analysed

for pollen, chemistry, stable isotopes, fossil pigments and Cladocera (Lotter & Boucherle, 1984; Lotter, 1985) (Fig. 3). A disturbance or hiatus in the sediment from the Younger Dryas prevents detailed studies of the transition to the Holocene, but it offers an interesting example of an early lake-level lowering in the European context (Gaillard, 1985). The transition from the Oldest Dryas to the Bölling, on the other hand, is very sharp for several parameters, and, like at Gerzensee, 12 km away, the $\delta^{18}O$ measured on precipitated carbonates (lake marl) rose rapidly with reforestation. This rapid rise can be interpreted as indicating a rise in summer temperature (Eicher & Siegenthaler, 1976; Eicher, 1987). Cladocera and fossil pigments mainly reflect changes in trophic state of the lake. Cladocera also show changes in relative sizes of the littoral and limnetic (pelagial) zones, also reflecting lake level changes. Such developments may or may not be triggered by climatic changes. The cladoceran fauna of Amsoldingersee shifts distinctly at the beginning of the Bölling. This can be

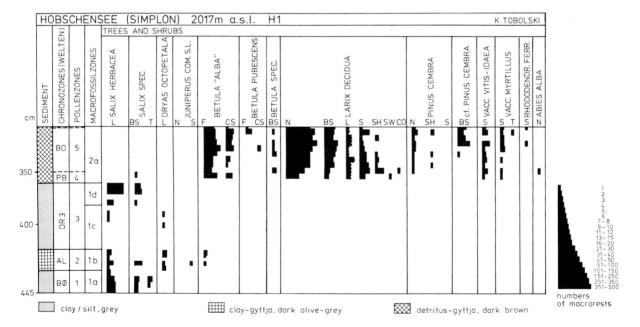

Fig. 2. Simplified macrofossil diagram from Hobschensee (after Lang & Tobolski, 1985), including selected taxa from the late-glacial and early Holocene, and showing the reforestation at the beginning of the Holocene. The scale on the right gives the number of macrorests recorded. BS = bud scale, CO = cone, CS = catkin scale, F = fruit, L = leaf, N = needle, S = seed, SH = short shoot, SW = seed wing, T = twig. BØ = Bölling, AL = Alleröd, DR3 = Younger Dryas, PB = Preboreal, BO = Boreal.

AMSOLDINGERSEE
Late-glacial and Early Holocene
after LOTTER & BOUCHERLE (1984)

conventional 14C-ages		Regional pollenzones FIRBAS 1949/1954 WELTEN 1982	Local PAZ	Vegetational history	Climate (temperature) $\delta^{18}O$	Sediment-type	Lake history (Cladocera, pigments)
regional (extrapolated) terrestrial plants (AMS)	locally measured whole sediment (decay)						
		V Boreal	A-8	Corylus, Ulmus + Quercus	temperature rising	lake marl	oligotrophic
10 000	10 280+120	IV Preboreal	A-7	Pinus-Betula forests + Corylus			
10 800		III Younger Dryas	A-6	clearance of Pinus forests	cold	fine detritus gyttja	uninterpretable
LST		II Alleröd	A-5	Pinus-Betula forests		chalk gyttja, LST	meso-eutrophic
12 000	13 020+170	Ib Bölling (s.l.)	A-4	open Betula forests	warm	chalk gyttja with fine detritus gyttja layers	oscillating between more or less eutrophic
			A-3	Juniperus-Hippophaë shrubs			
12 500	13 490+170	Ia Oldest Dryas	A-2	Betula+Salix dwarf shrubs		chalk gyttja	oscillating cond.*
16 000?			A-1	Steppe-Tundra	cold	glacial clay	cold/ice conditions
		Würm-glacial					covered by the Aare glacier

*between oligotrophic and mesotrophic

Fig. 3. Palaeoecology at Amsoldingersee (641 m) during the late-glacial and the early Holocene, according to Lotter & Boucherle (1984). All dates are given in conventional radiocarbon ages *sensu* Stuiver & Polach 1977, without calibration. The regional time scale follows a new dating series for the central Swiss Plateau worked out on terrestrial plant macrofossils (therefore without hard-water error). The locally measured samples of whole sediment have a hard water error. LST = Laachersee tephra.

interpreted as a shift from conditions oscillating between oligotrophic and mesotrophic to more or less eutrophic conditions, a change consistent with rising summer temperature (Boucherle, in Lotter (1985)). Time control is not easy to directly establish at Amsoldingersee because the three late-glacial ^{14}C–dates from the main core are affected by a hard water error (Fig. 3).

Lobsigensee on the central Swiss Plateau

During at least part of the Würm Glaciation the area northwest of Berne was covered by Rhone ice. Lobsigensee (Fig. 1), at 514 m in that area, is a small kettle hole lake presently with a surface of 2 ha and a maximum depth of 2.7 m. It has no surface inflow and only a small, sporadic overflow functional mainly during the period of snow melt. Multidisciplinary studies were carried out on littoral and profundal cores that were correlated by palynology (Ammann *et al.*, 1983; Ammann *et al.*, 1985).

Lobsigensee results

A transition from shrub tundra in the Oldest Dryas to forest in the Bölling is concluded not only from the pollen diagrams but also from size statistics on *Betula* pollen (Gaillard, 1983) and from *Betula* macrofossils (Tobolski, in Ammann & Tobolski (1983)). During the Oldest Dryas *Betula nana* was the dominant birch species. Since the initial juniper peak of the Bölling tree birches prevailed (Fig. 4). In the coleopteran and trichopteran faunas, important shifts from boreal and boreo-montane to temperate assemblages are recorded in the early Bölling. By using today's biogeographic ranges an increase in mean July temperature from 10–12 °C to 14–16 °C can be inferred from theses faunistic shifts (Elias & Wilkinson, 1983, 1985). Unfortunately the lowermost part of the Bölling (i.e., the juniper peak) did not contain any plant-independent beetles or caddisflies. In the chironomid and ceratopogonid faunas, Hofmann (1983, 1985a) found a major change coinciding with the beginning of the Bölling, i.e., the extinction of the cold-stenothermic fauna for the whole lake (littoral as well

127

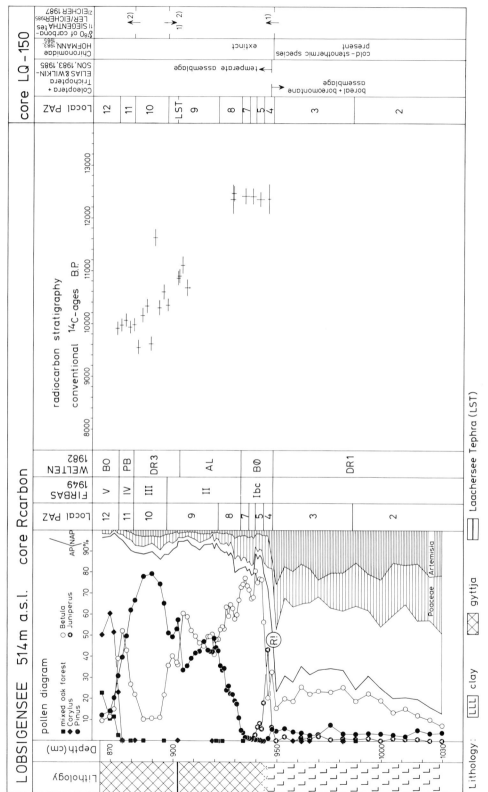

Fig. 4. Palaeoecological changes at Lobsigensee compared with the high-resolution radiocarbon-stratigraphy based on AMS datings of terrestrial plant macrofossils (Oeschger *et al.*, 1985; Andree *et al.*, 1986). Dates are presented in conventional, i.e. not calibrated radiocarbon years (Stuiver & Polach, 1977). The greatest environmental changes coincide with plateaus of constant age in the ^{14}C-stratigraphy. R! = reforestation.

as profundal). Among the Cladocera an abrupt faunal change was observed during the early part of the Bölling (during the juniper peak), namely the disappearance of subarctic species (Hofmann, 1985b). The Ostracoda developed a high species diversity during the Oldest Dryas but disappeared with the advent of the Bölling and didn't reappear until the late part of the Boreal, shifts that Löffler (1986) compared with results from lakes in Carinthia and interpreted to reflect the onset of meromixis at the beginning of the Bölling and of holomixis since the late Boreal. The Mollusca show a maximum number of individuals at the beginning of the Bölling and a somewhat gradual shift in species before this period (*Pisidium* spp. decrease; *Valvata piscinalis* increases (Chaix, 1983, 1985)). The stratigraphy of fossil algal and bacterial pigments displays a step change towards higher diversity and indicates more anoxic conditions at the beginning of the Bölling (Züllig, 1985, 1986). The stable isotopes $\delta^{18}O$ and $\delta^{13}C$ increase rapidly at the corresponding level (Siegenthaler & Eicher, 1985) (Fig. 4).

The transition from the Younger Dryas to the Preboreal is very sharp in the pollen diagrams (Ammann, 1985), and in the $\delta^{18}O$ stratigraphy (Eicher, 1987) (Fig. 4). Unfortunately the biostratigraphies of Coleoptera, Trichoptera and Mollusca could not be extended to this transition because in the littoral core used for these groups the transition is located in peat. In profundal cores (as the one presented in Fig. 4), Cladocera and fossil pigments were sampled at intervals that were too large, and Ostracoda were missing in this core section.

Interpretation of Lobsigensee results

The beginning of the Bölling biozone (*sensu* Welten, 1982; 12 500 yr B.P., as determined without hard water error, Andrée *et al.*, 1986), and the beginning of the Holocene (10 000 yr B.P.) are very distinct; each has synchronous shifts for several bio- and isotope stratigraphies. This is true for both: (1) the biota of terrestrial environments (pollen of terrestrial plants, terrestrial Coleoptera), and (2) the biota of lacustrine environments (pollen of aquatic plants, lacustrine Coleoptera, Trichoptera, Chironomidae, Ceratopogonidae, Ostracoda, Mollusca, algae and photosynthetic bacteria). This synchronism is higher than expected if we consider the predominantly individualistic behaviour of species (Iversen, 1954; Birks, 1981, 1986). Like Oberdorfer (1937), I interpret the rapidity of ecological change and the synchronism at the two major transitions to indicate external forcing, viz. rapid climatic warming (Atkinson *et al.*, 1987). No time-lag in the response of oxygen isotopes to the rise of summer temperature is expected. Some migrational lags of various lengths are recorded in the biostratigraphies. However, many biotic changes occur at the same stratigraphic levels. As an exemple: reforestation at the beginning of the Bölling could be interpreted to indicate the reaching and passing of a mean July temperature of 10 °C, although local differences in temperatures at timberlines add uncertainty to this interpretation (Friedel, 1967; Tranquillini, 1967; Hustich, 1983). The Coleoptera and Trichoptera point to a mean July temperature of 14–16 °C since the early Bölling. This would be warm enough for broadleaved deciduous trees like *Quercus* and *Corylus*, but these taxa do not arrive until 3000 years later. Thus, while response times may be very different for isotopes, plants and insects (Wright, 1984), if a climatic change is rapid enough and of great enough amplitude it may provoke a major stratigraphic synchronism (by 'telescoping together' different developments), e.g., as at the beginning of the Bölling.

How good a time scale do we have for the distinction between synchronism and metachronism? At Lobsigensee the AMS ^{14}C dates on terrestrial plant macrofossils (Oeschger *et al.*, 1985; Andrée *et al.*, 1986) enable us to reduce the sample thickness and refine chronostratigraphic resolution. Furthermore, by dating only terrestrial vascular plant remains we avoid the hard water error that may derive from dating whole sediment containing remains of aquatic biota or error from reworked carbon or penetrating rootlets. Sedimentary changes like carbonate layers in the gyttja (deposited since reforestation, see Fig. 4) would have introduced complications of changing sedimentation rates; we therefore chose two undisturbed one meter Livingstone core sections of pure fine detritus gyttja, linked them by the sharp isochronous volcanic ash layer from Laach (Boogard & Schmincke, 1984) and called this 'artificial core' Rcarbon. The striking feature in Fig. 4 is, that there are two plateaus of constant age, one around 12 500 yr B.P. and the other around 10 000 yr B.P. Thus during the two periods when the greatest environmental changes occur the high resolution radiocarbon dating does not help to establish a chronology for the assessment of rates of change.

A plateau of constant age could be the result of increased sedimentation rate, but here this can be excluded for two reasons: (1) throughout the core the sediment deposited since the reforestation is uniformly a fine detritus gyttja with very thin densely packed laminations (in the thin-section these laminations show a network of organic matter preventing the counting of possible annual layers), and (2) stratigraphically rapid changes like the reforestation or the expansion of *Corylus* would be even more rapid (which I consider unlikely) if I postulate increased sedimentation rate. Such considerations call for a reinvestigation in annually laminated sediment.

The geophysical reasons for a plateau of constant age can be, according to Andrée et al. (1986, p. 415): 'decreasing ^{14}C production rate or (by) dilution of the atmospheric ^{14}C with carbon of lower ^{14}C concentration. The drastic changes in the environmental system observed at this transition could, e.g., have accelerated ocean circulation, involving a reduction of the atmospheric ^{14}C level (e.g., Siegenthaler, Heimann & Oeschger 1980)'. For several reasons this explanation concerning the carbon cycle is favored by the geophysicists.

The palaeoecological consequences of plateaux of constant age are discussed in Ammann and Lotter (in prep.). There are several problems if these two plateaus prove to be real:

(1) high-resolution ^{14}C sampling will not improve chronological resolution,

(2) influx calculations or rates of change cannot be calculated if based on ^{14}C, and

(3) long distance correlations are strongly affected when based on radiocarbon dates around 12 500 yr and 10 000 yr B.P.

This last point raises the possibility that plateaus of constant age artificially sharpen transitions. We submit many samples for ^{14}C dating from sections with interesting stratigraphic changes and we therefore get many dates centered around these ages. But such plateaus function like traps for ^{14}C-ages and we possibly get more very similar dates than expected in calendar years. In the context of long distance correlation, this leads us to consider several events as synchronous which actually may be metachronous within the plateau.

Conclusion

There is a need for very detailed and multidisciplinary research concentrating on these rapid changes as recorded in varved lacustrine sediments and in ice cores (both to be correlated by δ^{18}O stratigraphy) to understand the leads and lags of these rapid climatic and environmental shifts. Such changes are obviously faster than (and superimposed on) the climatic changes resulting from orbital forcing. Of the following two key climatic parameters: atmorpheric CO_2 concentration and ocean circulation (deep water formation), the latter seems to be discernible in continental sediments and in tree rings (if Δ^{14}C is calculated), as well as in ocean cores. In addition, lake deposits offer a wide variety of biostratigraphies which may enable us to evaluate what future rapid climatic changes could possibly mean to ecosystems.

Acknowledgements

My cordial thanks to all who helped with this project: M. Andrée, M. M. Boucherle, L. Chaix, R. B. Davis, U. Eicher, S. A. Elias, M.-J. Gaillard, W. Hofmann, G. Lang, H. Löffler, A. Lotter, M. Moell, H. Oeschger, T. Riesen, C.

130

Scherrer, U. Siegenthaler, K. Tobolski, E; Venanzoni, B. Wilkinson, H. Zbinden, H. Züllig. The study was supported by the Swiss National Science Foundation.

References

Ammann, B., 1985. Introduction and Palynology: vegetational history and core correlation at Lobsigensee (Swiss Plateau). Diss. Bot. *87*: 127–134.

Ammann, B., L. Chaix, U. Eicher, S. A. Elias, M.-J. Gaillard, W. Hofmann, U. Siegenthaler, K. Tobolski & B. Wilkinson, 1983. Vegetation, Insects, Molluscs and Stable Isotopes from Late Würm deposits at Lobsigensee (Swiss Plateau). Studies in the Late Quaternary of Lobsigensee No. 7. Revue Paléobiologie 2: 221–277.

Ammann, B., M. Andrée, L. Chaix, U. Eicher, S. A. Elias, W. Hofmann, H. Oeschger, U. Siegenthaler, K. Tobolski, B. Wilkinson & H. Züllig, 1985. An attempt at a palaeoecological synthesis. Diss. Bot. 87: 165–170.

Ammann, B. & K. Tobolski, 1983. Vegetational Development During the Late-Würm at Lobsigensee (Swiss Plateau). Studies in the Late Quaternary of Lobsigensee No. 1. Revue Paléobiologie 2: 163–180.

Andrée, M., H. Oeschger, U. Siegenthaler, T. Riesen, M. Moell, B. Ammann & K. Tobolski, 1986. [14]C Dating of Plant Macrofossils in Lake Sediment. Radiocarbon 28: 411–416.

Atkinson, T. C., K. R. Briffa & G. R. Coope, 1987. Seasonal temperatures in Britain during the past 22 000 years, reconstructed using beetle remains. Nature 325: 587–592.

Birks, H. J. B., 1981. The use of pollen analysis in the reconstruction of past climates: a review. In T. M. L. Wigley, M. J. Ingram & G. Farmer, Climate and History. Cambridge University Press, Cambridge: 111–138.

Birks, H. J. B., 1986. Late-Quaternary biotic changes in terrestrial and lacustrine environments, with particular reference to north-west Europe. In B. E. Berglund (ed.), Handbook of Holocene Palaeoecology and Palaeohydrology. John Wiley, Chichester: 3–65.

Bogaard, P. v.d. & H.-U. Schminke, 1985. Laacher Tephra: a widespread isochronous late Quaternary tephra layer in central and northern Europe. Geol. Soc. Am. Bull. 96: 1554–1571.

Chaix, L., 1983. Malacofauna from the Late-Glacial Deposits of Lobsigensee (Swiss Plateau). Studies in the Late Quaternary of Lobsigensee No. 5. Revue Peléobiologie 2: 211–216.

Chaix, L., 1985. Malacofauna at Lobsigensee. Diss. Bot. 87: 148–150.

Eicher, U., 1987. Die spätglazialen sowie die frühpostglazialen Klimaverhältnisse im Bereich der Alpen: Sauerstoffisotopenkurven kalkhaltiger Sedimente. Geographica Helvetica 1987-2: 99–104.

Eicher, U. & U. Siegenthaler, 1976. Palynological and oxygen isotope investigations on Late-Glacial sediment cores from Swiss Lakes. Boreas 5: 109–117.

Elias, S. A. & B. Wilkinson, 1983. Lateglacial Insect fossil assemblages from Lobsigensee (Swiss Plateau). Studies in the Late Quaternary of Lobsigensee No. 3. Revue Paléobiologie 2: 189–204.

Elias, S. A. & B. Wilkison, 1985. Fossil assemblages of Coleoptera and Trichoptera at Lobsigensee. Diss. Bot. 87: 157–161.

Gaillard, M.-J., 1983. On the Occurrence of *Betula nana* L. Pollen Grains in the Late-Glacial Deposits of Lobsigensee (Swiss Plateau). Studies in the Late Quaternary of Lobsigensee No. 2. Revue Paléobiologie 2: 181–188.

Gaillard, M.-J., 1985. Postglacial palaeoclimatic changes in Scandinavia and Central Europe. A tentative correlation based on studies of lake level fluctuations. Ecologia Mediterranea 11: 159–175.

Hofmann, W., 1983. Stratigraphy of subfossil Chironomidae and Ceratopogonidae (Insecta: Diptera) in late-glacial littoral sediments from Lobsigensee (Swiss Plateau). Studies in the Late Quaternary of Lobsigensee No. 4. Revue Paléobiologie 2: 205–209.

Hofmann, W., 1985a. Developmental history of Lobsigensee: subfossil Chironomidae (Diptera). Diss. Bot. 87: 154–156.

Hofmann, W., 1985b. Developmental history of Lobsigensee: subfossil Cladocera (Crustacea) Diss. Bot 87: 150–153.

Friedel, H., 1967. Verlauf der alpinen Waldgrenze im Rahmen anliegender Gebirgsgelände. *In:* Oekologie der alpinen Waldgrenze, Mitt. Forstl. Bundes-Versuchsanstalt Wien 75: 81–172.

Hustich, I., 1983. Tree-line and tree-growth studies during 50 years: some subjective observations. *In:* Morisset, P. & S. Payette: Tree-line Ecology. Proceedings of the Northern Québec Tree-Line Conference, Collection Nordicana 47: 181–188.

Iversen, J., 1954. The Late-Glacial Flora of Denmark and its Relation to Climate and Soil. Danmarks Geol. Unders. II. Raekke 80: 87–119.

Keller, P., 1935. Pollenanalytische Untersuchungen an Mooren im Wallis. Vierteljahrsschrift Naturf. Ges. Zürich. 80: 17–74.

Küttel, M., 1979. Pollenanalytische Untersuchungen zur Vegetationsgeschichte und zum Gletscherrückzug in den westlichen Schweizer Alpen. Ber. Schweiz. Bot. Ges. 89: 9–62.

Lang, G., 1985. Palynologic and stratigraphic investigations of Swiss lake and mire deposits. A general view over a research programme. Diss. Bot. 87: 107–114.

Lang, G. & K. Tobolski, 1985. Hobschensee – Late-glacial and holocene environments of a lake at the timberline in the Central Swiss Alps. Diss. Bot. 87: 209–228.

Löffler, H., 1986. An early meromictic stage in Lobsigensee (Switzerland) as evidenced by ostracods and *Chaoborus*. Studies in the late Quaternary of Lobsigensee No. 12. Hydrobiologia 143: 309–314.

Lotter, A., 1985. Amsoldigersee – Late-glacial and holocene environments of a lake at the southern edge of the Swiss Plateau. Diss. Bot. 87: 185–208.

Lotter, A. & M. M. Boucherle, 1984. A Late-Glacial and Post-Glacial history of Amsoldingersee and vicinity, Switzerland. Schweiz. Z. Hydrol. 46: 192–209.

Müller, H.-N., 1984. Spätglaziale Gletscherschwankungen in den westlichen Schweizer Alpen und im nordisländischen Tröllaskagi-Gebirge. Küng, Näfels: 205 pp.

Oberdorfer, E., 1937. Zur spät- und nacheiszeitlichen Vegetationsgeschichte des Oberelsasses und der Vogesen. Z. Botanik 30: 513–57.

Oeschger, H., M. Andrée, M. Moell, T. Riesen, U. Siegenthaler, B. Ammann, K. Tobolski, B. Bonani, H. J. Hofmann, E. Morenzoni, M. Nessi, M. Suter & W. Wölfli, 1985. Radiocarbon chronology at Lobsigensee. Comparison of materials and methods. Diss. Bot. 87: 135–139.

Siegenthaler, U., M. Heimann & H. Oeschger, 1980. ^{14}C variations caused by changes in the global carbon cycle. Radiocarbon 22/2: 177–191.

Siegenthaler, U. & Eicher, 1985. Stable isotopes of carbon and oxygen in carbonate sediments of Lobsigensee. Diss. Bot. 87: 162–164.

Stuiver, M. & H. A. Polach, 1977. Reporting of ^{14}C data. Radiocarbon 19/3: 355–363.

Tranquillini, W., 1967. Ueber die physiologischen Ursachen der Wald- und Baumgrenze. In: Oekologie der alpinen Waldgrenze, Mitt. Forstl. Bundes-Versuchsanstalt Wien 75: 457–487.

Welten, M., 1982. Vegetationsgeschichtliche Untersuchungen in den westlichen Schweizer Alpen: Bern-Wallis; Denkschr. Schweiz. Naturf. Ges. 95: 1–104, 37 diagrams.

Wright, J. E., 1984. Sensitivity and response time of natural systems to climatic change in the Late Quaternary. Quaternary Science Reviews 3: 91–131.

Züllig, H., 985. Carotenoids from plankton and phototrophic bacteria in sediments as indicators of trophic changes: evidence from the Late-glacial and the early Holocene of Lobsigensee. Diss. Bot. 87: 143–147.

Züllig, H., 1986. Carotenoids from plankton and photosynthetic bacteria in sediments as indicators of trophic changes in Lake Lobsigen during the last 14000 years. Hydrobiologia 143: 315–319.

Littoral and offshore communities of diatoms, cladocerans and dipterous larvae, and their interpretation in paleolimnology

David G. Frey

Department of Biology, Indiana University, Bloomington, IN 47405, USA

Key words: offshore transport, incomplete integration, *Bosmina* response, *Tanytarsus lugens* community

Abstract

The remains of diatoms, cladocerans, and midges are usually the most abundant of freshwater organisms and to now have been most useful in interpreting past conditions in a lake. Each taxocene consists of two separate communities, one in the warm littoral zone and the other offshore. Remains of inshore organisms are moved offshore by wind-generated currents, the amount of transport varying with individual characteristics of the lakes. Nowhere do remains of the two communities become completely integrated numerically, although the remains of the littoral chydorid Cladocera become integrated by species before they are incorporated into the sediments. The taxa of the planktonic *Eubosmina* and of the offshore midges correspond to levels of productivity in present-day lakes, and hence changes in the fossil record are commonly regarded as indicating eutrophication over time. The deepwater midges respond to the concentration of dissolved oxygen in deep water, which may be controlled more by a decrease in volume of deep water through accumulation of sediments than by any real increase in edaphic productivity. While such changes are going on offshore during the Holocene, the littoral communities of cladocerans and midges are scarcely changing at all, suggesting a different response of the inshore from the offshore communities to longterm changes resulting from increasing productively or from other functions. Thus, considering these different responses of the two communities of organisms and their incomplete mixing, the remains of littoral and offshore taxa recovered from an offshore core of sediments must be tabulated separately and interpreted separately. For any studies involving accumulation rates, there must be an understanding of the integration of inshore and offshore remains, its variation over the lake bottom, and how it may have varied with marked fluctuations in water level.

Introduction

This report is concerned with three groups of organisms occurring in fresh waters – diatoms *(Bacillariophyceae)*, cladocerans *(Cladocera)*, and dipterous larvae (mainly *Chironomidae* = midges, plus *Ceratopogonidae* and *Chaoboridae)*. Except possibly for chrysophytes and remains of other groups of non-diatom algae, sometimes also ostracods and testaceous rhizopods, they provide the most abundant remains of aquatic organisms in offshore lake sediments, from which taxa can

Originally published in
Journal of Paleolimnology **1**: 179–191.

134

be recognized even to varieties and forms in the diatoms, species in the cladocerans, and genera, sometimes species, in the dipterous larvae. Most studies to date have been done on the diatoms, cladocerans, and midges, and these groups consequently are the best known. The questions asked about them, and by extrapolation about the other groups as well, are: 1) to what extent can these remains be used to determine past conditions in lakes and their controlling influences, and 2) for a given region are there any common patterns of change among lakes, or does each lake respond completely individualistically?

Some forcing variables are worldwide or regional, such as the warming near the end of the last glacial age, the longterm changes in temperature and precipitation in the Holocene, the recent general increase in the level of production in the lakes from human, agricultural, and industrial wastes, and, still more recently, acid deposition, heavy metals, and various toxic organics. Such major external changes should certainly have influenced the species of organisms present in the lakes and their abundance, and hence the occurrence of their remains in the sediments, but what about more minor changes in individual lakes? We recognize a lake as being a sensitive responder to what it receives from its watershed and the atmosphere, variably buffered by the characteristics of the lake and its watershed. Hence, unless the species are too euryoecious, they must have participated in the overall responses to these changes, although possibly only by readjustment of their absolute and relative abundances.

Another question is, to what extent is the material present in an offshore core representative of the lake as a whole? For purposes of the present discussion, a lake can be divided into three major habitats – a littoral benthic, an offshore benthic, and a planktonic habitat. Each of these has species occupying niches in it. A deepwater core to be representative of a lake must contain interpretable proportions of the species from all three zones. Does it? The deep water of stratified lakes experiences a decrease in dissolved oxygen from various causes and an increase in temperature as loss of stratification occurs through accumulation of sediments or reduction in water level. Organisms in the littoral zone and in the offshore water may well respond differently to changes in nutrient delivery from the watershed or to other influences. Thus, in any core study, the littoral and offshore species of a group must be separated if possible and interpretations made independently from them.

In the littoral zone there are many taxa of midges and cladocerans that do not occur in the plankton or the deepwater benthos (Fig. 1). When they molt or die their remains tend to stay in the sediments on site. They can be moved offshore by currents set up by the wind, but they do not become completely integrated numerically with remains produced offshore. Along a transect outward from shore there are progressive changes in the proportional representation of inshore and offshore taxa, and hence the question arises whether one core, or even several cores, from deepwater sediments accurately represent the overall composition of the biota in past times. Sediment focusing occurs with pollen and spores, and although it tends to occur in these two groups of aquatic animals as well, its effects are muted by the tendency toward persistence of littoral remains in shallow-water sediments. Many of these statements about littoral and offshore communities, the offshore transport of littoral remains, and their incomplete integration with remains produced offshore apply to the diatoms as well.

The chief purpose of this discussion is to document the incomplete offshore transport of remains of organisms living in shallow water and to point out the resulting problems of interpreting past conditions in a lake from remains recovered from an offshore core of sediments.

Diatoms

Many species of diatoms live in the littoral zone mostly attached to various substrates, others occur in the offshore plankton, and still others occur in wet soils of the watershed and in channels entering the lake. Frustules of the non-planktonic

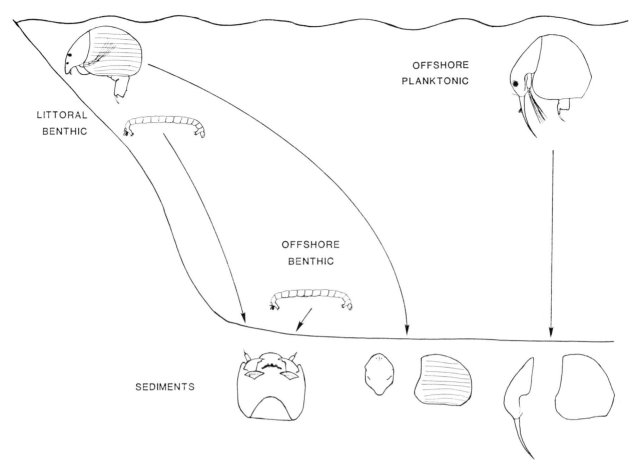

Fig. 1. Major habitats of Cladocera and Chironomidae (midges) and the accumulation of their remains in deepwater sediments. Chydorid Cladocera and most midges are benthic in the littoral zone, whereas *Bosmina* is planktonic in the open water. Some midge larvae also can occur in deepwater sediments, the species present being controlled by the oxygen concentration at the end of stratification. (Based in part on Hofmann, 1986[b]).

species (collectively called periphyton) become variably transported offshore and mixed with the remains of the planktonic species. Frustules of a hundred or more taxa typically can be recovered from a single lake, many in large numbers. Close-interval stratigraphies are constructed, like pollen diagrams.

Much of the present emphasis with diatoms concerns reconstruction of pH in past time. Hustedt (1939) distributed the diatom taxa among five categories of pH tolerance, ranging from acidobiontic to alkalibiontic. Subsequent attempts have been made to use these five categories through some sort of index for calculating

pH in past time. Nygaard's (1956) alpha index (the best of the three he proposed) was refined by Meriläinen (1967), and then further modified by Renberg & Hellberg (1982) to a form that is now used widely in Europe (Battarbee, 1986). Another approach not dependent on Hustedt's pH tolerance categories is an environmental calibration, which involves relating the taxa of diatoms and their abundance in the surficial sediments of a particular lake to the pH of that lake, repeated for a large number of lakes of different pH and other characteristics in a region. Equations for the pH relationship are generated by computer, which yield transfer coefficients that can be used to infer the pH in past time from the diatom assemblage recovered from sediments deposited in that time period. The inferred pH varies somewhat, depending on which particular procedure was used (Davis & Anderson, 1985; Davis, 1987). No single procedure has been accepted yet as the best one to approximate past pH.

Davis *et al.*, (1985) have attempted to ascertain what environmental parameters besides pH control the composition of the diatom taxocene. Total organic carbon (TOC), which in the acidic, oligotrophic lakes being studied is almost the same as dissolved organic carbon (DOC), was most important. It declines as pH declines below pH ~ 5.5, which means that at low pH fewer organic ligands will be available to bind with aluminium or other metallic ions and thereby reduce their toxicity. This is one partial explanation for the toxicity of acidified waters to fishes.

What happens with diatoms as the pH declines in the acidic range is that planktonic taxa tend to disappear (Battarbee, 1986). Hence, in studies of acidic waters the diatomist does not have to worry excessively about the separation of planktonic from periphytic taxa and the interpretation of their responses separately. But if we hope to develop a general science of paleolimnology, then lakes will be studied that have had near-neutral or alkaline pH throughout their existence, with no indication of any marked periods of acidity. In such instances planktonic taxa will be present along with the periphytic taxa, and in addition there will be a greater likelihood of dissolution of the frustules. Here is where problems might arise.

Sweets (1983) analyzed diatoms in the plankton, in sediment collectors, and in the sediments at four stations across Jellison Pond in Maine. He found that remains of the littoral taxa were moved down slope by resuspension and transport within a thin layer at the bottom, possibly only a few centimeters thick, but that they never became completely integrated numerically with remains of the offshore planktonic taxa. As Sweets emphasized, which is just as true for other groups of aquatic organisms as well, we must be concerned with the transport of frustules from their place of production to the place of sediment sampling, as

well as with losses from fragmentation and dissolution. Dixit (1986) came to much the same conclusion from his study and from those of earlier investigators that periphytic diatoms tend to be underrepresented in deepwater sediments, and that there is much habitat-dependent preferential accumulation in a lake. During transport, or during their being eaten by animals, the frustules can become fragmented, in which condition they seem to be more easily dissolvable. Dissolution occurs more commonly in alkaline pH, although other factors are also involved. Diatoms recovered from sediments can correspond almost precisely to their producing populations, as demonstrated nicely by Haworth's (1980) study, in which she was able to compare her closely dated sediment results with Lund's weekly phytoplankton samples, whereas in other lakes a very poor correspondence can result from a number of factors (Battarbee, 1986). In Lake Michigan, for example, as many as 70–90% of the frustules are dissolved before they reach the sediments at depths of 60 m or more (Parker *et al.*, 1978).

Thus, for the moment at least, diatomists are relatively secure because of concentrating their attention on the effects of acidity in poorly buffered waters. At low pH diatom frustules tend not to be redissolved. Planktonic taxa tend not to be present, and hence the integration of inshore with offshore taxa is a minor consideration. Replicate cores from deep water show intercore variability in numbers of frustules, but the percentages of the various taxa are often quite consistent from core to core (Battarbee, 1986). Accumulation rates, though, which free the taxa from the constraints of percentages (for studying fluxes), can be different from core to core, because of different sediment accumulation rates, differences in bioturbation, different concentrations of diatoms, etc. Such matters reflect the non-homogenity of sedimentation processes and accumulation of sediments over a lake bottom. These should be the concern of specialists in any group of organisms.

Cladocera

Cladocerans occur in much smaller number of species, roughly 100–150 for a continent, but their remains being chitinous are mostly resistant to microbial action and hence tend not to be destroyed over time, either in the water or by diagenesis in the sediments. The species can be recognized precisely from the various exoskeletal parts that occur – headshields, shells, postabdomens, claws, ephippia, and a few other parts for certain taxa. Apparently all components of the exoskeleton of the Bosminidae and Chydoridae preserve quantitatively, except possibly for the first instar, and hence they have been used most commonly, sometimes solely, for interpreting past conditions. However, all cladocerans in a local faunule leave remains in the sediments (Cotten, 1985), and hence, for some studies at least, all species should be tabulated and used in interpretations. Taxa of the other nine families are represented variously only by postabdomens, claws, ephippia, mandibles, and antennal segments, never by headshields and shells, except occasionally in the *Macrothricidae*. Each individual cladoceran molts an indefinite but relatively small number of times during its life, after only two or at best a few reproductive instars.

The cladocerans are organized into two distinct communites – planktonic and meiobenthic. The planktonic taxa consists mainly of species in the genus *Bosmina*, which are well preserved, and in *Daphnia, Ceriodaphnia,* and *Diaphanosoma*, which are not. Hofmann (1986b) has studied a number of lakes in northern and southern Germany, Switzerland, and the Baltic Sea east of Bornholm where in late-glacial time and the early Holocene, *Bosmina longispina* was present as the only taxon of the subgenus *Eubosmina*, then *B. coregoni* f. *kessleri* invaded, and finally *B. coregoni* f. *coregoni* occurred. The latter sometimes is the only *Eubosmina* present now. This sequence parallels the occurrence of *Eubosmina* taxa in present-day European lakes according to their trophic state, and also parallels changes in the deepwater midge fauna accompanying progressive decrease in hypolimnetic oxygen concentration. Hence the

Bosmina shift is considered to reflect changes in productivity over time – from oligotrophic at the beginning, through mesotrophic, to more eutrophic in recent time.

A shift from *Bosmina longispina* (sometimes incorrectly called *B. coregoni* in the literature) to *B. longirostris*, which is in a different subgenus, has been documented in a number of studies and has been considered a consequence of eutrophication. In Swiss lakes with annually laminated recent sediments, the rapidity of this species replacement is striking, sometimes seeming to occur in just one or a couple of years (Boucherle & Züllig, 1983). As the lakes recovered from eutrophication through removal of phosphorus from influents, *B. longispina* reappeared and *B. longirostris* dropped accordingly. The work that Hofmann (1986a, 1987) has been doing recently, though, shows that *B. longirostris* often occurred in a lake throughout the time that the succession from *B. longispina* to *B. coregoni* was occurring. Thus, the significance of the sudden shift from only one oligotrophic taxon of *Eubosmina* originally to the supposedly eutrophic *B. longirostris* after some threshold had been crossed needs reconsideration.

Chydorid presence and abundance in cores and some attempted interpretation of them have been reported by a fair number of persons over the past three decades, although realization of the presence of cladoceran remains in sediments goes back to the last century (Frey, 1964). Chydorids are meiobenthic (Frey, 1987), mainly in the littoral zone, except for *Chydorus* gr. *sphaericus*, which becomes planktonic during periods of high productivity. Actually, instead of being truly planktonic it seems to use the larger colonies of cyanobacteria as a substrate. Mueller (1964) from transects of three lakes in northern Indiana (with areas of 3.4, 27, and 204 ha) determined that the concentration of chydorid remains in the sediments decreases from the littoral zone outward into deeper water, and that the concentrations of *Bosmina* remains are greatest at depths corresponding to the upper part of the hypolimnion and then decline shoreward and lakeward from here (Fig. 2). The planktonic and meiobenthic taxa nowhere become integrated in concentration, but significantly the chydorid remains in the sediments become integrated by relative abundance of species regardless of their origin in diverse microhabitats inshore and at different times of the year (DeCosta, 1968; Goulden, 1969a). Thus, a sample of surficial sediments from almost anywhere in a lake will represent the total chydorid community of the lake integrated over microhabitat and time (at least one year, possibly several where bioturbation and resuspension at overturn are important). Consequent-

ly, percentage composition of chydorids is a good descriptor of present and past populations in the entire lake.

In late-glacial time only a few species of chydorids were present in the lakes studied (e.g., Frey, 1958; Goulden, 1964; Hofmann, 1986a) none of them cold stenotherms and all of them capable of continuing in the lake in relatively high abundance throughout the Holocene. We do not know of any truly cold stenothermal chydorids, although *Chydorus arcticus*, recently described by Røen (1987) may be one, and possibly also the several species of cladocerans on the subantarctic islands (Frey, 1988b). Hofmann's study (1986a) on Schöhsee and Grosser Plöner See in northern Germany shows a low number of species (not shown in the figure) and a low H' diversity in late-glacial time (Fig. 3), followed by a gradual increase in number of taxa and diversity in post-

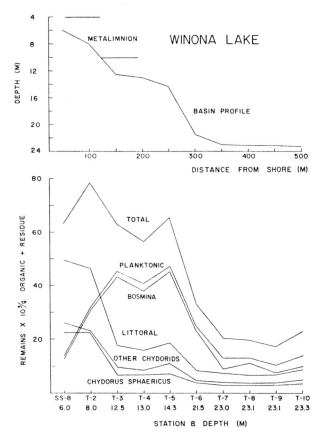

Fig. 2. Concentration of the remains of littoral and planktonic Cladocera in sediments of Winona Lake, Indiana, along a transect from the littoral zone offshore into the profundal. Nowhere do the littoral and planktonic remains become completely integrated with one another by concentration. The vertical axis of the lower figure is the number of remains per dry weight of sediment, minus the $CaCO_3$ calculated from the loss of CO_2 during ashing at 925 °C. (From Mueller, 1964.)

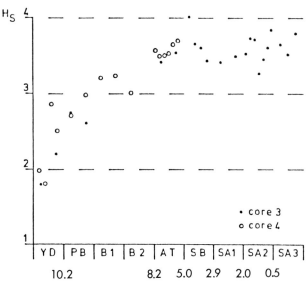

Fig. 3. Progressive increase in the Shannon-Wiener diversity index (H_s) for sedimentary remains of Cladocera over time in Lake Schöhsee in Holstein. The magnitude of the index is controlled by the number of taxa and their relative abundance. In this study the number of taxa present (although not shown) increased significantly from the Younger Dryas into Atlantic time. Letters along the horizontal axis are for the standardized pollen (time) zones of European sequences: YD = Younger Dryas, PB = Preboreal, B = Boreal, AT = Atlantic, SB = Subboreal, SA = Subatlantic. At the bottom are radiocarbon ages in thousands of years. (After Hofmann, 1986a)

glacial time to maxima in the Atlantic Period, which have persisted to the present. Goulden (1969b) had previously demonstrated a similar relationship in Lake Lacawac, PA. As the number of cladoceran species increased in postglacial time, the H' index tracked between the minimum and maximum H' values for any given number of species, swinging strongly toward the maximum diversity limit with higher species numbers in more recent time. Hofmann believes this increase most likely resulted from amelioration in climate, although it certainly is associated with maturation and stabilization of the lakes as well. In his two lakes the *Eubosmina* and midge taxa exhibited the species-replacement response considered typical of eutrophication, or at least reduction in deepwater oxygen concentration, but the chydorids did not. The latter found conditions satisfactory for them to remain at good numbers throughout the Holocene. Hofmann considers the ups and downs in percentage composition to be 'normal' variance in the data, although the taxocene obviously was responding to environmental changes not yet interpretable.

What Hofmann and others, perhaps beginning with Whiteside (1970), have attempted to do is classify the chydorids into habitat groups – macrophyte species, sand species, mud species, clear-water species, and polluted-water species. Such results show that the macrophyte and bare-bottom species maintained about the same relative relationship in Schöhsee throughout the Holocene. They did the same in Grosser Plöner See, except that here there was some disturbance at the top, where species that respond positively to pollution became more abundant. Yet during the Holocent both lakes were becoming overall more productive, as indicated by the *Bosmina* and the midges. Thus, the chydorids in the littoral zone seem not to respond to changes associated with increasing productivity, until the changes become great enough to affect the littoral habitat itself, as by the elimination of macrophytes through blooms of algae (Whiteside, 1969). Parenthetically, an increase in very fine inorganic turbidity that reduces light penetration could possibly bring about the same change. When conditions are stable and well adjusted, the frequency distribution of chydorid species corresponds almost precisely to the MacArthur broken-stick distribution (Goulden, 1966). Disturbance from whatever cause (agriculture, climate, volcanism: Frey, 1976; other causes: Whiteside & Swindoll, 1988) results in a disruption of this configuration, with the commonest species now much more common than expected from the model.

Other attempts have been made to use cladocerans in the same way as diatoms to calculate past pH, and with some success according to Steinberg et al., (1984) and Krause-Dellin and Steinberg (1987). Hofmann (1986a), however, has re-examined their calculations and the assignment of taxa to particular groups, and has concluded that the calculations are erroneous in a number of respects, at least for non-acidic lakes. Some of the chydorids were assigned to improper pH categories, and the pH for Schöhsee and Grosser Plöner See according to the formula used was controlled more by the circumneutral taxa than by the acidophilous ones. The ecology of the species is presently too poorly known to establish a meaningful range of environmental requirements.

Most scientists are still using percentage diagrams to interpret results and are still attempting to assess from them such gross environmental changes as eutrophication and acidification. But the changes in relative, or absolute, abundance of chydorids over time cannot be just variance in a highly variable system, controlled in part by the vagaries of sedimentation. The entire microscopic community must be responding to changes in deliveries of various substances from watershed and atmosphere as well as to changes in climate. We have to determine if the chydorids themselves can tell us anything about past conditions in a lake, and, by interference, the external influences that could have brought them about. Cladocera are part of a complex ecosystem, but we hope that they will give some primary information about past conditions and not merely confirm conditions suggested by other evidence.

It looks as if this may be possible through consideration of the entire cladoceran faunule of a

140

lake and by the use of sophisticated computer analysis. Cotten (1985) studied the cladoceran remains in the surficial sediments of 46 lakes in the 150 in eastern Finland that Meriläinen and his group are studying. She obtained a total of 71 species, including several new ones. Her three-dimensional data matrix consists of 46 lakes versus the percentages of the various taxa present versus the 21 measured environmental variables for the lakes and their watersheds. She used de-trended correspondence analysis for ordination, minimum variance clustering for classification of assemblages, and multiple discriminant analysis for checking various results.

She found five distinct lake groups well separated from one another (Fig. 4). The diatomists (Huttunen & Meriläinen, 1983) using about the same number of lakes from the same population but only partially overlapping those used by Cotten had previously found five groups based only on diatoms. Analyses by Cotten (1985) and by Huttunen *et al.*, (1988) showed that the two sets of five groups reflect the same limnological conditions almost precisely, and hence that the two groups of organisms should be almost equally sensitive in estimating parameters in the lakes in past time, the diatoms being a little better at the moment, possibly from their much greater number of taxa.

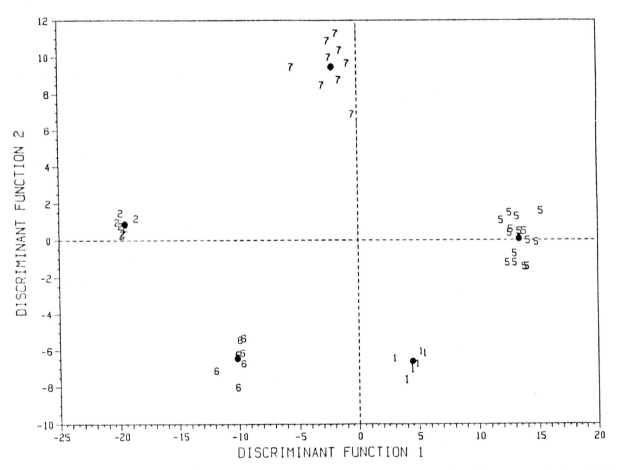

Fig. 4. Results of stepwise, multiple discriminant analysis of the remains of the 39 most abundant cladoceran taxa in the surficial sediments of 46 lakes in eastern Finland. Each group of numbers represent a distinct lake type based on cladoceran assemblages. Each number is a separate lake. The closed circles are the group means. The first discriminant function, accounting for 76.5% of the variance, is correlated primarily with nutrients and pH, the second, accounting for an additional 16.8% of the variance, with water transparency. (From Cotten, 1985).

Thus, the cladocerans seem to respond just as sensitively as diatoms to environmental conditions that affect processes in lakes. This is really not surprising, because the relative abundance of 25 species of chydorids in one lake in the Eemian Interglacial of Denmark was almost the same as in the whole country today (Frey, 1962; Whiteside, 1970), suggesting that their ecology has not changed appreciably over more than 100000 years. And since the Cladocera are just one part of a microscopic community in the littoral, which is much the same all over the World, one can assume with confidence that the other components, including the diatoms, did not change much either.

Dipterous larvae

Aquatic dipterous larvae consist of members of the families Chironomidae, Ceratopogonidae, and Chaoboridae, with the Chironomidae (midges) being most abundant as to individuals and in having the most species. In Europe there are estimated to be about 1400 described species of midges (Fittkau & Reiss, 1978), of which the number in a single lake can easily exceed 100 [115 in Grosser Plöner See (by Hofmann, ms), 150 in Sweden (Brundin, 1949)]. Although the number of species of chironomids in a lake is about twice as great as the number of chydorid Cladocera, the number of remains per unit of sediment is normally only about 1/100 as great, or even much less. Only the head capsules of the larvae occur in sediments, and occasionally the hypopygia of adult males. The latter can be important for determining species. The profundal benthic taxa can usually be identified to species from their larval head capsules, whereas the much more diverse littoral taxa can often be identified only to genus, and sometimes only to subfamily.

Thienemann (1920) set up a successional series of lake types based on the chironomid taxa in their offshore benthos. At one end was a *Tanytarsus* community considered to represent oligotrophic lakes, and at the other end a *Chironomus* community, with a *Stictochironomus* community in between. This change has been associated with and is probably controlled by the dissolved oxygen content of the deep water. As a lake becomes more productive (= oxygen demand), as the volume of the hypolimnion decreases by accumulation of sediments or reduction in water level (= oxygen supply), or as the rate of utilization of oxygen increases because of higher temperature, the oxygen concentration decreases. Thus, the oxygen concentration in deep water toward the end of stratification and the particular species of benthic midges that can survive this condition are controlled by other influences than just nutrient inputs and their utilization. The task of the paleolimnologist is to evaluate these various potential influences and decide which were most important in a particular instance. Quite likely the species composition of the deepwater midges is controlled more by increasing morphometric eutrophy over time from sediment accumulation than by edaphic eutrophy.

One of the difficulties with this system is that *Tanytarsus* is a genus occurring primarily in shallow water, with just a couple of species in deep water under special conditions. The oligotrophic community of Europe is now defined as a *Tanytarsus lugens* community, all the species of which are obligate cold stenotherms with a high oxygen demand. *Sergentia coracina* and *Stictochironomus rosenschoeldi* are considered to represent an intermediate condition. The meroplanktonic predator *Chaoborus*, which can tolerate even lower levels of oxygen than *Chironomus*, typically overlaps it in temporal sequence as oxygen reduction becomes more stressful.

In late-glacial time when water temperatures were cold, members of the *Tanytarsus lugens* community seemingly occupied the entire lake bottom. Low temperatures restricted the occurrence of other taxa. As the climate warmed, the *T. lugens* species were eliminated from shallow water and became restricted to the colder deep water of stratified lakes. In unstratified lakes they were eliminated rapidly by temperatures above their tolerance limits, even as early as Bølling time (Hofmann, 1983). Warming of the epilimnion in the Holocene allowed taxa requiring higher temperatures to invade the littoral zone, forming a diverse community, which has largely persisted to the present. What happened to the cold stenothermal offshore community is dependent on what the decline in oxygen concentration in deep water was over the Holocene. The community persisted if oxygen conditions remained satisfactory, or it was eliminated and replaced by *Chironomus* species if oxygen concentration declined sufficiently. There may or may not have been an in-between *Sergentia/Stictochironomus* stage.

Because most of the littoral species never move into the profundal at all, possibly because of too low temperatures, their remains recovered from the profundal sediments bear no relationship to oxygen concentration in deep water. This applies particularly to shallow-water species of *Tanytarsus*. The extent to which chironomid remains are moved offshore is still not completely resolved. In his various studies, Hofmann (1986b, for a summary) has found that littoral remains occur quite commonly offshore. Iovino (1975) concluded, however, that they tend to remain where the larvae lived. Iovino studied the seasonal occurrence and abundance of 53 species of midges along four transects in Pretty Lake and Crooked Lake, Indiana, and also the remains in the surficial sediments at each station sampled. The abundance of remains reflected almost precisely the annual abundance of the species living at each particular station (Fig. 5), from which Iovino concluded that offshore transport of littoral remains was minimal. In three larger lakes in Italy and Denmark he found almost no littoral remains in deep water, whereas in a shallower lake, as in Florida, the fossil assemblage in mid-lake was mixed. The offshore transport of head capsules is strongly influenced by surface area of the lake, as affecting wind disturbance, and the volume of deep water, and hence its success declines sharply with depth and distance from shore. In the middle of a large, deep lake there should be only occasional head capsules of littoral species in the sediments.

Like the littoral cladocerans, the littoral midges do not seem to display any relationship to eutrophication until the littoral zone itself is affected by the changes. Many species are present, occupying a great variety of niches. Little is known about their ecology, and few attempts have been made to predict values for any environmental parameter or changes in the waterbodies from the remains recovered from the sediments.

Discussion and summary

Thus, all three groups of organisms are strongly differentiated into inshore and offshore communities. The inshore taxa can be transported offshore by currents, the extent varying in part with size of the lake, but they never become completely integrated numerically with the offshore taxa. Changes in the offshore midges and in the planktonic *Eubosmina* have roughly paralleled eutrophication during the Holocene. The buildup of the chydorid and midge communities in the littoral zone and the restriction of cold-stenothermal midges to deep water near the beginning of the Holocene are in part responses to the increase in temperature at this time. The littoral communities of cladocerans and midges seem not to have responded to increasing productivity (eutrophication) of the lakes until the littoral zones themselves were affected.

For any group of organisms we need to understand the various factors of lake morphometry and of climate that influence the offshore transport of remains of littoral taxa and the extent of their integration with remains of organisms living offshore. For each group of organisms we also need some understanding of the persistence over time of the various types of remains, and the conditions that lead either to their dissolution or else fragmentation to particles too small to be handled on a quantitative basis. What is found in the sediments is only the end product of all the various processes that have been acting on the remains since the producing organisms were alive. For persons concerned only with the last couple of centuries or millennia, such changes in the quality of the remains may be of little real importance, but for any studies of paleolimnology farther back in time, this can be a very real concern.

The ultimate objective is to learn how these

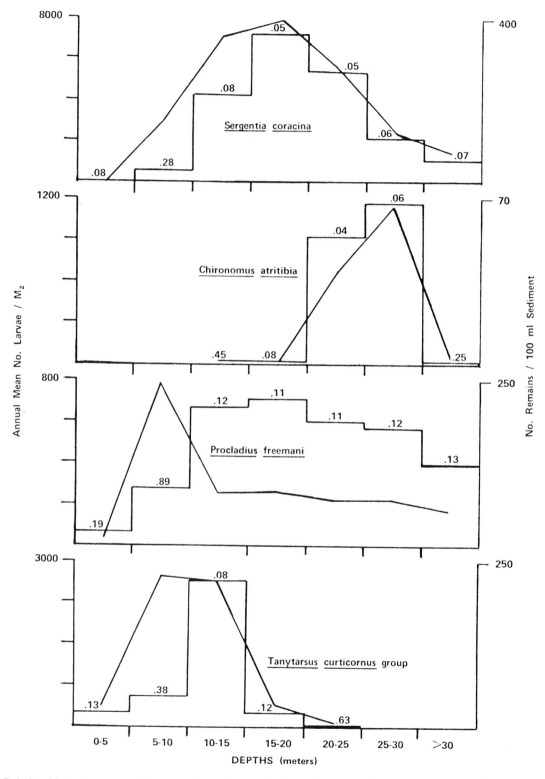

Fig. 5. Relationship between annual means of live chironomid larvae (histograms) and concentrations of their remains in surficial sediments (solid lines) in Crooked Lake, Indiana. The decimal numbers are the ratio between them. Similar diagrams for shallow-water species likewise show a marked concordance between the distribution of living animals and their remains, indicating that few remains of the littoral taxa are transported offshore into deep water. (From Iovino, 1975.)

144

various biotic components of the ecosystem were functioning in past time, not merely their response to major changes in environmental parameters but also to more minor changes. How, for example, did they respond to a volcanic ash fall, to a series of avalanches in the watershed, to slumping of deltaic deposits, to changing deliveries of cations and anions from the watershed or of dissolved and particulate organic matter? How did they respond to changes in temperature before and after the thermal maximum, to marked variation in lake level and volume from prolonged periods of reduced precipitation or higher temperature, to the invasion of planktivorous fishes during the Holocene, and to many other possible influences, all of which, for example, have helped produce the high variance in percentage composition of the chydorids through postglacial time? Obviously the magnitude of response to any of these perturbations will vary from lake to lake, because each lake has a variable buffering component in its system. Understanding these responses will probably come only from an analysis of the total community, not just the few species or groups that are easiest to identify and work with. The species present and their relative abundance will indicate past conditions through transfer coefficients worked out from a calibration series of diverse lakes and of their watershed influences. Because lakes are so highly individualistic, it is safe to predict that the fine response of each lake will also be individualistic to the particular influences on it. Only responses to regional or global changes in temperature and precipitation or to marked changes in nutrient inputs will be found to be more or less similar in all the lakes, although probably not contemporaneous.

Acknowledgements

I am indebted to W. Hofmann of West Germany for information from several of his papers on cladocerans and midges, which at the time of first writing this paper were still unpublished, and to R. W. Battarbee, R. B. Davis, and E. S. Deevey for their critical reading of the original manuscript, which led to substantial clarification of presentation.

References

Battarbee, R. W., 1986. Diatom analysis, pp. 527–570, in: Handbook of Holocene Palaeoecology and Palaeohydrology, B. E. Berglund (ed.). John Wiley: Chichester: i–xxiv, 1–869 pp.

Boucherle, M. M. & H. Züllig, 1983. Cladoceran remains as evidence of change in trophic state in three Swiss lakes. Hydrobiologia 103: 141–146.

Brundin, L., 1949. Chironomiden und andere Bodentiere der Südschwedischen Urgebirgsseen. Rep. Inst. Freshwat. Res. Drottningholm 30: 1–914.

Cotten, C. A., 1985. Cladoceran assemblages related to lake conditions in eastern Finland. Ph. D. dissertation, Indiana University, Bloomington. viii, 96 pp.

Davis, R. B., 1987. Paleolimnological diatom studies of acidification of lakes by acid rain: an application of Quaternary science. Quatern. Sci. Rev. 6: 147–163.

Davis, R. B. & D. S. Anderson, 1985. Methods of pH calibration of sedimentary diatom remains for reconstructing history of pH in lakes. Hydrobiologia 120: 69–87.

Davis, R. B., D. S. Anderson & F. Berge, 1985. Paleolimnological evidence that lake acidification is accompanied by loss of organic matter. Nature 316: 436–438.

DeCosta, J. 1968. Species diversity of chydorid fossil communities in the Mississippi Valley. Hydrobiologia 32: 497–512.

Dixit, S. S. 1986. Algal microfossils and geochemical reconstructions of Sudbury lakes: a test of the paleo-indicator potential of diatoms and chrysophytes. Ph. D. dissertation, Queen's Univ., Kingston, Ont.

Fittkau, E. J. & F. Reiss, 1978. Chironomidae, pp. 404–440 in: Limnofauna Europaea, J. Illies (ed.). Fischer: Stuttgart.

Frey, D. G., 1958. The late-glacial Cladocera of a small lake. Arch. Hydrobiol. 54: 209–275.

Frey, D. G., 1962. Cladocera from the Eemian Interglacial of Denmark. J. Paleontol. 36: 1133–1154.

Frey, D. G., 1964. Remains of animals in Quaternary lake and bog sediments and their interpretation. Ergebn. Limnol. 2:i–ii, 1–114.

Frey, D. G.,, 1976. Interpretation of Quaternary paleoecology from Cladocera and midges, and prognosis regarding usability of other organisms. Can. J. Zool. 54: 2208–2226.

Frey, D. G., 1988a. Cladocera. Chapter in 'Introduction to the study of meiofauna', R. F. Higgins and Hjalmar Thiel (eds.). Smithsonian Press. In Press.

Frey, D. G., 1988b. The Alona on the subantarctic islands is A. weinecki not A. rectangula (Chydoridae, Cladocera). Chapter in book honoring W. T. Edmondson. J. T. Lehman and N. Hairston, Jr. (eds.). Limnol. Oceanogr. In press.

Goulden, C. E., 1964. The history of the cladoceran fauna of Esthwaite Water (England) and its limnological significance. Arch. Hydrobiol. 60: 1–52.

Goulden, C. E., 1966. La Aguada de Santa Ana Vieja: an interpretative study of the cladoceran microfossils. Arch. Hydrobiol. 62: 373–404.

Goulden, C. E., 1969a. Interpretative studies of cladoceran microfossils in lake sediments. Mitt. Internat. Verein. Limnol. 17: 43–55.

Goulden, C. E., 1969b. Developmental phases of the biocoenosis. Proc. Nat. Acad. Sci. 62: 1066–1073.

Haworth, E. Y., 1980. Comparison of continuous phytoplankton records with the diatom stratigraphy in the recent sediments of Blelham Tarn. Limnol. Oceanogr. 25: 1093–1103.

Hofmann, W., 1983. Stratigraphy of subfossil Chironomidae and Ceratopogonidae (Insecta: Diptera) in late glacial littoral sediments from Lobsigensee (Swiss Plateau). Studies in the Late Quaternary of Lobsigensee 4. Rev. Paleobiol. 2: 205–209.

Hofmann, W., 1986a. Developmental history of the Grosser Plöner See and the Schöhsee (north Germany): cladoceran analysis, with special reference to eutrophication. Arch. Hydrobiol., Suppl. Bd. 74: 259–287.

Hofmann, W., 1986b. Analyse tierische Mikrofossilien in Seesedimenten: langfristige Veränderungen der limnische Fauna im Zusammenhang mit der Entwicklung ihres Lebensraum. Habilitationsschrift eingericht der Hohen Mathematisch-Naturwissenschaftliche Fakultät der Christian-Albrechts-Universität zu Kiel.

Hofmann, W., 1987. Stratigraphy of Cladocera (Crustacea) and Chironomidae (Insecta: Diptera) in three sediment cores from the Central Baltic Sea as related to paleosalinity. Int. Rev. ges. Hydrobiol. 72: 97–106.

Hofmann, W., ms. On the significance of chironomid analyses (Insecta: Diptera) for paleolimnological research. 18 pp.

Hustedt, F., 1939. Systematische und ökologische Untersuchungen über die Diatomeenflora von Java, Bali und Sumatra nach dem Material der Deutschen Limnologischen Sunda-Expedition. III. Die ökologische Faktoren und ihr Einfluss auf die Diatomeenflora. Arch. Hydrobiol. Suppl. 16: 274–394.

Huttunen, P. & J. Meriläinen, 1983. Interpretation of lake quality from contemporary diatom assemblages. Hydrobiologia 103: 91–97.

Huttunen, P., J. Meriläinen & C. A. Cotten. 1988. Rough title: Comparison of diatoms and Cladocera in predicting past conditions in lakes, in: Method for the investigation of lake deposits: palaeoecological and palaeoclimatological aspects, IGCP Project 158, Soviet Working Group. Vilnius, Lithuania.

Iovino, A. J., 1975: Extant chironomid larval populations and the representativeness and nature of their remains in lake sediments. Ph. D. dissertation, Indiana University, Bloomington. 54 pp. + appendix.

Krause-Dellin, D. & C. Steinberg, 1987. Cladoceran remains as indicators of lake acidification. Hydrobiologia 143: 129–143.

Meriläinen, J., 1967. The diatom flora and the hydrogen ion concentration of water. Ann. Bot. Fenn. 4: 51–58.

Mueller, W. P., 1964. The distribution of cladoceran remains in surficial sediments from three northern Indiana Lakes. Invest. Indiana Lakes & Streams 6: 1–63.

Nygaard, G., 1956. Ancient and recent flora of diatoms and Chrysophyceae in Lake Gribsø. Fol. Limnol. Scand. 8: 32–94.

Parker, J. K., H. L. Conway & D. N. Edgington, 1978. Dissolution of diatom frustules and silicon cycling in Lake Michigan, USA. Verh. Internat. Verein. Limnol. 20: 336–340.

Renberg, I. & T. Hellberg, 1982. The pH history of lakes in southwestern Sweden, as calculated from the subfossil diatom flora of the sediments. Ambio 11: 30–33.

Røen, U., 1987. *Chydorus arcticus* n.sp., a new cladoceran crustacean (Chydoridae: Chydorinae) from North Atlantic arctic and subarctic areas. Hydrobiologia 145: 125–130.

Steinberg, C., K. Arzet & D. Krause-Dellin, 1984. Gewässerversäuerung in der Bundesrepublik Deutschland im Lichte paläontologischer Studien. Naturwissenschaften 71: 631–633.

Sweets, P. R., 1983. Differential deposition of diatom frustules in Jellison Hill Pond, Maine. MS dissertation, University of Maine, Orono. ix, 261 pp.

Thienemann, A., 1920. Untersuchungen über die Beziehungen zwischen dem Sauerstoffgehalt des Wassers und der Zusammensetzung der Fauna in norddeutschen Seen. Arch. Hydrobiol. 12: 1–65.

Whiteside, M. S., 1969. Chydorid (Cladocera) remains in surficial sediments of Danish lakes and their significance to paleolimnological interpretations. Mitt. Internat. Verein. Limnol. 17: 193–201.

Whiteside, M. C., 1970. Danish chydorid Cladocera: modern ecology and core studies. Ecol. Monogr. 40: 79–118.

Whiteside, M. C. and M. R. Swindoll. 1988. Guidelines and limitations to cladoceran paleoecological interpretations. Palaeogr., Palaeoclimat., Palaeoecol. 62: 40–412.

Early postglacial chironomid succession in southwestern British Columbia, Canada, and its paleoenvironmental significance

Ian R. Walker [1] & Rolf W. Mathewes
Department of Biological Sciences, and Institute for Quaternary Research, Simon Fraser University, Burnaby, British Columbia, V5A 1S6, Canada; [1] Present address: Department of Biology, Queen's University, Kingston, Ontario, K7L 3N6, Canada

Key words: British Columbia, chironomids, Pacific Northwest, paleoclimate, paleolimnology

Abstract

Chironomids typical of cold, well-oxygenated, oligotrophic environments are common in late-Pleistocene deposits, but these taxa are rare in Holocene sediments of most small temperate lakes. Hypotheses to explain the demise of these taxa include variations in climate, lake trophic state, lake levels, terrestrial vegetation, and/or sediment composition. In southwestern British Columbia, this demise correlates with palynological evidence for a lodgepole pine decline, and for rapid climatic amelioration, at about 10 000 yr B.P. Faunal changes are poorly correlated with lithological boundaries. The similar timing of the declines among lakes suggests that a regional influence, climate, has possibly been the principal determinant of early chironomid faunal succession.

Introduction

Although pioneering investigations of chironomid stratigraphy suffered from small sample sizes, poor taxonomic resolution or both, recent studies in North America and Europe (e.g. Brodin, 1986; Günther, 1983; Hofmann, 1971, 1978, 1983a, 1983b; Lawrenz, 1975; Schakau & Frank, 1984; Walker & Mathewes, 1987a; Walker & Paterson, 1983) have provided significant information regarding lake development. In most postglacial stratigraphic profiles a distinct pattern of chironomid faunal development is apparent. During the late-Pleistocene chironomids typical of cold, oligotrophic environments were common. Most

of these taxa disappeared from small temperate lakes prior to Holocene time.

No consensus has emerged as to why the late-Pleistocene fauna disappeared. Although Walker & Mathewes (1987a) have suggested a direct or indirect climatic determinant, this hypothesis has stimulated vigorous debate (Walker & Mathewes, 1987b; Warner & Hann, 1987). It has also been suggested that changes in sediment composition, terrestrial vegetation, lake trophic state, or lake levels might be more important than climate in determining this pattern of chironomid faunal development.

Data included in the present paper provide information pertinent to these hypotheses. Sedi-

Originally published in
Journal of Paleolimnology 2: 1–14.

148

ment cores have been examined from two lakes, Mike and Misty Lakes, in southwestern British Columbia (B.C.) for comparison with results obtained previously at Marion Lake, B.C. (Walker & Mathewes, 1987a). Although Mike Lake lies in close proximity to Marion Lake, Misty Lake is 360 km northwest of Marion and Mike Lakes.

Study sites

Mike Lake

Mike lake (225 m a.s.l.; 49° 16.5′ N, 122° 32.3′ W) is located 3 km south of Marion Lake. Because it lies farther from the mountains than Marion Lake and at lower elevation (Fig. 1), the climate at Mike Lake is probably slightly

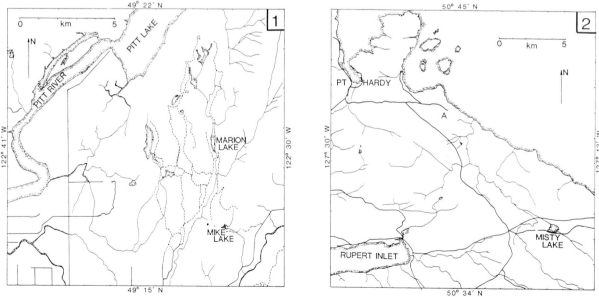

Fig. 1. Locations of Mike and Misty Lakes in southwestern British Columbia, Canada (1 = Mike Lake area, 2 = Misty Lake area).

149

Table 1. Climatic summaries (1951–1980) for Loon Lake (49° 18′ N, 122° 35′ W; 354 m elev.), and Administration (49° 16′ N, 122° 34′ W; 143 m elev.), Univ. British Columbia Research Forest, at Haney, and for Port Hardy Airport (50° 41′ N, 127° 22′ W; 22 m elev.), northern Vancouver Island, B.C. (Environment Canada, 1982)

| | Univ. B.C. Res. For. | | Port Hardy |
	Loon Lk	Administr.	Airport
Mean Daily Temperature			
Coldest Month (Jan)	0.5	1.4	2.4 °C
Warmest Month (Jul)	16.3	16.8	13.8
Precipitation			
Rain: Annual	2459.1	2059.3	1705.8 mm
Wettest Month (Dec)	343.3	306.3	260.3
Driest Month (Jul)	86.7	65.5	52.0
Snow: Annual	195.2	81.6	72.1 cm
Frost-free Period	199	198	177 d
Degree-days			
Above 0 °C	3092.7	3445.7	2931.2 °C
Above 5 °C	1633.7	1882.1	1350.0

warmer and drier. Forests surrounding the lake are placed within the drier subzone of the Coastal Western Hemlock zone (Klinka, 1976). Climatic differences between Mike and Marion Lakes are expected to be slight, however. Although two University of British Columbia Research Forest weather stations near the lakes (Table 1) are separated by 211 m elevation, mean temperatures differ by less than 1 °C. Annual precipitation is about 16% less at the lower site. Consequently, a very similar climatic regime exists at Marion and Mike Lakes today, and this similarity would also have existed in the past.

Mike Lake has a surface area of 4.5 ha and maximum depth of 6.5 m. The lake's catchment extends to about 200 m above lake level. Owing to the limited, 1.7 km² catchment, inflowing streams are small. During summer a distinct thermal stratification is apparent (Fig. 2). In late summer, temperatures between 21 and 27.5 °C were recorded for surface water. During the same period, bottom water varied from 10 to 13 °C. Water below 4.5 m contained less than 5.0 mg/l O₂.

Mike Lake is surrounded, and presumably underlain by a thick morainal blanket (Klinka, 1976). Bedrock beneath the lake and catchment

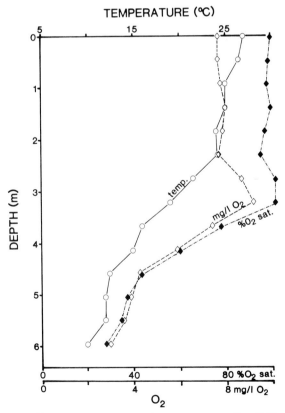

Fig. 2. Late summer oxygen and temperature profile of Mike Lake, B.C. (late afternoon September 7, 1987).

150

consists of base-poor crystalline plutonic rocks, diorite, of the coast mountain complex (Roddick, 1966). The surrounding forests are similar to those at Marion Lake.

Misty Lake

Misty Lake (70 m elev.; 50° 36.3′ N, 127° 15.7′ W) is situated 360 km northwest of Marion and Mike Lakes, near Port Hardy on northern Vancouver Island (Fig. 1). Despite the great distance separating this site from Marion and Mike Lakes, the similar vegetation and climate also place this site (Farley, 1979) within the Coastal Western Hemlock biogeoclimatic unit (wetter subzone). Port Hardy (Table 1) receives less rain and snow than the southern stations. Although this area is warmer in winter, the summers are cooler than those recorded near Marion and Mike Lakes. Differences in the forest cover are evident. Although local forests are dominated by western hemlock (*Tsuga heterophylla*) and western red cedar (*Thuja plicata*), the low relief and cool summer climate have allowed extensive paludification (Hebda, 1983). Thus peatland-forest complexes are prominent throughout the area. According to Hebda (1983), Douglas-fir (*Pseudotsuga menziesii*) is uncommon, restricted to dry sites, and trees typical of higher elevations near Vancouver (e.g., yellow cedar: *Chamaecyparis nootkatensis*) are more widespread. Shore pine (*Pinus contorta* ssp. *contorta*) is also more common near Misty Lake.

Misty Lake has a surface area of 36 ha and maximum depth of 5.2 m. This lake's catchment extends to approximately 100 m above lake level, encompassing 10 km². Since an extensive stream system enters the lake, the core was taken near the centre, distant from the inflow. The lake and catchment are underlain by Mesozoic rocks. To the northeast side are Cretaceous sedimentary rocks consisting largely of shales, sandstones, siltstones, and conglomerates, with some coal. Southwestward, Triassic rocks, including both sedimentary (limestone and dolomite) and volcanic units (andesite, basalt, and rhyolite), are exposed (Prov. of B.C., undated).

Methods

The methods used in the stratigraphic study of Mike and Misty Lakes differ little from those described for Marion Lake (Walker & Mathewes, 1987a). A 5-cm-diameter sediment core 6.43-m-long was obtained from the centre of Mike Lake at a water depth of 6.47 m. At Misty Lake, 7.53 m of sediment were removed near the lake centre, in 5.2 m of water. For Mike Lake, the 1.0-m-long piston core segments were stored intact, but Misty Lake sediments were bagged as smaller units. For Misty Lake, the upper 7.00 m of sediment were cut into 0.10 m sections, which were individually sealed in plastic bags. Below 7.00 m sediment was packaged as 0.05 m slices. During analysis, sediment subsamples of 1.0 to 2.0 ml were examined at 0.80 m intervals throughout most of each core. Closer sampling was necessary to characterize expected changes within late-glacial sediments and, for Mike Lake, near the Mazama volcanic ash.

The sediment subsamples were deflocculated in warm 6% KOH and then sieved (0.075 mm mesh). The coarse matter retained was later manually sorted, at 50X magnification in Bogorov counting trays. The yield from each subsample varied from 40.5 to 276.5 head capsules.

Fossil chironomids were mounted in Permount® and identified, with reference principally to Hamilton (1965) and Wiederholm (1983). Diagnostic features used for identification of specific taxa are reported elsewhere (Walker, 1988). Percentage diagrams were plotted using the computer program MICHIGRANA developed at University of Michigan by R. Futyma & C. Meachum.

Head capsule accumulation rates (HCAR) were calculated assuming a uniform rate of sedimentation between radiocarbon-dated levels. Thus,

$$HCAR = (N/V) \cdot SAR$$

where N is the number of head capsules recovered from a sample, V is the sample volume (cm³), and SAR is the rate of sediment accumulation ($cm \cdot yr^{-1}$).

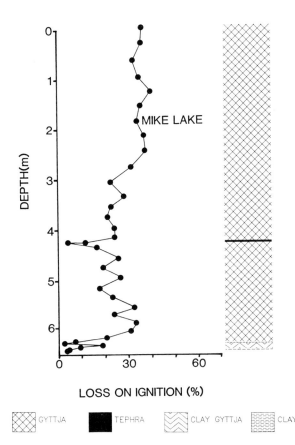

LOSS ON IGNITION (%)

GYTTJA   TEPHRA   CLAY GYTTJA   CLAY

Late summer temperature and dissolved oxygen profiles (Fig. 2) were determined at Mike Lake using a Yellow Springs Instruments O_2 meter (model #54) and probe (model #5739).

Results: Mike Lake

The basal sediment (6.40–6.43 m) of Mike Lake is inorganic (Fig. 3), composed mostly of grey clay with little, if any, sand or coarser matter. A mottled grey-brown clay-gyttja was subsequently deposited (6.32–6.40 m), grading into the organic gyttja above 6.32 m. This progression to more organic-rich sediments is interrupted by a thin compact clay layer between 6.26 and 6.275 m. Subsequent sediments, above 6.26 m, consist of a rather uniform gyttja, except for the Mazama ash at 4.25 to 4.28 m. Although organic matter and water compose much of the sediment bulk,

◀ *Fig. 3.* Sediment lithology and loss on ignition diagram for dry sediments of Mike Lake, B.C. Thin bands of clay are indicated by the dashed lines.

Table 2. Radiocarbon age for Mike and Misty Lake sediments, British Columbia, Canada.

Sample depth	Material dated	Laboratory reference No.	Age* (yr B.P.)	Accumulation rate
Mike Lake†				
425 cm	Sediment	RIDDL-647	6860 ± 60	0.062 cm/yr
428	Sediment	RIDDL-648	7080 ± 60	0.014
589	Sediment	RIDDL-649	10060 ± 100	0.054
598	Sediment	RIDDL-650	10610 ± 110	0.016
628	Sediment	RIDDL-651	11720 ± 80	0.027
640	Sediment	RIDDL-653	12600 ± 100	0.014
Misty Lake				
90–100	Sediment	BETA-16582	1760 ± 80	0.054
290–300	Sediment	BETA-16583	2860 ± 80	0.182
490–500	Sediment	BETA-16584	5720 ± 90	0.070
590–600	Sediment	BETA-16585	6960 ± 110	0.081
705–710	Sediment	BETA-16586	10180 ± 130	0.035
735–740	Sediment	GSC-4029	12100 ± 130	0.016

* $\delta^{13}C$ of −25.0 assumed.

† Accelerator Mass Spectrometry dates on Mike Lake sediments, following KOH, HCl, and HF treatments.

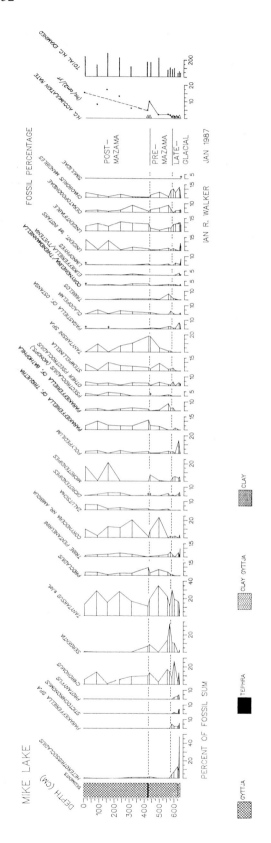

MIKE LAKE

PERCENT OF FOSSIL SUM

Fig. 4. Percentage diagram representing chironomid stratigraphy at Mike Lake, B.C. ['+'–indicates presence in small numbers; *Parakiefferiella*? cf. *triquetra* and *P.* cf. *bathophila* respectively correspond to *P. nigra* and *P.* sp.B of Walker & Mathewes (1987a).] The positions of radiocarbon dates are indicated in the head capsule accumulation rate column by open triangles. Since no dates are available on the post-Mazama sediments, the post-Mazama data points (solid circles) have not been connected in this column.

mineral matter constitutes, by weight, approximately 70 to 80% of the dry residue from 6.1 to 3.0 m. Above 3.0 m sediments are only slightly less inorganic (*ca.* 60%).

Accelerator radiocarbon dates were obtained from the Radio-isotope Direct Detection Laboratory (RIDDL) at McMaster University, Hamilton, Ontario, on organic residue from the lower portion of the Mike Lake core, as summarized in Table 2. Basal organic-rich sediment at 640 cm dated to 12 600 ± 100 yr B.P. Thus the timing of deglaciation at Mike Lake is very similar to that at Marion Lake. Dates of 10 610 ± 110 and 10 060 ± 100 yr B.P., on sediments near 5.9 m, approximately define the Pleistocene/Holocene boundary.

Although a detailed palynological investigation of Mike Lake's sediments is not yet complete, preliminary data suggest a vegetation history similar to that at Marion Lake. As at Marion Lake, the basal sediments (> 6.40 m; > 12 000 yr B.P.) are a clay in which pollen of two shrubs, willow (*Salix*) and soapberry (*Shepherdia canadensis*), is prominent, as well as pine (*Pinus contorta* type). Subsequent forest establishment is marked by a sharp increase in sediment organic content and a preponderance of lodgepole pine (*Pinus contorta*) pollen. Pollen also suggests the presence of fir (*Abies*), spruce (*Picea*), and poplar or cottonwood (*Populus*). The thin clay band between 6.26 and 6.275 m (Fig. 3) is apparently not distinguished by a distinctive fossil spectrum. Lodgepole pine pollen continues to dominate the sediments through late-glacial time (6.40 to *ca.* 5.90 m; *ca.* 12 000 to 10 000 yr B.P.) with the proportions of fir (*Abies*), spruce, and alder (*Alnus*) pollen increasing above the clay band.

Pollen of western hemlock and mountain hemlock (*Tsuga mertensiana*) are relatively abundant near the Pleistocene/Holocene boundary (*ca.* 5.90 m; 10 000 yr B.P.). Early Holocene sediments (above 5.85 m) include a high proportion of Douglas-fir pollen, suggesting the beginning of a xerothermic interval (Mathewes, 1985). However, the renewed abundance of western hemlock pollen (above *ca.* 5.2 m), beginning of enough pollen of western red cedar at *ca.* 3.5 m to indicate local arrival of the trees, and corresponding decline in Douglas-fir pollen indicate a Holocene shift to the moist climate extant in the lower Fraser Valley.

The chironomid record (Fig. 4) for Mike Lake is clearly comparable to that at Marion Lake (Walker & Mathewes, 1987a), also allowing discussion in terms of three intervals. As at Marion Lake, the lowermost interval (6.43 to 5.90 m) encompasses late-glacial sediments deposited prior to 10 000 yr B.P. The second, pre-Mazama interval (5.90 to 4.28 m) was deposited between *ca.* 10 000 yr B.P. and 6800 yr B.P. The third interval, comprising sediments above the Mazama ash (4.25 to 0.0 m), spans the period from 6800 yr B.P. to the present.

Late-glacial assemblages
The rate of head capsule accumulation was low, *ca.* 1.0 hc cm^{-2} yr^{-1} (hc = head capsules). Prominent late-glacial taxa (Fig. 4) at Mike Lake included each of the oligotrophic, cold-stenothermous elements (*Heterotrissocladius, Parakiefferiella* sp.A, *Protanypus,* & *Stictochironomus*) recorded at Marion Lake, apart from *Pseudodiamesa.* [*Pseudodiamesa* has been found in the late-glacial sediments of Marion Lake (Walker, 1988) since our earlier publication: Walker & Mathewes, 1987a.] Many other taxa (e.g., *Chironomus, Corynocera* nr. *ambigua, Microtendipes, Pagastiella* cf. *ostansa, Psectrocladius, Sergentia, Tanytarsus* s.lat.) are also represented. Although, as compared to Marion Lake, the cold-stenothermous, oligotrophic taxa at Mike Lake constitute a smaller proportion of the total fauna, the late-glacial trend is distinctly similar. This faunal element persists throughout the late-glacial

to essentially disappear at 5.7 to 5.85 m, near the Pleistocene/Holocene boundary. *Heterotrissocladius* is the only genus of this group represented in later sediments.

Holocene assemblages
The Holocene is characterized by gradually increasing head capsule accumulation rates, rising from near 1.0 hc cm^{-2} yr^{-1} during the earliest Holocene to 17.0 hc cm^{-2} yr^{-1} for modern sediments. A marked separation between early and late Holocene faunas is not evident (Fig. 4).

Heterotrissocladius is rare, and is the only late-glacial taxon characteristic of cold, oligotrophic waters represented in the Holocene sequence. *Sergentia* is relatively very abundant in the earliest Holocene sediments, rapidly declining in later deposits. Most other taxa (e.g. *Chironomus, Corynocera* nr. *ambigua, Tanytarsus* s.lat.) occur throughout Mike Lake's Holocene sediments. The fauna does not change markedly with deposition of the Mazama Ash, but an abrupt increase in *Microtendipes* abundance is apparent later in Holocene time. Rheophilous chironomids are rare in both the late-glacial and Holocene sediments.

Results: Misty Lake

Inorganic sediments were also encountered at the base of the Misty Lake core on northern Vancouver Island (Fig. 5). This clay deposit, extending from 7.53 to 7.40 m includes sand and pebbles as minor constituents. Thereafter, throughout the remaining late-glacial and Holocene deposits, the sediment is a uniform dark brown dy, averaging *ca.* 55% mineral matter on a dry weight basis (Fig. 5).

Radiocarbon dates have been obtained from Beta Analytic Laboratories (BETA) and the Geological Survey of Canada (GSC), as summarized in Table 2. Basal organic-rich sediments (7.35 to 7.40 m) date to 12 100 ± 130 yr B.P. A date of 10 180 ± 130 yr B.P. at 7.05–7.10 m approximately defines the Pleistocene/Holocene boundary. This indicates only 0.45 m of late-

154

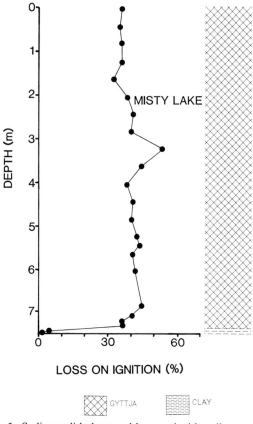

Fig. 5. Sediment lithology and loss on ignition diagram for dry sediments of Misty Lake, B.C.

glacial deposition. Slow sedimentation continued through the early Holocene, but increased towards the present.

A detailed palynological record is, as yet, unavailable for Misty Lake, but preliminary notes on major changes are available. Lodgepole pine pollen dominates throughout late-glacial sediments. Fir, spruce, and mountain hemlock pollen also occur. Western hemlock pollen is first evident at 7.20–7.25 m (*ca.* 11 000 yr B.P.). The beginning of the Holocene is marked by the first ocurrence of Douglas-fir pollen (6.90–7.00 m).

In contrast to Marion and Mike Lakes,

Fig. 6. Percentage diagram representing chironomid stratigraphy at Misty Lake, B.C. (' + ' – indicates presence in small numbers). The positions of radiocarbon dates are indicated in the head capsule accumulation rate column by open triangles. ▶

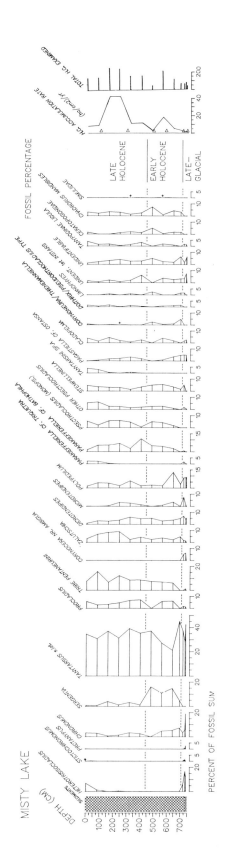

Douglas-fir pollen is not abundant during the early Holocene. Instead, western hemlock and spruce pollen prevail from 7.0 to *ca.* 4.0 m (*ca.* 10 000 to *ca.* 4000 yr B.P.). This suggests a wetter and perhaps cooler early Holocene climate than existed at the two southerly sites. The mid-Holocene shift to a cooler and/or wetter climate, and paludification of adjacent forests is marked by the prevalence of skunk cabbage (*Lysichiton americanum*) pollen above 6.0 m, and later occurrence of burnet (*Sanguisorba* sp.) and Douglas' Gentian (*Gentiana douglasiana* pollen). Above 4.0 m, western hemlock and Cupressaceae pollen (probably western red cedar) dominate, as they do today.

The major chironomid changes at Misty Lake are also best described in terms of three intervals, the late-glacial, early Holocene, and late Holocene (Fig. 6). The late-glacial (≥ 7.00 m) is represented by the lowermost 0.45 m. A division between early and late Holocene deposits is possible at a marked decline in *Sergentia* abundance, *ca.* 4.40 m (about 5500 yr B.P.).

Late-glacial assemblages
At Misty Lake, the influx of chironomid head capsules was initially low, about 1.0 hc cm^{-2} yr^{-1}. The oligotrophic, cold-stenothermous element is represented by *Heterotrissocladius*, *Protanypus*, and *Stictochironomus* (Fig. 6). The two latter taxa are present in very small numbers, and only in the two lowermost samples. As at Mike Lake, the oligotrophic, cold-stenothermous elements have essentially disappeared by 10 000 yr B.P.

Several other taxa (e.g. *Chironomus, Corynocera* nr. *ambigua, Dicrotendipes*) also occur in the late-Pleistocene sediments. Apart from the lowermost sample, as many taxa occur in the late-glacial samples as in the Holocene collections.

Holocene assemblages
Head capsule accumulation rates at Misty Lake gradually increased throughout much of the Holocene. Peak influx exceeded 40 hc cm^{-2} yr^{-1}

(Fig. 5) at 1800 to 2900 yr B.P. The Holocene faunal changes at Misty Lake illustrate few trends. Most striking is the early Holocene prominence of *Sergentia*, which abruptly decreases *ca.* 5000 yr B.P.

In the uppermost sediments, three taxa which had been present during the late-glacial again appear, *Heterotrissocladius, Parakiefferiella* cf. *triquetra*, and *Stictochironomus*. Two of these taxa, *Heterotrissocladius* and *Stictochironomus* seem to be associated with cool, oligotrophic habitats in British Columbia (Walker, 1988). As at Mike Lake, few rheophilous chironomids (e.g. *Doithrix, Pseudorthocladius, Rheocricotopus*) were identified from the core.

Discussion

The cores from Mike and Misty Lakes provide new information regarding late-glacial and early postglacial chironomid faunal succession in coastal British Columbia. The early decline of the '*Heterotrissocladius* fauna' is the most conspicuous event recorded. This decline might be linked to terrestrial vegetation changes, lake shallowing, sedimentary changes, climatic amelioration, or succession attributable to the chemical or trophic ontogeny of the lakes (Walker & Mathewes, 1987a, 1987b; Warner & Hann, 1987). Much of the following discussion addresses these possibilities.

The late-glacial fauna
The late-glacial '*Heterotrissocladius* fauna' of British Columbia includes at least 4 taxa – *Heterotrissocladius, Parakiefferiella* sp.A, *Protanypus*, and *Stictochironomus*. The association of these taxa is evident in both extant and fossil communities (Donald *et al.*, 1985; Mayhood & Anderson, 1976; Walker, 1988; Walker & Mathewes, 1987a). This fauna is probably a regional analogue to that inhabiting Brundin's (1958) *Heterotrissocladius subpilosus* lakes. Although *Paracladius* has not been recorded from late-glacial sediments of British Columbia lakes, it frequently occurs with

the '*Heterotrissocladius* fauna' in present day communities in western Canada (Donald *et al.*, 1985; Mayhood & Anderson, 1976; Walker, 1988).

All species of *Heterotrissocladius* are considered to be cold-stenothermous, and occur in well-oxygenated, oligotrophic waters (Sæther, 1975a). Individual species differ significantly as to their limits of tolerance. Species of the *Heterotrissocladius subpilosus* group have the narrowest limits of tolerance, and consequently are the best indicators of cold, oligotrophic conditions. Larvae of the *Heterotrissocladius marcidus* group are more eurytopic, ocurring in mesotrophic waters (Sæther, 1979) and at temperatures at least as warm as 18 °C (Sæther, 1975a). Unfortunately, we are not confident that *H. subpilosus* group fossils can be reliably distinguished from the *H. marcidus* group fossils. Although *Heterotrissocladius* appears to be most common at high elevations and in the cold profundal waters of the largest and deepest lakes of British Columbia, it does occur in some small low-elevation lakes, including Marion Lake (Walker, 1988).

Parakiefferiella sp.A appears to be more stenotopic than *Heterotrissocladius*, and thus is a better indicator of past environmental conditions. The larvae are easily identifiable (Sæther, 1970: as 'genus near Trissocladius') and have a distribution in the Canadian Cordillera limited to cold, high-elevation lakes, and the profundal of the deepest and most oligotrophic lakes at low-elevations (Donald *et al.*, 1985; Mayhood & Anderson, 1976; Sæther, 1970; Walker, 1988). It is also recorded from Lake Huron (Ontario), Manicouagan Reservoir (Québec), and the North American arctic (Hare, 1976; Walker, 1988).

Larvae of *Protanypus* are also typical of cold, well-oxygenated, oligotrophic waters (Sæther, 1975b). Surface mud samples from the Canadian Cordillera demonstrate their abundance in cold, high-elevation lakes, and in the profundal of deep, oligotrophic lakes (Walker, 1988).

Although *Stictochironomus* species are widely-distributed in warm, mesotrophic waters (Sæther, 1979; Wiederholm, 1983), the larvae of this genus were rare in surface collections from British Columbia and Alberta, except at high elevations

(Walker, 1988). *Stictochironomus* is widely distributed in arctic habitats (Danks, 1981).

Although interpretations based upon single indicator species are always weak, the occurrence of several of the above taxa together in a fossil assemblage provides strong evidence of a cold, well-oxygenated, oligotrophic environment, such as that present in high-elevation lakes and the profundal of very large, deep, low-elevation lakes in British Columbia (Walker, 1988). Similar assemblages have been described in the arctic (Bliss, 1977; de March *et al.*, 1978; Hershey, 1985a, 1985b; Oliver, 1964, 1976).

Chironomid succession & lithological changes

Warwick's (1980) observations regarding chironomid sucession in the Bay of Quinte, Lake Ontario, have focussed much attention upon the role of sediment composition as a determinant of chironomid faunas. Several chironomid taxa occur principally in mineral substrata (Coffman & Ferrington, 1984). *Heterotanytarsus* and some Tanytarsini build sand cases (Coffman & Ferrington, 1984). *Heterotrissocladius* disappeared near a lithological boundary within sediments from Green Lake, Michigan (Lawrenz, 1975).

Despite these observations, the major early postglacial faunal changes in British Columbia are poorly correlated with lithological changes. At Marion, Mike, and Misty Lakes, deposition of organic-rich gyttja began *ca.* 12 000 yr B.P. (Fig. 3 & 5, Table 2; Walker & Mathewes, 1987a; Wainman & Mathewes, 1987). Although *Protanypus* and *Stictochironomus* may have disappeared about this time at Misty Lake, *Heterotrissocladius* was present until *ca.* 10 000 yr B.P. (Fig. 6). *Heterotrissocladius*, *Parakiefferiella* sp.A, *Protanypus*, and *Stictochironomus* were also prominent until *ca.* 10 000 yr B.P. at Marion (Walker & Mathewes, 1987a) and Mike Lakes (Fig. 4). At this time there is no indication of a change in sediment composition.

This poor correspondence suggests that changes in sediment composition were not responsible for the decline of the late-glacial, oligotrophic, cold-stenothermal fauna in British Columbia. We also note that deposition of the Mazama ash did not permit the distinctive late-glacial fauna to become temporarily re-established at Mike (Fig. 4) or Marion Lakes (Walker & Mathewes, 1987a). Walker (1988) found *Cryptotendipes* and *Paralauterborniella* to be the only taxa with strong affinities to mineral substrata in low-elevation British Columbia lakes. Neither of these taxa was noted in the cores.

Links to terrestrial vegetation

The decline of the '*Heterotrissocladius* fauna' correlates closely with the decline of lodgepole pine in British Columbia's coastal forests. Although this correspondence is suggestive of a link between forest vegetation and lake faunas, similar chironomid faunal changes elsewhere in North America are either poorly correlated with vegetation changes, or are associated with a different kind of vegetation change (Walker & Mathewes, 1987b). In eastern North America the faunal changes occur either prior to, or about the time forests became established (Lawrenz, 1975; Walker & Paterson, 1983). In British Columbia, the change in fauna is associated with a shift in coniferous forest type. Major Holocene changes in forest vegetation at Portey Pond in New Brunswick are not accompanied by chironomid faunal changes (Walker & Paterson, 1983). Although further study is required, we doubt that the declines in late-glacial fauna were related to changing forest composition.

We have previously suggested that climate might directly or indirectly influence chironomid faunas (Walker & Mathewes, 1987a, 1987b). If the chironomid faunas were responding to early postglacial climatic adjustments, the terrestrial and aquatic environments could be responding to the same ultimate cause.

Lake depth & faunal succession

The importance of lake depth to chironomid communities has been stressed by Sæther (1980). Marion, Mike, and Misty Lakes all exceeded 10 m depth during the late-glacial, and have gradually shallowed with the progressive accumulation of sediments. In addition, the surface level of each lake would have responded to changing climate. Surface levels may have been lowest during the warm and dry early Holocene interval of British Columbia (9000 to 7000 yr B.P.), and highest during the late-glacial and late-Holocene periods (King, 1980; Mathewes, 1985). These surface level fluctuations would probably have been modest, being limited by the level of the outflowing stream. Water levels may also have fluctuated with local events, including beaver activity.

As expressed earlier, the records from post-glacial chironomid profiles across North America illustrate a similar late-glacial successional pattern (Lawrenz, 1975; Walker & Mathewes, 1987a; Walker & Paterson, 1983). The timing of the *Heterotrissocladius* decline may have varied by a thousand or more years across North America (Walker & Mathewes, 1987a), but within southwestern British Columbia, the decline appears to have ended at about 10 000 yr B.P. (It can be difficult however to precisely define the decline's end, and a significant error may be inherent to radiocarbon dating of whole lake sediments.)

If changes in lake depth were responsible for the late-glacial faunal changes, then the explanation must also account for the similar timing of the declines, not only within southwestern British Columbia, but between North American and European lakes as well. If parallel lake level fluctuations did occur in North America and Europe, the ultimate cause must be climatic. We have not, however, noted any evidence to suggest that the British Columbia study lakes were ever closed basins. Thus, water level fluctuations were probably small. The late-glacial depth of other study sites probably varied, from site to site, from about 4 m (Walker & Paterson, 1983) to over 30 m (Hofmann, 1971). Despite this range of depths, similar results have been obtained. Thus, water levels cannot have been the principal mechanism for late-glacial faunal change. The progressive shallowing of the British Columbia lakes may, however, explain many of the gradual Holocene changes within our study lakes (e.g., the *Sergentia* decline).

Variations in climate & lake trophic state

Chironomids are widely recognized by limnologists as indicators of lake trophic state. Their fossil remains have proven valuable as indicators of human impact, including the eutrophication and acidification of lakes (e.g. Brodin, 1986; Henrikson *et al.*, 1982; Kansanen, 1985; Warwick, 1980). The faunal changes occurring at the close of the Pleistocene are very similar to those expected as a result of increasing lake productivity. All of the late-glacial taxa are dependent upon cold, oligotrophic and oxygenated environments. Marion, Mike, and Misty Lakes must therefore have been unproductive during late-glacial time.

Many limnologists (e.g. Deevey, 1953) have considered eutrophication to be a natural consequence of lake aging, i.e. an autogenic/ontogenic process, yet recent data are contradictary (Walker, 1987). Early postglacial increases in lake productivity could, instead, be related to climatic amelioration (Livingstone, 1957). A satisfactory 'lake ontogenic/eutrophication' hypothesis would have to accommodate the similar timing of the faunal changes among lakes in southwestern British Columbia. If the rate of eutrophication were dependent upon soil leaching rates, then eutrophication rates should be dependent upon climate and surficial geology. The *Heterotrissocladius* declines occur at about the same time at each of our southwestern British Columbia study sites, despite differences in soil parent material and topography.

The late-glacial period is widely recognized as an interval of rapid climatic change. Pollen – climate transfer functions suggest a climatic warming in British Columbia *ca.* 10 000 yr B.P. (Mathewes & Heusser, 1981). Since this amelioration correlates closely with the disappearance of most cold-stenothermous/oligotrophic chironomids in the study lakes, these climatic and faunistic changes might be related. Nevertheless, refuge for cold-stenothermous chironomids should have persisted within profundal regions of the study lakes. Thus, direct climatic effects cannot be solely responsible for the complete disappearance of cold-stenothermous taxa. These taxa would be eliminated from the warmer and more productive littoral habitats, however.

Indirect climatic effects may have eliminated these taxa from profundal habitats. Since the characteristic late-glacial chironomid taxa are restricted to oligotrophic lakes, we propose that their decline was linked to increasing lake productivity. As Livingstone (1957) proposed, increased Holocene productivity may have begun with late-glacial climatic amelioration. Although this provides a plausible explanation for the decline of the oligotrophic/cold-stenothermous chironomids characteristic of the late-Pleistocene, does it have a more general significance? Lake productivity and chironomid faunas are influenced by many factors apart from climate. Although major fluctuations in climate are likely to be reflected by changes in the chironomid record, many similar changes in chironomid faunas will be unrelated to climate. Changes that relate to climate should be apparent on a regional scale.

We believe that direct and indirect climatic effects provide the most plausible explanation of the observed late-glacial faunal changes, but further research is necessary to test each of the hypotheses presented. Other hypotheses may also exist. Brodin (1986) and Hofmann (1983a, 1988) have also considered climatic change to be a significant determinant of chironomid faunal changes.

Complete postglacial chironomid records are available for only seven North American lakes (Lawrenz, 1975; Walker & Mathewes, 1987a, 1988); Walker & Paterson, 1983), including Mike and Misty lakes. These studies illustrate prominent trends in the early development of lacustrine chironomid faunas. Although a distinct late-glacial fauna is apparent in most postglacial records, much can be learned from the unusual records that exist, including that reported by Walker & Mathewes (1988).

Acknowledgements

We thank A. Furnell, C. E. Mehling, and B. Walker for assistance while coring Mike and Misty Lakes. Many of the radiocarbon dates were provided courtesy of the Geological Survey of Canada (GSC) or with the courtesy of D. E. Nelson (Department of Archaeology, Simon Fraser University) using the RIDDL facilities at McMaster University, Hamilton, Ontario. Permission to core Mike Lake, provided by the B.C. Provincial Park administration, is gratefully acknowledged. Laboratory help was provided by S. Chow and K. Dixon. This research was supported by Simon Fraser University with a scholarship to I.R.W. and by the Natural Sciences and Engineering Research Council of Canada through grant #A3835 to R.W.M., and a postgraduate scholarship to I.R.W.

References

Bliss, L. C. (ed.), 1977. Truelove Lowland, Devon Island, Canada: A high arctic ecosystem. University of Alberta Press, Edmonton.

Brodin, Y., 1986. The postglacial history of Lake Flarken, southern Sweden, interpreted from subfossil insect remains. Int. Revue ges. Hydrobiol. 71: 371–432.

Brundin, L., 1958. The bottom faunistical lake type system and its application to the southern hemisphere. Moreover a theory of glacial erosion as a factor of productivity in lakes and oceans. Verh. int. Ver. Limnol. 13: 288–297.

Coffmann, W. P. & L. C. Ferrington, Jr., 1984. Chironomidae. In R. W. Merritt & K. W. Cummins (eds.), An Introduction to the Aquatic Insects of North America. 2nd ed. Kendall/Hunt Publ., Dubuque, Iowa: 551–652.

Danks, H. V., 1981. Arctic Arthropods: A review of systematics and ecology with particular reference to the North American fauna. Entomol. Soc. Can., Ottawa.

de March, L., de March, B. & Eddy, W., 1978. Limnological, fisheries, and stream zoobenthic studies at Stanwell-Fletcher Lake, a large high arctic lake. Arctic Islands Pipelines Program, Preliminary Report 1977. Department of Indian and Northern Affairs Publication No. QS-8160-004-EE-A1, Ottawa.

Deevey, E. S., Jr., 1953. Paleolimnology and climate. In H. Shapley (ed.), Climatic change: Evidence, causes, and effects. Harvard Univ. Press: 273–318.

Donald, D. B., D. J. Alger & G. A. Antoniuk, 1985. Limnological studies in Jasper National Park. Part ten: The north boundary lakes. Can. Wildlife Serv., Edmonton.

Environment Canada, 1982. Canadian climate normals (1951–1980), 2, 3, 4 & 6. Atmos. Environ. Serv., Ottawa, Can.

Farley, A. L., 1979. Atlas of British Columbia – People, environment and resource use. Univ. B.C. Press, Vancouver.

Günther, J., 1983. Development of Grossensee (Holstein, Germany): variations in trophic status from the analysis of subfossil microfauna. Hydrobiologia 103: 231–234.

Hamilton, A. L., 1965. An Analysis of a Freshwater Benthic Community with special reference to the Chironomidae. Ph.D. thesis, Univ. of B.C., Vancouver, B.C.

Hare, R. L., 1976. The macroscopic zoobenthos of Parry Sound, Georgian Bay. M.Sc. thesis, University of Waterloo, Waterloo, Canada.

Hebda, R. J., 1983. Late-glacial and postglacial vegetation history at Bear Cove Bog, northeast Vancouver Island, British Columbia. Can. J. Bot. 61: 3172–3192.

Henrikson, L., J. B. Olofsson & H. G. Oscarson, 1982. The impact of acidification on Chironomidae (Diptera) as indicated by subfossil stratification. Hydrobiologia 86: 223–229.

Hershey, A. E., 1985a. Littoral chironomid communities in an arctic Alaskan lake. Holarct. Ecol. 8: 39–48.

Hershey, A. E., 1985b. Effects of predatory sculpin on the chironomid communities in an arctic lake. Ecology 66: 1131–1138.

Hofmann, W., 1971. Die postglaziale Entwicklung der Chironomiden und Chaoborus-Fauna (Dipt.) des Schöhsees. Arch. Hydrobiol. Suppl. 40.

Hofmann, W., 1978. Analysis of animal microfossils from the Großer Segeberger See (F.R.G.). Arch. Hydrobiol. 82: 316–346.

Hofmann, W., 1983a Stratigraphy of Cladocera and Chironomidae in a core from a shallow North German lake. Hydrobiologia 103: 235–239.

Hofmann, W., 1983b. Stratigraphy of subfossil Chironomidae and Ceratopogonidae (Insecta: Diptera) in late glacial littoral sediments from Lobsigensee (Swiss Plateau). Studies in the late Quaternary of Lobsigensee 4. Revue Paleobiol. 2: 205–209.

Hofmann, W., 1988. The significance of chironomid analysis (Insecta: Diptera) for paleolimnological research. Palaeogeogr. Palaeoclimatol. Palaeoecol. 62: 501–510.

Kansanen, P. H. 1985. Assessment of pollution history from recent sediments in Lake Vanajavesi, southern Finland. II. Changes in the Chironomidae, Chaoboridae and Ceratopogonidae (Diptera) fauna. Ann. zool. fenn. 22: 57–90.

King, M., 1980. Palynological and macrofossil analyses of lake sediment from the Lillooet area, British Columbia. M.Sc. thesis, Simon Fraser University, Burnaby, British Columbia.

Klinka, K., 1976. Ecosystem units – their classification, interpretation and mapping in the University of British

160

Columbia Research Forest. Ph.D. thesis, Univ. B.C., Vancouver.

Lawrenz, R. W., 1975. The developmental paleoecology of Green Lake, Antrim County, Michigan. M.S. thesis, Central Michigan University, Michigan, U.S.A.

Livingstone, D. A., 1957. On the sigmoid growth phase in the history of Linsley Pond. Am. J. Sci. 225: 364–373.

Mathewes, R. W., 1985. Paleobotanical evidence for climatic change in southwestern British Columbia during late-glacial and Holocene time. Syllogeus 55: 397–422.

Mathewes, R. W. & L. E. Heusser, 1981. A 12 000 year palynological record of temperature and precipitation trends in southwestern British Columbia. Can. J. Bot. 59: 707–710.

Mayhood, D. W. & R. S. Anderson, 1976. Limnological survey of the Lake Louise area, Banff National Park. Part 2 – The lakes. Canadian Wildlife Service, Environment Canada, Calgary.

Oliver, D. R., 1964. A limnological investigation of a large arctic lake, Nettilling Lake, Baffin Island. Arctic 17: 69–83.

Oliver, D. R., 1976. Chironomidae (Diptera) of Char Lake, Cornwallis Island, N.W.T., with descriptions of two new species. Can. Ent. 108: 1053–1064.

Province of British Columbia, undated. British Columbia Geological Highway Map. B.C. Ministry Energy Mines Petroleum Resour., Victoria.

Roddick, J. A., 1966. Vancouver North, Coquitlam, and Pitt Lake map-areas, British Columbia with special emphasis on the evolution of the plutonic rocks. Geol. Surv. Can, Mem. 335.

Sæther, O. A., 1970. A survey of the bottom fauna in lakes of the Okanagan Valley, British Columbia. Fish. Res. Bd Can. Techn. Rep. 196: 1–29.

Sæther, O. A., 1975a. Nearctic and Palaearctic *Heterotrissocladius* (Diptera: Chironomidae). Bull. Fish. Res. Bd Can. 193: 1–67.

Sæther, O. A., 1975b. Two new species of *Protanypus* Kieffer, with keys to Nearctic and Palaearctic species of the genus (Diptera: Chironomidae). J. Fish. Res. Bd Can. 32: 367–388.

Sæther, O. A., 1979. Chironomid communities as water quality indicators. Holarct. Ecol. 2: 65–74.

Sæther, O. A., 1980. The influence of eutrophication on deep lake benthic invertebrate communities. Prog. Water Technol. 12: 161–180.

Schakau, B. & C. Frank, 1984. Die Entwicklung der Chironomiden-Fauna (Diptera) des Tegeler Sees im Spät- und Postglazial. Verh. Ges. Okol. 12: 375–382.

Wainman, N. & R. W. Mathewes, 1987. Forest history of the last 12 000 years based on plant macrofossil analysis of sediment from Marion Lake, southwestern British Columbia. Can. J. Bot. 65: 2179–2187.

Walker, I. R., 1987. Chironomidae (Diptera) in paleoecology. Quat. Sci. Rev. 6: 29–40.

Walker, I. R., 1988. Late-Quaternary palaeoecology of Chironomidae (Diptera: Insecta) from lake sediments in British Columbia. Ph.D. thesis, Simon Fraser Univ., Burnaby, B.C.

Walker, I. R. & R. W. Mathewes, 1987a. Chironomidae (Diptera) and postglacial climate at Marion Lake, British Columbia, Canada. Quat. Res. 27: 89–102.

Walker, I. R. & R. W. Mathewes, 1987b. Chironomids, lake trophic status, and climate. Quat. Res. 28: 431–437.

Walker, I. R. & R. W. Mathewes, 1988. Late-Quaternary fossil Chironomidae (Diptera) from Hippa Lake, Queen Charlotte Islands, British Columbia, with special reference to *Corynocera* Zett. Can. Ent. 120: 739–751.

Walker, I. R. & C. G. Paterson, 1983. Post-glacial chironomid succession in two small humic lakes in the New Brunswick – Nova Scotia (Canada) border area. Freshwater Invertebr. Biol. 2: 61–73.

Warner, B. G. & B. Hann, 1987. Aquatic invertebrates as paleoclimatic indicators? Quat. Res. 28: 427–430.

Warwick, W. F., 1980. Palaeolimnology of the Bay of Quinte, Lake Ontario: 2800 years of cultural influence. Can. Bull. Fish. Aquat. Sci. 206: 1–117.

Wiederholm, T. (ed.), 1983. Chironomidae of the Holarctic region. Keys and diagnoses. Part 1 – Larvae. Entomol. scand. Suppl. 19.

Chironomids, lake development and climate: a commentary

William F. Warwick
National Hydrology Research Institute, Aquatic Ecology Division, 11 Innovation Blvd., Saskatoon, Saskatchewan, S7N 3H5, Canada

Key words: Chironomidae, palaeolimnology, sedimentation effects

The controversy concerning chironomid remains as indicators of palaeoclimate (Warner & Hann, 1987; Walker & Mathewes, 1987a) originates largely from the emphasis placed on climate as the driving force behind the decline in the late-glacial *Heterotrissocladius* assemblage in Marion Lake (Walker & Mathewes, 1987b) and in Mike and Misty lakes (Walker & Mathewes, 1989). Although passing reference was made to trophic change in the Marion Lake paper, Walker & Mathewes argued that the principle factor leading to decline in the cold stenothermous fauna was climatic amelioration (= warming) towards the close of the Pleistocene. More consideration is given in their 1989 paper to other factors, but emphasis still remains largely on temperature. Climate undoubtedly has some influence on chironomid assemblages in lakes, but how close is the connection? Does it include a direct temperature component as Walker and Mathewes imply?

Water depth is a major factor influencing chironomid communities in lakes. During the late-glacial period, each of the lakes was comparatively deep – between 12–13 m in Mike and Misty lakes and 14–15 m in Marion Lake. If present-day Mike Lake stratifies at 6.5 m depth, then stratification almost certainly occurred at these greater depths. As Walker & Mathewes (1987b) state, 'deep-water chironomids of stratified lakes are isolated from the direct effects of changing air temperature…'. Thus, water depth almost certainly rules out the direct influence of climatic warming.

Substrate also exerts a strong influence on chironomid communities. Walker & Mathewes (1989) state that early postglacial faunal changes correlate poorly with lithological changes. And yet there is clear evidence of sediment/fauna interactions, the most obvious of which are associated with the Mazama ash layer. Taxa like *Heterotrissocladius*, *Procladius*, *Zalutschia*, *Parakiefferiella* cf. *bathyphila* (sub *P. nigra* in Marion Lake?), *Parakiefferiella* cf. *triquetra* (sub *P.* sp. *B* in Marion Lake?), *Stempellinella* and *Limnophyes* responded positively to the injection of volcanic ash in Mike and Marion lakes. Conversely, taxa like *Chironomus*, *Sergentia*, *Tanytarsus* and *Corynocera* cf. *ambigua* responded negatively.

These responses provide an important key for interpreting the late-glacial sediments. The species of *Heterotrissocladius* found in these sediments most likely was *H. latilaminus*. According to Saether (1975), it occurs at all depths in Marion Lake and can tolerate a wide range of biotopes from littoral to profundal and oligotrophic to slightly eutrophic. From its association with the Mazama ash layer, it also appears to be sediphilic*. This is not surprizing as other species of the genus bear the same proclivity. *Heterotrissocladius marcidus*, which occurs in small num-

---

* I define *sediphilic* as those taxa responding positively to the accumulation of mineral sediments.

Originally published in
Journal of Paleolimnology 2: 15–17.

bers in the littoral zone of Marion Lake, lives equally well on organic or mineral substrates (Saether, 1975). *Heterotrissocladius changi* Saeth. responded positively to the accumulation of mineral sediments in the Bay of Quinte (Warwick, 1980a). As a eurytopic sediphile, *H. latilaminus* would hardly fit the role of an oligotrophic, cold-stenothermous element.

The second most important taxon from the point of view of sediment interpretation was *Parakiefferiella* sp. *A*. Although not present at the Mazama ash layer, supplementary information on the taxon comes from Warwick (1980b). The abundance of *Parakiefferiella* sp. *A* (sub. Gen. nr. *Heterotrissocladius*) increased sharply during the early stages of the 1852-53 influx of mineral sediments, but reached peak abundance in the aftermath of maximum deposition. It was followed by *Heterotrissocladius* which reached maximum abundance slightly later when accumulation rates had declined. While both prefer unstable, highly mineral sediments, *Parakiefferiella* sp. *A* appears to be slightly more tolerant.

These observations are important when applied to the late-glacial faunal changes. In Mike Lake, *Heterotrissocladius* was most abundant initially in the clay-gyttja sediments at ca. 6.35 m. The taxon began to decline in abundance (ca. 6.30 m) as the sediments changed to gyttja and organic concentrations neared 20%. The deposition of a thin compact layer of clay around 6.27 m sharply depressed organic concentrations to <5% and *Heterotrissocladius* increased in abundance at the expense of sediphobic species like *Chironomus*. The reversal was short-lived, however, and as organic concentrations again climbed above 20%, *Heterotrissocladius* declined once more to disappear above 5.80 m as organic concentrations exceeded 30%.

Although not described by Walker & Mathewes (1989), a second sedimentary event is suggested by an increase in *Parakiefferiella* sp. *A* and a decrease in *Chironomus* at ca. 5.80 m. Organic concentrations remained above 20%, however, and there was no indication of a revitalized *Heterotrissocladius* population.

In Misty Lake, the scenario is simpler. *Hetero-trissocladius* was most abundant in the sediments immediately overlying the basal clay sediments at ca. 7.40 m. The taxon then declined steadily as organic concentrations increased and was phased out above 7.00 m as these exceed 30%.

In Marion Lake, similar sequences are apparent even though ignition data are lacking. The abundance of *Parakiefferiella* sp. *A* peaked twice in the late-glacial sediments at ca. 8.80 m and 8.45 m to mark two accumulation episodes as noted for Mike Lake. Both episodes were further marked by concurrent increases in the abundance of *Heterotrissocladius* at ca. 8.65 m and 8.25 m. As in the Bay of Quinte, *Heterotrissocladius* here too seems to be slightly less sediphilic than *Parakiefferiella* sp. *A*.

Walker & Mathewes (1987b, 1989) argue that the disappearance of late-glacial *Heterotrisso-cladius* communities from Europe and North American lakes was climatically driven. But the changes noted here are clearly linked to sedimentary processes. The question then becomes what kind of sedimentary process could apply synchronously, across a wide geographic area and in such a variety of biotopes? The only single common denominator seems to be the manner in which maturation of bottom sediments takes place. Fauna colonizing the relatively sterile mineral sediments left after deglaciation would necessarily be tolerant of unstable bottom conditions. However, as beachlines and surrounding catchment areas stabilize and ground cover develops, the ratio between organic and mineral increments to the sediments change and with them the composition of their resident communities. As organic concentrations increase, sediphilic taxa like *Heterotrissocladius* can no longer compete and are phased out in favor of species like *Chironomus* which require stable bottom sediments and readily accessible food resources.

The importance of the ratio between organic and mineral sediment fractions is shown by the response of *Heterotrissocladius* to the Mazama ash layer. In Mike Lake, the injection of ash at 4.25–4.28 m reduced organic concentrations below 5% and *Heterotrissocladius* reappeared.

The taxon then persisted in the sediments, albeit in small numbers, as long as organic concentrations remained below 30%, but as soon as this was exceeded, *Heterotrissocladius* was phased out once more.

The same sequence was repeated in Marion Lake. *Heterotrissocladius*, which persisted in small numbers up to 6.10 m, increased sharply in abundance following the injection of Mazama ash and remained relatively abundant up to 4.1 m before entering a decline phase at ca. 3.00 m.

The persistence of *Heterotrissocladius* in the post-Mazama sediments of Mike and Marion lakes coincides with the western hemlock maximum during which cooler and wetter conditions prevailed. In view of their responsiveness to the sedimentary environment, it is not hard to envisage increased allochthonous inputs washed into the lakes as the driving force maintaining these fauna. Mathewes (1973) acknowledged that sedimentation rates in Marion Lake were quite variable and that much of the material was 'allochthonous in origin, brought in by the stream and perhaps slope wash'. The four-fold difference in catchment size almost undoubtedly accounts for the greater sediment depth in Marion Lake. It may be that this factor, too, accounts for the difference in size of the *Heterotrissocladius* communities in the two lakes.

If cooler, moist conditions were responsible for these relatively strong *Heterotrissocladius* communities through the increased accumulation of allochthonous materials, it is not hard to envisage the effect of the drier, warmer climate of the xerothermic interval on these communities. With little input of allochthonous materials from outside sources, *Heterotrissocladius* disappeared altogether from Mike Lake sediments and was very much reduced in Marion Lake. Although climatically driven in the broadest sense, these changes were wrought by reduced rainfall or runoff rather than temperature *per se*.

The driving force behind chironomid community changes in these lakes appears to be the interplay between mineral sediment accumulation and the availability, both in qualitative and quantitative terms, of organic food resources rather than direct temperature change. It is unfortunate that the sediment properties of Marion Lake were not fully characterized and the effects of sediments and lake trophy analyzed in Walker & Mathewes' (1987b) initial study; fortunately the sediments of nearby Mike Lake can serve as a surrogate for interpretation purposes.

References

Mathewes, R. W., 1973. A palynological study of postglacial vegetation changes in the University Research Forest, southwestern British Columbia. Can. J. Bot. 51: 2085–2103.

Saether, O. A., 1975. Nearctic and Palaearctic *Heterotrissocladius* (Diptera: Chironomidae). Bull. Fish. Res. Bd Can. 193: 67 p.

Walker, I. R. & R. W. Mathewes, 1987a. Chironomids, lake trophic status, and climate. Quat. Res. 28: 431–437.

Walker, I. R. & R. W. Mathewes, 1987b. Chironomidae (Diptera) and postglacial climate at Marion Lake, British Columbia, Canada. Quat. Res. 27: 89–102.

Walker, I. R. & R. W. Mathewes, 1989. Early postglacial chironomid succession in southwestern British Columbia, Canada, and its paleoenvironmental significance. J. Paleolimnology.

Warner, B. C. & B. J. Hann, 1987. Aquatic invertebrates as paleoclimatic indicators. Quat. Res. 28: 427–430.

Warwick, W. F., 1980a. Palaeolimnology of the Bay of Quinte, Lake Ontario: 2800 years of cultural influence. Can. Bull. J. Fish. Aquat. Sci. 206: 117 p.

Warwick, W. F., 1980b. Chironomidae (Diptera) responses to 2800 years of cultural influence: a palaeolimnological study with special reference to sedimentation, eutrophication, and contamination processes. Can. Entomol. 112: 1193–1238.

Much ado about dead diptera

Ian R. Walker[1] & R.W. Mathewes

Department Biol. Sci., Simon Fraser University, Burnaby, British Columbia V5A 1S6, Canada; [1]Present address: Dept. of Biology, Queen's University, Kingston, Ontario K7L 3N6, Canada

Warwick's (1989) concerns with the effect of sediment composition and sedimentation rates upon chironomid faunal composition are drawn largely from his paleolimnological investigation of anthropogenic eutrophication processes in the Bay of Quinte, Lake Ontario (Warwick, 1980a, 1980b). Thus, a brief review of this literature is pertinent to our discussion. The chironomid fauna of the Bay of Quinte had changed little for two to three thousand years prior to *ca.* 1850, when the influx of inorganic sediments increased abruptly. The relative abundance of many chironomid taxa changed at or about the same time. It is obvious that European colonists, through their exploitation of land-based resources, had dramatically altered the bay.

Warwick (1980a, 1980b, 1989) attributes many of the chironomid faunal changes to direct effects of mineral sediment influx. He suggests, for example, that the profundal taxa *Parakiefferiella* sp.A (= genus near *Heterotrissocladius* of Warwick, 1980a, 1980b) and *Heterotrissocladius* responded positively to increased mineral sedimentation. It is obvious however that many other characteristics of the bay had also been altered. Significantly, Warwick (1980b: p. 71–73) has noted a coincident expansion of profundal habitat.

Subsequent diatom studies revealed that benthic diatom taxa decreased greatly relative to planktonic forms during this interval (Stoermer *et al.*, 1985). This diatom change results from decreased water transparency, most likely due to increased sediment load. Such changes in water transparency can significantly alter the temperature profile of a lake (Yan, 1983), thus influencing chironomid faunal composition (Uutala, 1986: p. 69). Absorption of solar energy in the shallow postdisturbance photic zone would promote an upward displacement of the thermocline. A warmer, shallower epilimnion and deeper, cold profundal region would result. It is possible that other limnological processes also influenced chironomid faunal changes at this time.

Although both hypotheses seem consistent with Warwick's data, the environmental changes which occurred were complex, and it is unclear which effects actually contributed to the altered fauna. If mineral sedimentation was the principal influence, we may inquire why those chironomids, which responded positively to these changes, were almost entirely profundal taxa.

Warwick (1989) considers both *Heterotrissocladius* and *Parakiefferiella* sp.A to have an affinity for regions of high mineral sedimentation (= sediphilic *sensu* Warwick, 1989), but these taxa were common in pre-disturbance time when sedimentation rates were low, and sediment organic content high. These taxa were extirpated in the late 19th century as the Bay of Quinte became increasingly eutrophic. We believe the 'sediphilic' designation of Warwick (1989) is premature. These taxa are common in clear, cold waters as well as the turbid waters of glacial lakes (Hare, 1976; Walker, 1988; Walker & Mathewes, 1987a, 1987b, 1989). Their late-glacial presence in lakes of southwestern British Columbia correlates with very low sediment accumulation rates (Walker & Mathewes, 1989).

Warwick (1989) calls upon our data (Walker &

Originally published in
Journal of Paleolimnology 2: 19–22.

Table 1. Chironomid head capsule count data (= number of head capsules per sample) for pre-Mazama and post-Mazama sediments in Marion and Mike Lakes. 'Mazama-positive' (= abundant after Mazama Ash) and 'Mazama-negative' (= abundant prior to Mazama Ash) designations follow Warwick (1989). *Parakiefferiella* cf. *bathophila* and *P.* cf. *triquetra* respectively correspond to *P.* sp.B and *P. nigra* of Walker & Mathewes (1987a).

Depth (m)	Marion Lake				Mike Lake	
	pre-Mazama		post-Mazama		pre-Mazama	post-Mazama
	6.17	6.10	6.05*	6.00	4.35	4.24*
'Mazama-positive'						
Heterotrissocladius	2.0	1.0	3.5	5.5	2.5	1.5
Limnophyes	1.0	–	1.0	–	–	2.0
Parakiefferiella cf. *bathophila*	–	1.0	–	1.0	–	3.0
Parakiefferiella cf. *triquetra*	–	2.0	5.0	–	8.5	19.5
Procladius	7.0	5.0	–	3.0	5.0	16.5
Stempellinella	1.0	1.0	–	–	4.0	7.0
Zalutschia	–	0.5	–	0.5	0.5	2.5
'Mazama-negative'						
Chironomus	–	–	3.0	1.0	22.0	17.0
Corynocera nr. *ambigua*	3.0	–	–	–	13.0	2.0
Sergentia	10.0	3.0	3.0	0.5	14.5	12.0
Tanytarsus s.lat.	31.0	7.0	14.5	13.0	31.5	19.5
other chironomids	47.5	10.5	17.5	32.5	63.5	71.0
Total number of head capsules	102.5	31.0	47.5	57.0	165.0	173.5

* Sample taken at upper surface of Mazama Ash.

Mathewes, 1987a, 1989) to confirm the importance of mineral sedimentation. He considers minor peaks for several taxa following deposition of the Mazama Ash to be an important key to interpretation. Unfortunately, he did not have access to the raw count data (Table 1), and thus could not assess their statistical significance. Since several of the increases and decreases actually precede the Mazama Ash, his analysis also reflects the difficulty of discerning the precise position of samples in the chironomid diagram of Walker & Mathewes (1987a).

We also note that most of the taxa which

Warwick interpreted as increasing in the post-Mazama interval, including *Limnophyes*, *Parakiefferiella* cf. *bathophila*, *P.* cf. *triquetra* (= *Paracladius* cf. *triquetra* of Warwick, 1980a, 1980b), *Zalutschia*, and most *Tanypodinae* either declined or failed to respond as mineral sedimentation increased in the Bay of Quinte (Warwick, 1980a, 1980b). Warwick (1989) considers the late-glacial environment to have been favourable for sediphiles, but *Heterotrissocladius* is the only 'late-glacial' taxon which Warwick identifies as increasing after the Mazama Ash. Although *Heterotrissocladius* increases in abundance in post-

Mazama sediments of Marion Lake, it does not increase at nearby Mike Lake (Table 1). Also, at Marion Lake, *Heterotrissocladius* abundance did not decline once organic sedimentation had resumed (Fig. 1). Thus mineral sedimentation cannot adequately explain these temporal distributions.

As further support for the mineral sedimentation hypothesis Warwick (1989) provides a detailed analysis of late-glacial events in our cores. We disagree with his interpretations. The apparent decline of *Heterotrissocladius* abundance *ca.* 6.30 m in Mike Lake occurred as the supply of cold, clay-laden glacial meltwater to Mike Lake is believed to have stopped. Since many chironomid taxa are unable to complete their life cycles in very cold water (Danks & Oliver, 1972), the supposed decline of *Heterotrissocladius* in response to the changing sediment composition is confounded by the immigration of new thermophilous chironomids. Since the data are presented as relative abundances (%), an apparent decrease in one taxon may be attributable to increases in the relative abundance of other taxa.

The minor peak in the relative abundance of *Heterotrissocladius* with deposition of the thin clay band *ca.* 6.27 m in Mike Lake cannot be considered significant. Forty-one head capsules were recovered from the clay and sixty from the underlying clay-gyttja. Of these fossils, 5.5 *Heterotrissocladius* were recovered from the clay, and 5 from the underlying clay-gyttja.

Over-interpretation of our data is evident in many of Warwick's arguments. He notes several minor peaks in the abundance of *Heterotrissocladius* in the Marion Lake profile (Fig. 1), but draws little attention to the fact that these minor peaks are completely out of phase with those of *Parakiefferiella* sp.A (a taxon which we and Warwick agree has similar ecological requirements to *Heterotrissocladius*). Despite the importance of *Protanypus* and *Stictochironomus* to our interpretations, Warwick has completely ignored these taxa. We note that *Stictochironomus* increases in abundance throughout the late-glacial (Fig. 1) as sediment organic content increases. Nevertheless, *Stictochironomus* abruptly disappears at the close of the late-glacial.

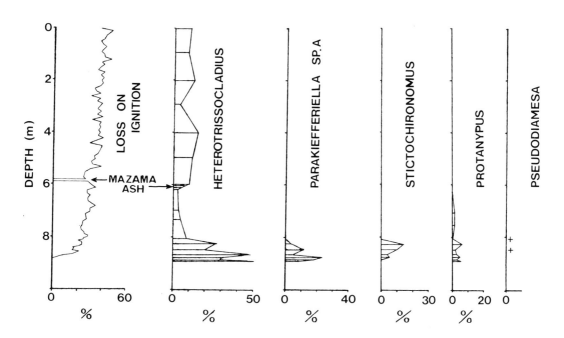

Fig. 1. Sediment loss on ignition and chironomid relative abundance diagrams (of taxa most important to interpretation of late-glacial events) for Marion Lake, British Columbia. Data from Wainman & Mathewes (1987), and Walker (1988).

168

Warwick has found an intriguing correlation between low sediment organic content and occurrence of *Heterotrissocladius* at Mike Lake, but this correlation cannot explain the increased prevalence of *Heterotrissocladius* in the more organic post-Mazama sediments of Marion Lake (Fig. 1).

Warwick (1989) has limited his analysis of postglacial events to our cores from British Columbia lakes, despite the relevance of other North American and European profiles. Although Lawrenz (1975) has placed a lithological boundary near the Green Lake *Heterotrissocladius* decline, the change in sediment composition is so slight it cannot be discerned on the loss on ignition profile. Warwick rules out the possibility of direct temperature effects on Chironomidae, despite the many littoral taxa which are excluded from arctic and alpine regions (Walker, 1988). Complex interactions are probable among various environmental factors (including oxygen concentrations, lake productivity, nutrient concentrations, mineral sedimentation, and temperature) and chironomid fauna, but we believe that Warwick has not convincingly demonstrated the importance of mineral sedimentation in our data or his study. We continue to favour the climatic hypothesis as an explanation for the late-glacial demise of oligotrophic, cold-stenothermous Chironomidae.

References

Danks, H. V. & D. R. Oliver, 1972. Seasonal emergence of some high arctic Chironomidae (Diptera). Can. Ent. 104: 661–686.

Hare, R. L., 1976. The Macroscopic Zoobenthos of Parry Sound, Georgian Bay. M.Sc. thesis, Univ. of Waterloo, Waterloo, Ont., Canada.

Lawrenz, R. W., 1975. The Developmental Paleoecology of Green Lake, Antrim County, Michigan. M.S. thesis, Central Michigan Univ., Mt. Pleasant, Mich., U.S.A.

Stoermer, E. F., J. A. Wolin, C. L. Schelske & D. J. Conley, 1985. Postsettlement diatom succession in the Bay of Quinte, Lake Ontario. Can. J. Fish. aquat. Sci. 42: 754–767.

Uutala, A. J., 1986. Paleolimnological assessment of the effects of lake acidification on Chironomidae (Diptera) assemblages in the Adirondack region of New York. Ph.D. thesis, State Univ. of N.Y., Syracuse, N.Y., U.S.A.

Wainman, N. & R. W. Mathewes, 1987. Forest history of the last 12 000 years based on plant macrofossil analysis of sediment from Marion Lake, southwestern British Columbia. Can. J. Bot. 65: 2179–2187.

Walker, I. R., 1988. Late-Quaternary Palaeoecology of Chironomidae (Diptera: Insecta) from Lake Sediments in British Columbia. Ph.D. thesis, Simon Fraser Univ., Burnaby, B.C., Canada.

Walker, I. R. & R. W. Mathewes, 1987a. Chironomidae (Diptera) and postglacial climate at Marion Lake, British Columbia, Canada. Quat. Res. 27: 89–102.

Walker, I. R. & R. W. Mathewes, 1987b. Chironomids, lake trophic status, and climate. Quat. Res. 28: 431–437.

Walker, I. R. & R. W. Mathewes, 1989. Early postglacial chironomid succession in southwestern British Columbia, Canada, and its paleoenvironmental significance. J. Paleolim. 2: 1–14.

Warwick, W. F., 1980a. Chironomidae (Diptera) responses to 2800 years of cultural influence: a palaeolimnoligical study with special reference to sedimentation, eutrophication, and contamination processes. Can. Ent. 112: 1193–1238.

Warwick, W. F., 1980b. Paleolimnology of the Bay of Quinte, Lake Ontario: 2800 years of cultural influence. Can Bull. Fish. aquat. Sci. 206: 1–117.

Warwick, W. F., 1989. Chironomids, lake development and climate: a commentary. J. Paleolim. 2: 15–17.

Yan, N. D., 1983. Effects of changes in pH on transparency and thermal regime of Lohi Lake, near Sudbury, Ontario. Can. J. Fish. aquat. Sci. 40: 621–626.

The developmental history of Adirondack (N.Y.) Lakes

Donald R. Whitehead,[1] Donald F. Charles,[1,2] Stephen T. Jackson,[3] John P. Smol,[4] &
Daniel R. Engstrom[5]
[1] *Department of Biology, Indiana University, Bloomington, IN 47405, USA;* [2] *Environmental Research
Laboratory, U. S. Environmental Protection Agency, Corvallis, OR 97333, USA;* [3] *Department of
Geological Sciences, Brown University, Providence, RI 02912, USA;* [4] *Department of Biology, Queen's
University, Kingston, Ontario, Canada K7L 3N6;* [5] *Limnological Research Center, University of Minnesota,
Minneapolis, MN 55455, USA*

Key words: Paleolimnology, lake developmental history, watershed-lake interactions, lakewater chemistry, Adirondacks

Abstract

We utilized paleoecological techniques to reconstruct long-term changes in lake-water chemistry, lake
trophic state, and watershed vegetation and soils for three lakes located on an elevational gradient
(661–1150 m) in the High Peaks region of the Adirondack Mountains of New York State (U.S.A.).
Diatoms were used to reconstruct pH and trophic state. Sedimentary chrysophytes, chlorophylls and
carotenoids supplied corroborating evidence. Pollen, plant macrofossils, and metals provided information
on watershed vegetation, soils, and biogeochemical processes. All three lakes were slightly alkaline
pH 7–8) and more productive in the late-glacial. They acidified and became less productive at the end
of the late-glacial and in the early Holocene. pH stabilized 8000–9000 yr B.P. at the two higher sites and
by 6000 yr B.P. at the lowest. An elevational gradient in pH existed throughout the Holocene. The highest
site had a mean Holocene pH close to or below 5; the lowest site fluctuated around a mean of 6. The
higher pH and trophic state of the late-glacial was controlled by leaching of base cations from fresh
unweathered till, a process accelerated by the development of histosols in the watersheds as spruce-
dominated woodlands replaced tundra. An apparent pulse of lake productivity at the late-glacial-
Holocene boundary is correlated with a transient, but significant, expansion of alder (*Alnus crispa*)
populations. The alder phase had a significant impact on watershed (and hence lake) biogeochemistry.
The limnological changes of the Holocene and the differences between lakes were a function of an
elevational gradient in temperature, hydrology (higher precipitation and lower evapotranspiration at
higher elevation), soil thickness (thinner tills at higher elevation), soil type (histosols at higher elevation),
vegetation (northern hardwoods at lower elevation, spruce-fir at higher), and different Holocene vegetational sequences in the three watersheds.

Originally published in
Journal of Paleolimnology **2**: 185–206.

Introduction

Although it was once suggested that the developmental history of temperate lakes should follow a unidirectional pathway from oligotrophy to eutrophy, current knowledge of late-glacial and Holocene vegetational and climatic history and modern ecosystem biogeochemistry suggests that more complex and variable trajectories should be expected. This has been confirmed by many paleoecological studies. For example, the late-glacial and Holocene vegetational changes in mountainous regions of northeastern North America are now reasonably well understood. In general there appears to have been a progression from tundra to spruce-dominated boreal woodlands, to mixed conifer-hardwood forest with white pine, to hardwoods with much hemlock, to hardwoods with little hemlock, to hardwoods with some spruce (e.g., Whitehead et al., 1986; Whitehead & Jackson, 1988; Davis, 1983; Webb, 1981; Overpeck, 1985; Davis & Jacobson, 1985; Webb et al., 1983; Davis et al., 1980; Gaudreau & Webb, 1988). There have been major regional and local changes in vegetation, in some cases involving shifts from conifer domination to hardwoods and back to conifers once again. Some of these changes were driven by climate, some by differential migration rates (and pathways) (Davis, 1981b, 1983), some by biogenic events (Davis, 1981a). It is also evident that there were significant vegetational responses to the climatic changes of the mid-Holocene (Jackson, 1983, 1988; Spear, 1981; Davis et al., 1980). These climatic and/or vegetational changes should have resulted in continually shifting biogeochemical characteristics of watersheds. The altered biogeochemistry would, in turn, influence lakewater chemistry and lake biota.

The characteristics of past watershed-lake systems can be inferred from the extensive recent literature on the biogeochemistry of terrestrial ecosystems in the Northeast (e.g., Likens et al., 1977; Bormann & Likens, 1979; Cronan, 1980; Olson et al., 1981; Lovett et al., 1982; Vitousek, 1977). From this work it is apparent that soil, stream, and lake water chemistry are controlled by the complex interaction of soil and bedrock characteristics, weathering rate, hydrology, and vegetation composition and dynamics. Accordingly, it should be possible to infer the developmental history of watershed-lake ecosystems if one has adequate information on vegetational/climatic history and the biogeochemistry of modern forest systems in the region. It should also be possible to infer variations that might be expected given differences in watershed size, elevation, soil thickness, and bedrock type.

The present paper presents the results of an interdisciplinary paleoecological investigation of the developmental history of three watershed-lake systems located on an elevational gradient (661–1150 m) in the High Peaks region of the Adirondack Mountains in New York (Fig. 1). The objectives of the study are to (1) reconstruct the vegetational history of the region and of each watershed, (2) reconstruct the chemical and biological history of each lake, and (3) elucidate the factors that have controlled the development of each system. The analyses have included pollen, plant macrofossils, diatoms, chrysophytes, sedimentary pigments, total metals, and ^{14}C dating. Much of this study, in particular the late-glacial and Holocene acidity changes experienced by these lakes, has been described elsewhere (Whitehead et al., 1986; Whitehead & Jackson, 1988; Jackson, 1988).

The Adirondacks: geology, climate, vegetation

The Adirondack Mountains (Fig. 1) are underlain by a variety of Proterozoic metamorphic and igneous rocks (Buddington, 1953). Elevations range from less than 300 m to over 1500 m within the High Peaks region. The soils are predominantly histosols and spodosols (Witty, 1968).

The available data suggest that the area was covered by continental ice following the St. Pierre Interstade or the Mackinaw Interstade and that the High Peaks were probably ice-free before 12 500 B.P. (Connally & Sirkin, 1971, 1973; Denny, 1974; Craft, 1976, 1979).

The present climate of the Adirondack region

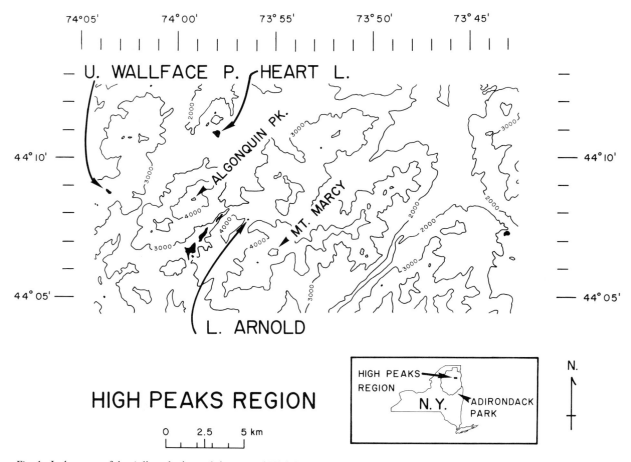

Fig. 1. Index map of the Adirondacks and the central High Peaks region. Locations of three lake sites are noted (from Whitehead *et al.*, 1986).

is characterized by cold snowy winters and cool wet summers (Mordoff, 1949). Precipitation is distributed relatively evenly throughout the year (Dethier, 1966). Temperature and length of the frost-free season decrease with elevation while precipitation increases. Cloud cover, wind velocity, and cloud droplet deposition also increase significantly with elevation (Reiners & Lang, 1979; Schlesinger & Reiners, 1974; Holroyd, 1970).

The vegetation of the Adirondack Mountains consists of a series of altitudinally defined integrating vegetation zones (Jackson, 1988). The zones are, in order of increasing elevation: (1) temperate hardwoods-northern hardwoods-hemlock-pine forest zone (in the Champlain Valley); (2) mixed conifer-northern hardwoods

zone; (3) subalpine spruce-fir-paper birch zone; (4) subalpine fir forest zone; and (5) alpine tundra zone. There are also important secondary patterns resulting from variations in soil texture, soil moisture, microclimate, slope, aspect, and disturbance history (Braun, 1950; DiNunzio, 1972; Heimberger, 1934; Holway *et al.*, 1979; Reiners & Lang, 1979, Scott & Holway, 1969; Sprugel, 1976; Young, 1934).

The lake sites

The three lake sites in this study are within 10 km of each other in the High Peaks region of the Adirondacks (Fig. 1). They differ substantially in elevation, watershed vegetation, and water chemistry. Morphometric, watershed, hydrological, water chemistry, and vegetation characteristics are summarized in Table 1.

The lowest site, Heart Lake (661 m), has the longest mean hydraulic retention time, the highest pH, and the highest alkalinity. The watershed is entirely within the mixed conifer-northern hardwoods forest zone.

In contrast, the highest site, Lake Arnold (1150 m), has the shortest mean hydraulic retention time, the lowest pH, the lowest alkalinity, and the highest aluminum concentrations. Watershed vegetation consists of a complex mosaic of mature red spruce-balsam fir stands and dense stands of young balsam fir. Paper birch occurs locally in the more mature stands.

The third site, Upper Wallface Pond, is at an intermediate elevation (948 m) and is intermediate in almost all characteristics except vegetation. The watershed is entirely within the subalpine spruce-fir-paper birch forest zone. Most of the watershed is vegetated by relatively young fir and paper birch, although there are local stands of mature red spruce and fir.

Methods

Sediment cores were taken with a square-rod piston sampler (Cushing & Wright, 1965). The primary cores for pollen, diatoms, chrysophytes, sediment chemistry, and ^{14}C dating were taken from the deepest portion of each lake. Additional cores for plant macrofossil analyses were taken in shallower water.

Radiocarbon dates were provided by the Radiocarbon Laboratory of the Smithsonian Institution and the Geochron Laboratories Division of Krueger Enterprises, Inc. All dates given in this paper are in uncorrected ^{14}C-years.

The methods used for analysis of pollen, plant macrofossils, and diatoms are described in Whitehead et al. (1986). Diatom taxa were assigned to one of Hustedt's (1939) pH categories based on literature references (e.g., Lowe, 1974; Cholnoky, 1968; Mariläinen, 1967; Renberg, 1976) and independently based on the present distribution of the taxa in 38 Adirondack lakes (Charles, 1982,

1985). The techniques used for analysis of chrysophytes are described by Smol (1986) and Christie & Smol (1986).

Sediment samples for chemical analysis were digested with concentrated HF, HNO_3, and HCl in a teflon bomb and analyzed by atomic absorption for a suite of six elements (K, Mg, Ca, Fe, Mn, Al). Mn and Al were not measured on Heart Lake sediments.

Pigments were extracted from wet sediments in 100 ml of acetone using techniques modified from Sanger & Gorham (1972). Chlorophyll derivatives were measured as the absorbance of the raw acetone extract at 665 nm and expressed in relative units where one unit is equal to an absorbance of 1.0 in a 10 cm cell. Carotenoids were extracted from the acetone mixture in petroleum ether after saponification in KOH and methanol. The absorbance of the carotenoids at 448 nm was then expressed in units analogous to those for chlorophyll.

Watershed-lake History

Regional vegetational trends

The major aspects of vegetational change in the High Peaks region since deglaciation can be inferred from the pollen diagrams. Pollen sequences from all three sites are similar (Whitehead & Jackson, 1988), so we will present only the diagram for Heart Lake (Fig. 2). Pollen data are described in greater detail by Whitehead & Jackson (1988).

The vegetation was essentially treeless prior to 11 000 yr B.P. Open spruce-dominated woodlands developed between 11 500 to 9 700 yr B.P. (Jackson, 1988; Whitehead & Jackson, 1988).

Spruce pollen declines sharply beginning about 10 500 years ago. This decline is accompanied by a transient maximum of alder (especially Alnus crispa) pollen. The alder maximum is especially prominent and lasts longer at the higher elevation sites.

Between 9700–7000 yr B.P., spruce decreased

Table 1. Location, morphometric, chemical, and watershed characteristics of Heart Lake, Upper Wallface Pond, and Lake Arnold. Soils in the Heart Lake watershed are developed in shallow to deep sandy glacial till; organic horizons range in depth from 0–10 cm. In the Upper Wallface Pond and Lake Arnold watersheds, soils typically consist of 10–50 cm of organic layer over weathered bedrock; till deposits appear to be lacking.

Lake	Latitude/ Longitude	Elev. (m)	Surface area (ha)	Max. depth (m)	Volume (10⁴ m³)	Watershed area (ha)	Mean* annual runoff (cm/a)	Mean hydraulic retention time (a)	pH⁺ range	Alkalinity† μeq l⁻¹	Total§ Al μg l⁻¹	Vegetation zone
Heart Lake	44° 10′ 50″ N 73° 58′ 03″ W	661	11.2	14.0	58.8	55.9	47	1.9	6.4–6.7	37	65	Northern hardwoods
Upper Wallface Pond	44° 08′ 47″ N 74° 03′ 15″ W	948	5.5	9.0	21.0	58.3	78	0.4	4.9–5.0	– 5	40	Spruce-fir-paper birch
Lake Arnold	44° 07′ 53″ N 74° 11′ 38″ W	1150	0.4	2.5	0.4	13.7	103	0.03	4.8–4.9	–11	570	Spruce-fir-paper birch

* Determined by subtracting evapotranspiration, calculated using method of Thornthwaite & Mather (1957), from precipita on (Knox & Nordenson, 1955).
⁺ Values for aerated lake-average samples taken monthly from May-August, 1978 and 1979, for Heart and Upper Wallface Pond, and in May and August, 1979, for Lake Arnold.
† Determined using Gran plot technique (Stumm & Morgan, 1970); measurements made on same dates as pH.
§ Measured using method of Dougan & Wilson (1974). Maximum possible eq l⁻¹ of Al, calculated using Cronan's (1978) method increases consistently with increasing elevation.

HEART LAKE
SELECTED POLLEN TYPES

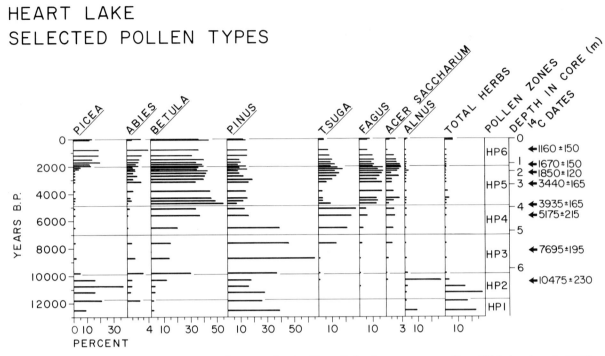

Fig. 2. Percentage diagram of dominant pollen types for the Heart Lake deep-water core (from Whitehead *et al.*, 1986).

markedly, mixed conifer-hardwood forests were established at lower elevations, and fir-birch forests at higher elevations. The mixed conifer-hardwood forests contained white pine, fir, birch, ash, sugar maple, hornbeam, and possibly oak.

Hemlock-dominated northern-hardwoods forests developed at lower elevations while fir-birch forests persisted at higher elevations between ~7000 and ~4800 yr B.P. Hemlock populations expanded rapidly and it became the dominant tree at lower elevations. White pine, birch (both paper and yellow), beech, sugar maple, and fir were also constituents of the mixed forests.

At ~4800 yr B.P., hemlock underwent a dramatic and extremely rapid population decline, apparently biogenically induced (Davis, 1981a; Allison *et al.*, 1986). Hemlock was replaced by birch, beech, and maple. Thus, there was an abrupt shift from conifer (hemlock) to hardwood (beech, pale, birch) dominance. Hemlock began a recovery at lower elevations at about 3400 yr B.P.

The past 2000 years were characterized by establishment of the existing vegetational zones. The major vegetational change was the expansion

of spruce populations at all elevations below 1300 m. Fir expanded slightly and hemlock, beech, and maple decreased in importance.

Localized watershed vegetational trends

The macrofossil data from the three lake sites confirm the regional trends outlined above and permit a reconstruction of the vegetational changes in each of the watersheds. Complete macrofossil profiles for all three lakes are presented and discussed in Jackson (1983, 1988).

At Heart Lake (Fig. 3), macrofossil remains (>12 000 yr B.P.), are indicative of tundra prior to 12 000 B.P., although there were localized spruce trees (Jackson, 1983, 1988).

The remainder of the profile for spruce needles roughly parallels the pollen curve for spruce. Fir appeared about 10 600 years ago and was represented consistently, but in low numbers, throughout the Holocene. White pine appeared in the Heart Lake watershed about 9000 years ago and was an important constituent of the mixed

175

HEART LAKE
SELECTED PLANT MACROFOSSIL TAXA

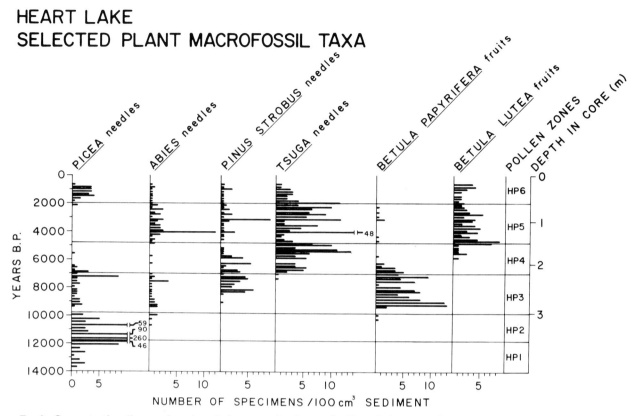

Fig. 3. Concentration diagram for selected plan macrofossil types for Heart Lake. Macrofossil data from a core taken in shallow water (from Whitehead *et al.*, 1986).

conifer-hardwood forests between 8400 and 6000 years B.P.

Hemlock became established in the Heart Lake basin at about 7400 years ago and expanded rapidly. It was abundant until the decline at 4800 yr B.P. and expanded again in the watershed beginning at about 3400 yr B.P. Needle frequency decreased steadily from 2000 years ago to the present.

Paper birch became established in the watershed about 10000 years ago and was abundant from 9000 to ~6000 yr B.P. Yellow birch did not appear in the record until 6000 years ago. It appears to have expanded rapidly immediately following the decline of hemlock.

The macrofossil record for Upper Wallface Pond (Jackson, 1983, 1988) indicates that the vegetational changes in this watershed have been different from those of Heart Lake. The record suggests that the forests near Upper Wallface

Pond were dominated by fir and paper birch from 10000 to 2000 years ago (shifting to spruce-fir-birch only in the last 2000 years). The presence of yellow birch, white pine, and hemlock between 9000 and 3000 years ago is consistent with an upward shift in their upper elevational limit in response to the climatic changes of the Hypsithermal (Jackson, 1983, 1988; Spear, 1981; Davis *et al.*, 1980).

The Lake Arnold macrofossil record (Fig. 4) also confirms the general pattern seen in the pollen diagram (Jackson, 1988; Whitehead & Jackson, 1988). Spruce was the dominant tree in the open late-glacial woodlands. Fir and paper birch were abundantly represented by 9500 years ago. Fir and paper birch dominated the watershed from 9700 to 2000 years ago. Some white pine, hemlock, and yellow birch trees grew near Lake Arnold during the Hypsithermal. Spruce began a significant increase at about 2500 years ago. The

176

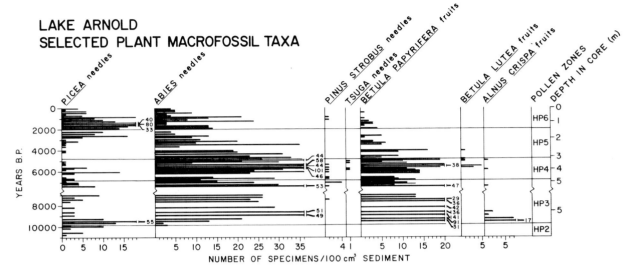

Fig. 4. Concentration diagram for selected plan macrofossil types for Lake Arnold. Data for pollen zones HP6 – upper HP3 from a macrofossil core taken near shore. Data for pollen zones HP2 and lower HP3 from the pollen/diatom core (from Whitehead *et al.*, 1986).

forests near Lake Arnold have been dominated by spruce-fir and, to a lesser extent, paper birch during the last 2000 years.

Diatom stratigraphy

A total of about 250 diatom taxa was identified from the cores of the three lakes. Fewer than 20–30 taxa were encountered in many of the late-glacial assemblages. In contrast, the Holocene diatom counts usually included at least 40 taxa. Species richness of the assemblages decreased with increasing lake elevation. This diversity gradient may be a function of decreasing lake size and diversity of aquatic habitat, independent of elevation. The complete diatom data for Heart Lake and Upper Wallface are in Reed (1982), and for Lake Arnold in Whitehead *et al.* (1986). Only the diagram for Heart lake is shown here (Fig. 5).

Heart Lake

The diatom data suggest that Heart Lake was initially slightly alkaline and underwent a late-glacial and early Holocene acidification. There were subtle changes in pH during the remainder

of the Holocene. Productivity may have been highest around 10000 yr B.P. (Reed, 1982). The Heart Lake diatom flora before that time was dominated by *Cyclotella stelligera*, and benthic alkaliphils were abundant (Fig. 5) (Brugam, 1983; Charles, 1985).

From 10000 yr B.P. to 4800 yr B.P. there were significant changes in both planktonic and benthic communities and steady increases of *Cyclotella comta* (Ehr.) Kutz, *Melosira distans* (Ehr.) Kutz, and *M. lirata* (Ehr.) Kutz, and varieties. Benthic acidophils were present in low concentrations.

The benthic acidophils increased significantly and the *Melosira* varieties decreased at 4800 yr B.P. (the hemlock decline). *Cyclotella comta* declined at about 3400 yr B.P. Most benthic acidophils continued to increase to about 1500 yr B.P. Among the benthic acidophils that increase after 9800 yr B.P. are *Anomoeneis serians* var. *brachysira* and *Frustulia rhomboides*, two taxa whose abundance is often positively correlated with DOC (Davis *et al.*, 1985).

Upper Wallface Pond. The late-glacial diatom assemblage from Upper Wallface Pond (Reed, 1982) is similar to that from Heart Lake. The euplanktonic forms, *Cyclotella comta*, *C. stelligera*, and *Fragilaria crotonensis* Kitton, occurred in

HEART LAKE
DOMINANT DIATOM TAXA

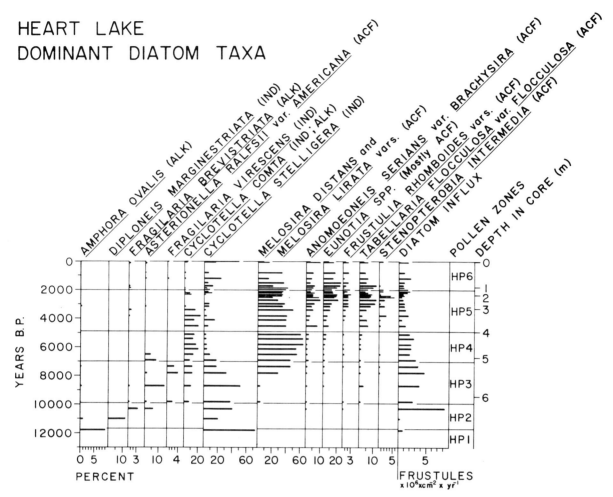

Fig. 5. Percentages of dominant diatoms in the Heart Lake deep-water core and influx rate of frustules. At least 50% of the counts for each level are represented. The pH category (Hustedt, 1939) to which each taxon was assigned is indicated in parentheses following the name: ACB = acidobiontic; ACF = acidophilic; IND = indifferent/circumneutral; ALK = alkaliphilic. The taxon designated *Cyclotella stelligera* includes forms that fit descriptions of both *C. stelligera* and *C. glomerata*. The two could not be reliably distinguished. Shape and markings were similar, and the diameter and striae count ranges overlapped. In general, forms found in the top half of the core were most similar to *C. glomerata*, while those in the bottom fit the description of *C. stelligera* best (from Whitehead *et al.*, 1986).

Upper Wallface Pond only during this period and a short time thereafter. Diatom accumulation rate was greatest durng this same interval. The ratio of *C. comta* to *C. stelligera* suggests an early acidification trend (Charles, 1985).

From 10 000 yr B.P. to the present there is a gradual increase in benthic acidophils, most of which also increased near the top of the Heart Lake core. These changes suggest an acidification trend.

Fragilaria acidobiontica Charles occurred in abundance only in the surface sediment sample. This may be indicative of an increased aluminum concentration or other changes resulting from increased acid deposition (Charles, 1986).

Lake Arnold. Benthic alkaliphils were common in the late-glacial and *Cyclotella stelligera* occurred only during this period. From 10 000 yr B.P. to the present, the diatom flora has been composed primarily of benthic acidophilic and acidobiontic taxa. Toward the surface of the core, the percentage of acidobiontic diatoms increases (Whitehead *et al.*, 1986).

General Considerations. The diatom stratigraphies in Heart Lake, Upper Wallface Pond, and Lake Arnold are different, though several taxa are common to all the cores. In a cluster analysis of diatom assemblages from all levels of all cores (no stratigraphic constraint), the late-glacial assemblages from the three lakes clustered together. In contrast, Holocene assemblages grouped mostly by lake. This suggests that conditions in the lakes were initially similar and then diverged during the Holocene.

In all lakes the greatest changes in diatom assemblages occurred around 10 000 yr B.P. The magnitude of change following this period varied among the lakes, but most changes were relatively slow. There are no clearly definable diatom zones after this period.

Changes in productivity of the lakes were assessed using both diatom accumulation rates and 'trophic preferences' of individual taxa. The former method assumes that net diatom frustule accumulation rates parallel overall lake productivity (e.g., Carney, 1982, Brugam, 1984). The latter method involves interpreting the diatom stratigraphy in terms of the trophic categories to which some individual taxa can be assigned (Lowe, 1974; Reed, 1982; Carney, 1982). Both approaches indicate that productivity was greatest during the late-glacial and lower during the Holocene. This trend to lower productivity has been termed meiotrophication by Quennerstedt (1955) and is probably related to the processes that result in natural acidification.

Chrysophyte stratigraphy

The developmental histories inferred from chrysophytes are similar to those inferred from diatoms, and strongly support our interpretation of more alkaline conditions during the early Holocene. The mallomonadacean trends are similar for Heart Lake (Fig. 6) and Upper Wallface Pond (Fig. 1 in Christie & Smol, 1986) and parallel the diatom trends. Lake Arnold is anomalous, as scales are absent to extremely rare below the 95 cm level (~750 yr B.P.).

There are very few scales in the late-glacial sediments of Heart Lake (Fig. 6). We do not believe that this is a preservational problem, as chrysophyte cysts are present in these sediments, but these cyst morphotypes are those characteristic of non-scaled species. Scaled chrysophytes appear not to have been part of the phytoplankton at that time. This is also true for the late-glacial sediments for Upper Wallface Pond (Christie & Smol, 1986) and Little Round Lake in Ontario (Smol & Boucherle, 1985).

At both Heart Lake and Upper Wallface Pond alkaliphilic mallomonads are common only in the early Holocene (e.g., *Mallomonas pseudocoronata* in Zone HP3 in Heart Lake, *Mallomonas caudata* and *M. elongata* in HP3 at Upper Wallface Pond). Much of the Holocene record at both sites is dominated by the generalist *M. crassisquama*. There is some indication of modest increases in acidophilic taxa during the Holocene (e.g. *M. acaroides* at Upper Wallface Pond). The oscillating decline of *M. allorgei* and the stratigraphic pattern for *Synura spinosa* and *S. uvella* in Heart Lake might also be correlated with slight Holocene acidification.

In short the data provided by mallomonadacean scales confirm the general trends inferred from the diatom stratigraphies. The lakes were more alkaline in the early Holocene and thereafter underwent acidification.

Diatom-inferred pH reconstructions

Lakewater pH trends were inferred using predictive equations derived from modern diatom assemblage and water chemistry data from 37 Adirondack lakes (Charles, 1985). The most accurate pH redictions for the Adirondacks are provided by index α (Nygaard, 1956), index B (Renberg & Hellberg, 1982), multiple regression (MR) of pH categories (Davis & Anderson, 1985; Charles, 1985), and multiple regression of the percentage of diatoms occurring in groups of taxa defined by cluster analysis (TC) (Charles, 1985).

179

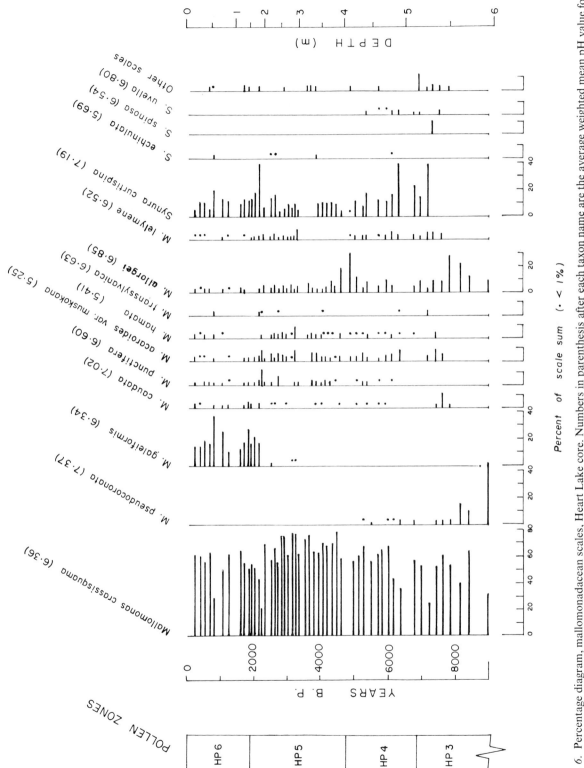

Fig. 6. Percentage diagram, mallomonadacean scales, Heart Lake core. Numbers in parenthesis after each taxon name are the average weighted mean pH value for each taxon in the Adirondack lake set.

180

Diatom-inferred pH values for Heart Lake, Upper Wallface Pond, and Lake Arnold correspond well and show similar trends (Whitehead *et al.*, 1986). The weighted average pH values were calculated for all three lakes and are presented in Fig. 7 so that trends among the three lakes can be compared more easily. The averages were calculated from only the MR, index α, and index B values. The pH values were inversely weighted by the standard errors of the predictive relationships used to calculate them. The weighted pH values are used in results described below.

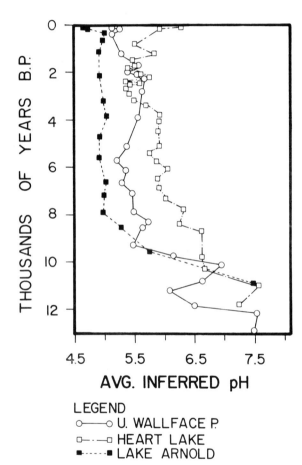

Fig. 7. Weighted average diatom inferred pH or Heart Lake, Upper Wallface Pond, and Lake Arnold. Averages were determined using the index α, index B, and multiple regression (MR) derived pH values shown in Figs 5, 8, and 11. Each pH value was weighted by the standard error of the predictive equation from which it was derived. Weights were 1.0, 1.0, and 1.18 for the α, B, and MR inferred pH values, respectively (from Whitehead *et al.*, 1986).

Heart Lake. The inferred pH of Heart Lake was above 7.0 before about 10 500 yr B.P. and declined steadily until ~6000 yr B.P. It remained at about 6.0 until 4000 yr B.P., then declined to 5.5 or less around 3000 yr B.P. Heart Lake was more acidic at this time than during any other period in its history. pH gradually increased from about 2000 yr B.P. to the present.

It is important to note that there may also have been increases in water color (and dissolved organic matter) during the Holocene. This is suggested by increases in *Anomoeoneis serians* var. *brachysira* and *Frustulia rhombiodes* late in the Holocene and increases in *Melosira distans* and *M. lirata* after 8000 yr B.P. The first two taxa are benthic acidophils whose abundance is positively correlated with humic-stained waters (Davis *et al.*, 1985). The *Melosira* species are also associated with more dystrophic situations (Camburn & Kingston, 1986). The changes in apparent color appear to be correlated with changes in hemlock populations in the watershed. This suggests that increases in humic acids in the soil (associated with increasing hemlock dominance) led to increases in water color.

Upper Wallface Pond. The inferred pH of Upper Wallface Pond was initially ~7.5, dropped sharply to a minimum of ~6.0 at ~11 500 yr B.P., increased to a maximum of about 7.0 at 10 000 years ago. pH then dropped to ~5.5

HEART LAKE

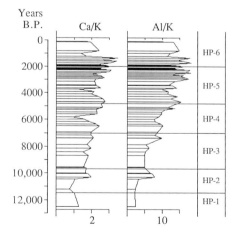

Fig. 8. Ca:K and Al:K ratios for Heart Lake core.

around 9000 yr B.P. (Fig. 9). The inferred pH was between ~5.7 and 5.3 for the remainder of the Holocene (Fig. 9).

The apparent rise in pH from 11 500 to 10 000 years ago is correlated with evidence of higher productivity (higher diatom accumulation rate and higher frequency of diatom taxa characteristic of productive lakes). However, because of the dominance by a few euplanktonic taxa, 'diversity' of the sedimentary assemblages is lower and the pH reconstruction is based on fewer taxa.

The dominant euplanktonic diatoms (*Cyclotella stelligera* and *C. comta*) disappeared from the assemblages around 9000 years ago, by which time the inferred pH had declined to ~5.5. Virtually no euplanktonic taxa occurred for the rest of the Holocene.

Lake Arnold. The inferred pH of Lake Arnold was between 7.5 to 6.5 during the late-glacial and decreased sharply to ~5.0 by 8000 yr B.P. (Fig. 7). It then remained close to 5.0 until very recently, when it declined slightly.

General Considerations. The diatom and chrysophyte stratigraphies and inferred pH data indicate that all three lakes were alkaline (pH > 7.0) immediately after deglaciation (~10 000–13 000 yr B.P.). Each lake acidified early, the rate slowing as relatively stable watershed conditions (soil, vegetation) were reached.

U. WALLFACE POND

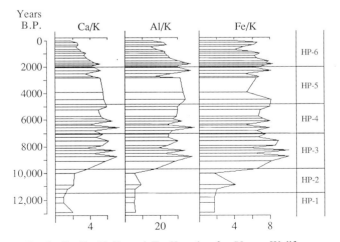

Fig. 9. Ca:K, Al:K, and Fe:K ratios for Upper Wallface Pond core.

The degree and rate of acidification increased with elevation (Fig. 7). The most rapid rate of pH decline was about 0.2–0.4 pH units in 100 years. Average Holocene (9000 yr B.P. to present) inferred pH for the lakes also followed the elevational gradient (Heart Lake = 5.9; Upper Wallface Pond = 5.5; Lake Arnold = 5.0). The average Holocene pH values for each lake are statistically different from each other at the 0.01 level based on t-test comparisons. The recent increase in *Fragilaria acidobiontica* (Charles, 1986) and the inferred pH decrease near the surface at Lake Arnold and Upper Wallface Pond (clearly evident in short cores) indicates acidification in historic times, apparently due to acid deposition from the atmosphere (Charles *et al.*, 1987; Charles & Norton, 1985). This is confirmed by chrysophyte data (Christie & Smol, 1986). This trend has been further documented for these two lakes by close-interval diatom and chrysophyte analysis of 0.5 m long cores of the top sediment Charles, unpublished data; Smol, unpublished data; Christie & Smol, 1986).

Geochemistry

The linkage between long-term vegetational change and the biogeochemistry of catchment soils is revealed by the chemical stratigraphy of lake sediments. Changes in soil composition are most clearly represented by the proportions of clastic (allogenic) and non-clastic (authigenic) components in the sediment. Elements such as K, Na, and Mg in Adirondack lake sediments are derived largely from the erosion of clastic silicates. Calcium, Al, Fe, and Mn, while often common in the clastic fraction, may also have substantial authigenic phases derived from solutional transport by organic acids (Mackereth, 1966; Engstrom & Wright, 1984). Because humus-rich soils enhance the flux of organic-bound metals and inhibit clastic erosion, ratios of allogenic to authigenic components, such as Fe:K, provide a direct index for Holocene soil development. Such ratios are desirable for the interpretation of bulk chemical analyses because they correct for the proportion of Ca, Al, and Fe in silicate lattices by normalizing to K, which is almost entirely in the allogenic fraction.

The ratios of Ca, Al, and Fe to K show distinct stratigraphic patterns at all three Adirondack sites (Figs 8–10). Ratios of the above elements to Mg (not illustrated) are similar. These ratios are lowest during the late-glacial and rise sharply at ~10 000 yr B.P. with the expansion of fir populations and development of closed forest at the two high altitude sites, Wallface and Arnold. At Heart Lake the increase is more gradual, beginning with the development of spruce woodlands ~10 500 yr B.P. and then rising in concert with hemlock populations between 7000 and 4500 yr B.P. A subsequent oscillation in the geochemical ratios follows the decline and recovery of hemlock at Heart Lake between 4500 and 2000 yr B.P. Ratios decrease with the reappearance of spruce in the last 2000 years. The drop in Ca:K at Heart Lake ~6000 yr B.P. and the decline in all ratios (Ca:K, Al:K, and Fe:K) between 8000 and 6000 yr B.P. at Lake Arnold have no clear correlates in the pollen or plant macrofossil record.

The geochemical profiles from the three Adirondack lakes confirm the pattern of Holocene soil development postulated for the High Peaks region by Whitehead *et al.* (1986). The clastic-rich sediments at the base of each core record the erosion of unweathered tundra inceptisols in a sparsely vegetated late-glacial tundra landscape. With the transformation of watershed vegetation to spruce woodland, soil organic content increased, impeding erosion and enhancing mineral weathering. Metals such as Fe and Al, eluviated by organic acids from upper soil horizons, were carried by surface drainage to the lakes (c.f. Pennington *et al.*, 1972; Engstrom & Hansen, 1985). This process continued with the subsequent replacement of woodland by a mixed conifer-hardwood forest at Heart Lake and fir-paper birch forest at higher elevations (Wallface and Arnold). The associated changes in sediment chemistry were more abrupt at Wallface and Arnold, however, implying that soil eluviation and humus accumulation occurred more rapidly under the high-elevation conifer forests. Soil solutions under conifer stands are typically richer in the humic compounds that eluviate base metals than those under hardwoods (Cronan & Aiken, 1985).

The gradual rise in Ca:K and Al:K in Heart Lake sediments between 7000 and 4800 yr B.P. indicates that low-elevation soils became increasingly spodic with the shift from hardwood to hemlock dominance during the mid-Holocene (7000 and 4800 years B.P.). The subsequent decrease in the geochemical ratios suggests that this trend in soil developement was interrupted by the replacement of hemlock dominated forests by northern hardwoods between 4800 and 3400 yr B.P.

The final decline in Ca:K, Al:K, and Fe:K at all sites after 2000 yr B.P. is more difficult to explain. The correlation of this trend with the reappearance of spruce and regional decline of hemlock suggests a linkage with vegetation, although the direction of chemical change is opposite to that noted during previous periods of conifer expansion. A decrease in the flux of authigenic metals from forest soils during zone HP2 is an unlikely cause given the soil-acidifying influence of conifer litter. An increase in clastic inputs, which could account for lower ratios, is equally improbable at the upper elevation sites because the inorganic content of the sediments declines during the late-glacial (Figs 11–13).

It is also unclear what caused the decline in geochemical ratios in the Lake Arnold core at 9000–6500 yr B.P., or the subsequent long-term

LAKE ARNOLD

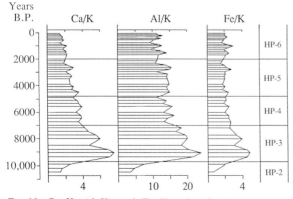

Fig. 10. Ca:K, Al:K, and Fe:K ratios for Lake Arnold core.

Heart Lake

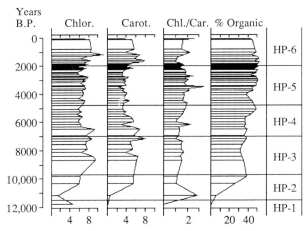

Fig. 11. Total Nitrogen, sedimentary pigments, chlorophyll/ carotenoid ratio, and per cent organic matter, Heart Lake core.

U. Wallface Pond

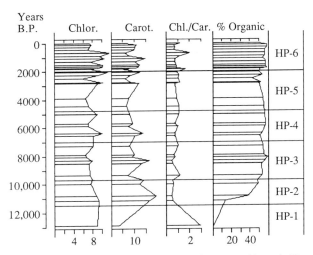

Fig. 12. Total Nitrogen, sedimentary pigments, chlorophyll/ carotenoid ratio, and per cent organic matter, Upper Wallface Pond core.

decrease in Ca:K. The decreases in ratios in this period B.P. are contemporaneous with a subtle drop in sedimentary organics (Fig. 13), but it is not correlated with any notable change in catchment vegetation that would imply edaphic control. The continuous decrease in Ca:K is not matched by a comparable decline in Al:K or Fe:K and thus may represent the long-term loss of soil-calcium through weathering.

Alternative mechanisms for these geochemical signals, such as change in redox conditions at the

sediment interface, are unlikely in light of the different controls on sedimentation of Ca, Al, and Fe. Although some forms of sedimentary Fe are redox sensitive, authigenic phases of Al and Ca occur largely as organometallic complexes and are not subject to redox cycling within the lake. The polymerization and sedimentation of dissolved humic materials is pH-dependent (Tipping & Ohnstad, 1984), however, so that increased accumulation of Ca, Al, and Fe may partially reflect decreasing lakewater pH. Such pH

Lake Arnold

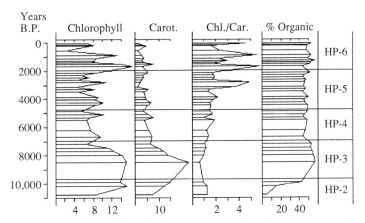

Fig. 13. Total Nitrogen, sedimentary pigments, chlorophyll/ carotenoid ratio, and per cent organic matter Lake Arnold core.

changes are consistent with the soil development sequence envisioned above. Long-term changes in clastic weathering might also contribute to the observed geochemical trends. The preferential leaching of K from silicate lattices, for example, would have the same effect on bulk sediment composition as increasing inputs of Ca, Al, and Fe. However, highly similar trends among the three elements (except Ca:K at Lake Arnold), and the fact that ratios with Mg as the denominator show the same pattern argue against differential weathering.

Pigments

The pigment data (Figs 11–13) show (1) an early rise in carotenoid concentrations, and (2) a strong correlation of high carotenoids and low chlorophyll/carotenoid ratios with the presence of a euplanktonic diatom flora.

The carotenoid peak is most pronounced at Lake Arnold between 10 000 and 8000 yr B.P. The peak is smaller and slightly earlier at Upper Wallface Pond, and is essentially absent at Heart Lake. This spatial and temporal pattern corresponds to that of the alder maximum, which is most prominent and lasts longest at high elevations (Whitehead *et al.*, 1986; Whitehead & Jackson, 1989). We suggest that the higher carotenoids may indicate a period of higher lake productivity, possibly associated with increased nutrient flux from the nitrogen-fixing alder.

High sedimentary carotenoids and low carotenoid/chlorophyll ratios are generally associated with more eutrophic lakes (Swain, 1985), although conditions for pigment preservation may be an important factor.

The correspondence between pigment and diatom profiles is most evident at Heart Lake where euplanktonic taxa are prominent in diatom assemblages throughout the Holocene. Pigment concentrations are highest (and carotenoid: chlorophyll lowest) from 10 500 to 6500 yr B.P., a period characterized by peak abundance of the plankter, *Cyclotella stelligera*. Both pigment concentrations and *C. stelligera* percentages are lowest from 6000 to 2000 yr B.P., after which time both profiles show a slight increase. This pigment-plankton correlation in weaker at Wallface Pond where euplanktonic diatoms disappeared in late-glacial times, and is absent at Lake Arnold where planktonic taxa have never been important. The correspondence of high pigment concentration and a maximum of planktonic diatoms suggests that Heart Lake was more productive in the early Holocene than later during the period of hemlock dominance in the watershed. At Wallface Pond an early Holocene peak in lake productivity was short-lived, and at Lake Arnold it may have been associated with algae other than diatoms. The corresponding shift in the carotenoid/chlorophyll ratios at all three sites can be attributed to differences in the relative production of these two pigment groups by different algal assemblages (Engstrom *et al.*, 1985; Swain, 1985).

Discussion

The correspondence between pollen, plant macrofossil, diatom, chrysophyte, and chemical stratigraphy suggests a tight linkage between vegetation and soils, and suggests that vegetation was a major factor controlling the biogeochemistry changes in these Adirondack watershed-lake systems. The alkaline and circumneutral lakewater conditions of the late-glacial probably resulted from the rapid leaching of base cations from fresh till and the transport of these solutes into the lakes. Edaphic changes associated with the development of spruce woodlands, including accelerated weathering, humus buildup, increased cation exchange capacity (CEC), lowered soil pH, and the increased availability or organic acids (Crocker & Major, 1955; Jacobson & Birks, 1980), led to a pronounced chemical transformation of lakewater in the late-glacial and early Holocene. Initially the flux of essential nutrients to the lakes remained high and this was reponsible for the higher levels of productivity at the late-glacial/Holocene transition. Gradually the export of base cations decreased, the flux of organic acids increased, and the pH of lake waters declined (Nihlgård, 1970; Brady, 1974; Cronan et al., 1978).

The shift from spruce woodland to mixed conifer-hardwood (fir dominated forests at higher elevations) at the late-glacial/Holocene boundary was marked by the transient expansion of alder which had important biogeochemical consequences, especially at the higher sites. Alder is an important colonizer in terrestrial boreal ecosystems and its presence results in dramatic increases in soil nitrogen (and nitrification) and decreases in soil pH (van Cleve et al., 1971; Crocker & Major, 1955; Binkley, 1982). This nitrogen pulse may have contributed to the trophic maximum inferred for the lakes between 11 000 and 9000 yr B.P.

The marked lakewater acidification trend of the late-glacial and early Holocene is a logical consequence of watershed biogeochemical events. Accelerated weathering and leaching would result in a rapid depletion of the readily available base cations. Hence, soil water concentrations of these cations would decline. The soil trends (continuing development of histosols and spodosols, increasing CEC, increasing production of organic acids, decreasing base saturation) would decrease both soil and pH, thus contributing further to a lowering of lake water pH (Chandler, 1939; Nihlgård, 1970; Brady, 1974; Cronan et al., 1978). This trend would be augmented by acidification of throughfall under the canopy of the developing spruce woodlands. The increasing abundance of fir would contribute to this as well (Eaton et al., 1973; Cronan, 1984; Raynal et al., 1983).

The probable increase in standing plant biomass from the late-glacial into the early Holocene (transition from tundra to spruce woodland to mixed conifer-hardwoods) may also have contributed to acidification. A net uptake (and accumulation in plant biomass) of cations would accompany the biomass increase. Cation uptake would be accompanied by below-ground proton release (Nilsson et al., 1982; van Breemen et al., 1984). In the presence of organic anions produced in the soil, this proton release would contribute to groundwater acidification.

Patterns of landscape development follow dif-ferent trajectories at the three sites. Diatom stratigraphy shows that limnological change was more abrupt at Wallface Pond and Lake Arnold than at Heart Lake. The pH of the upper two sites declined rapidly during the early Holocene and reached low stable values by 8000–9000 yr B.P. The acidification was prolonged at Heart Lake, continuing until about 6000 yr B.P. This contrast is readily explained by the differences in thickness of mineral soil and the divergent vegetational and edaphic trends manifest in the pollen and chemical records. Heart Lake has had a diverse vegetational history beginning with the transformation of spruce-woodland to northern hardwood forest, the rise and fall of hemlock dominance, and a final increase in fir and spruce in the watershed. A gradual drop in lake pH and a concomitant shift in the diatom assemblages might be expected at this site because of a slower transformation of catchment soils. By contrast, the rapid pH change at the upper elevation sites, Wallface and Arnold, corresponds with the early development of fir-dominated forests (which have occupied these watersheds to the present day). Spodosols and histosols (humus-rich soils) under fir and spruce stands should have the most pronounced acidifying impact on watershed drainage water (Nihlgård, 1971; Cronan et al., 1978; Cronan Aiken, 1985).

The pronounced Holocene elevational pH gradient fits expectations because higher elevation watersheds would have: (1) thinner mineral soils (hence a smaller potential reservoir of base cations), (2) lower weathering rates, (3) greater dilution of cations (due to higher precipitation, increased fog deposition, and lower evapotranspiration), (4) increasing dominance of spruce and fir resulting in acidified throughfall and higher organic acid production, and (5) extensive areas of histosols with higher CEC and lower base saturation (Witty, 1968; Vitousek, 1977; Cronan et al., 1978; Cronan, 1980; Olson et al., 1981; Lovett et al., 1982).

There were subtle mid- and late-Holocene changes in pH/alkalinity for all three lakes. For example, at Lake Arnold inferred pH appears to drop to about 4.7 in the uppermost levels. This undoubtedly reflects changes due to acid deposition and possibly indicates increasing aluminum levels in lake water (Whitehead *et al.*, 1986; Charles, 1985).

At Upper Wallface Pond there was: 1) a gradual decline of pH from >5.5 to ~5.2 between 9000 and 6000 yr B.P., (2) a slight increase in pH (~5.2 to ~5.7) between 6000 and 2500 years ago, and (3) a decline in pH in the last 2000 years. Watershed vegetational changes may have contributed. The expansion of hemlock in the watershed between 6500 and 5000 yr B.P. may have contributed to the earlier Holocene acidification trend. The expansion of spruce beginning just 2000 years ago probably contributed to the later acidification trend.

At Heart Lake, inferred pH remained close to 6.0 from 7000 to after 4000 yr B.P., dipped to a minimum between 3800 and 2700 yr B.P. (pH apparently dropped below 5.5 at this time) and increased over the past 1500 years. No short-term changes in lake water chemistry are evident at 4800 yr B.P., although inputs from the watershed should have undergone significant changes at this point due to the hemlock decline. Such changes have been detected elsewhere in the northeast (Whitehead *et al.*, 1973; Whitehead & Crisman, 1978; Crisman & Whitehead, 1978; Smol, 1982; Boucherle *et al.*, 1986). It is possible that sample intervals are too widely spaced to detect the trends or that the replacement of hemlock by birch, beech, and maple was sufficiently rapid to mask any biogeochemical changes caused by the hemlock die-off. However, some changes did occur in diatom stratigraphy (Fig. 5) that may be causally related.

The drop in pH between 3800 and 2700 yr B.P. may be associated with the re-establishment and expansion of hemlock in the watershed (fir also appears to expand in the watershed during this interval). The steady decline of hemlock over the last 2000 years and an increased frequency of fires in the region (E. Kurtz, personal communication) may have contributed to the most recent acidification trend at Heart Lake. The decreasing frequency of hemlock and increasing dominance of some hardwoods could cause an increase in groundwater pH. In addition, fires would mineralize some base cations and increase discharge to the lake. However, it is important to note that spruce increased in the watershed in the last 2000 years and we would expect this to counteract the geochemical impact of declining hemlock.

The importance of hemlock in controlling catchment biogeochemistry is emphasized by the strong correlation between the Ca:K and Al:K ratios and hemlock pollen and macrofossils (Fig. 10). This suggests a strong control of soil leaching processes.

Diatom stratigraphy provides some evidence for the increased export of humic compounds implied by the geochemical stratigraphy. For example, *Anomoeoneis serians* var. *brachysira* and *Frustulia rhomboides* increase in abundance at all three sites in concert with increases in the geochemical ratios (Ca:K, Al:K, and Fe:K) and the establishment of conifer-dominated forests. At Heart Lake the rise in these taxa is gradual, reaching peak values around 3000 yr B.P. while at Upper Wallface Pond and Lake Arnold the increase is abrupt and much earlier (9000–10 000 yr B.P.). The implied increase in water color, although time-transgressive among sites, is closely linked to the presence of humus–rich soils as inferred from chemical and pollen stratigraphy. Lakes receiving drainage from peatlands or forests with deep organic soils are typically the most highly colored (Rapp *et al.*, 1985; Engstrom, 1987). A greater influx of dissolved humic compounds would have increased the acidity of Adirondack lakes and may have lowered primary production through sequestering of nutrients (Schindler & Jackson, 1976) or by increased light attenuation and shading (Shapiro, 1957). Our study lakes are only moderately colored today (11–20 Pt-units), but may have been more so in the recent past, having lost dissolved organic carbon as a result of acid precipitation.

In conclusion, our paleoecological study of three Adirondack lakes provides a well-integrated picture of developmental history and watershed-lake interactions. The three lakes have undergone comparable developmental changes during the late-glacial and early Holocene which were controlled by a complex interaction of soils, changing climate (and hydrological budget), and watershed vegetation. These lakes were all initially alkaline and experienced acidification and a transient increase in productivity close to the transition from the late-glacial to the Holocene. The acidification trend persisted well into the Holocene. These early developmental events were driven in part by cation leaching stimulated by climatic amelioration and the development of coniferous vegetation (and the organic soils typical of such vegetation) and in part by transient establishment of alder (a symbiotic nitrogen-fixer) at about 10 000 yr B.P.

We suggest that the early developmental history of most softwater drainage lakes in northeastern United States should be similar to that of these Adirondack lakes. However, significant differences in bedrock and mineral soil might dictate a different developmental pathway (e.g., Ford, 1984).

The Holocene trajectories of the three lakes show different patterns reflecting differences in soil thickness, hydrological budget, weathering rate, and watershed vegetation that were related to elevation. The higher sites acidified more rapidly and stabilized at a lower pH than the low elevation sites. As this elevational gradient in response pattern and the late-glacial changes are predictable from modern watershed biogeochemical models (Vitousek, 1977), this study illustrates the utility of paleoecology for the verification of such models.

Subtle changes in water chemistry during the late Holocene at Heart Lake appear to have been influenced by major changes in the proportion of hemlock (vs. hardwoods) in the watershed. This illustrates how important a single species of tree can be in influencing the biogeochemistry of a watershed-lake system. As the hemlock decline of 4800 yr B.P. was an extremely abrupt event, the study provides some insight concerning the response of both watershed and lake systems to a 'catastrophic' watershed disturbance.

Lastly, this study provides important information concerning natural rates of acidification in watershed-lake systems. These rates contrast dramatically with those derived from paleoecological studies of acid-sensitive lakes in the Adirondacks that have been impacted by acid deposition (e.g. Big Moose Lake: Charles *et al.*, 1987). The natural acidification of the late-glacial and Holocene was slow, on the order of a thousand or more years for a decline of one pH unit. In contrast, anthropogenic acidification generated a comparable decline in pH at Big Moose Lake in twenty years.

Acknowledgements

This work was supported by NSF Grants DEB-77–03907, DEB-79–12210, and BSR-86–17622. We thank the Adirondack Park Agency, the Adirondack Mountain Club, and the New York State Department of Environmental Conservation for help with site selection and logistics. We owe an enormous debt of gratitude to Mark C. Sheehan and Robert J. Wise for their assistance with virtually every phase of field and laboratory work. We thank Dennis E. Matthews for his analysis of the sediment chemistry data and Susan E. Reed for the diatom counts. We are deeply indebted to the numerous field assistants who back-packed coring equipment to Lake Arnold and Upper Wallface Pond, often under difficult field conditions. We are grateful to R. B. Davis, S. A. Norton, J. Ford and several anonymous reviewers for critical comments on this study.

References

Allison, T. D., R. E. Moeller & M. B. Davis, 1986. Pollen in laminated sediments provides evidence for a mid-Holocene forest pathogen outbreak. Ecology 67: 1101–1105.

Binkley, D., 1982. Nitrogen fixation and net primary production in a young Sitka alder stand. Can. J. Bot. 60: 281–284.

Bormann, F. H. & G. E. Likens, 1979. Pattern and Process in a Forested Ecosystem. Springer-Verlag, New York.

Boucherle, M. M., J. P. Smol, T. C. Oliver, S. R. Brown & R. McNeely, 1986. Limnological consequences of the decline in hemlock 4800 years ago in three southern Ontario lakes. Hydrobiologia 143: 17–225.

188

Brady, N. C., 1974. The Nature and Properties of Soils. 8th Ed. McMillan, N.Y., 639 pp.

Braun, E. L., 1950. Deciduous Forests of Eastern North America. MacMillan, New York, N.Y., 596 pp.

van Breemen, N., C. T. Driscoll & J. Mulder, 1984. Acidic deposition and internal proton sources in acidification of soils and waters. Nature 307: 599–604.

Brugam, R. B., 1983. The relationship between fossil diatom assemblages and limnological conditions. Hydrobiologia 98: 223–235.

Brugam, R. B., 1984. Holocene paleolimnology. In H. E. Wright, Jr. (ed.). Late Quaternary Environments of the United States, 2. The Holocene. University of Minnesota Press, Minneapolis, 208–221.

Buddington, A. F., 1953. Geology of the Saranac Quadrangle, New York. N.Y. State Museum Bull. Number 346. The University of the State of New York, Albany, N.Y., 100 pp.

Camburn, K. E. & J. C. Kingston, 1986. The genus *Melosira* from soft-water lakes with special reference to northern Michigan, Wisconsin, and Minnesota. In J. P. Smol, R. W. Battarbee, R. B. Davis & J. Mariläinen (eds). Diatoms and Lake Acidity. W. Junk, Dordrecht, The Netherlands, 17–34.

Carney, H. J., 1982. Algal dynamics and trophic interactions in the recent history of Frains Lake, Michigan. Ecology 63: 814–1826.

Chandler, Jr., R. F., 1939. Cation exchange properties of certain forest soils in the Adirondack section. J. Agric. Res. 59: 491–505.

Charles, D. F., 1982. Studies of Adirondack Mountain (N.Y.) lakes: limnological characteristics and sediment diatom-water chemistry relationships. Doctoral dissertation, Indiana University, Bloomington, Indiana, USA.

Charles, D. F., 1985. Relationships between surface sediment diatom assemblages and lakewater characteristics in Adirondack lakes. Ecology 66: 994–1011.

Charles, D. F., 1986. A new diatom species, *Fragilaria acidobiontica*, from acidic lakes in northeastern North America. In J. P. Smol, R. W. Battarbee, R. B. Davis & J. Meriläinen (eds.). Diatoms and Lake Aciditiy. W. Junk, Dordrecht, The Netherlands, 35–44.

Charles, D. F., D. R. Whitehead, D. R. Engstrom, B. D. Fry, R. A. Hites, S. A. Norton, J. S. Owen, L. A. Roll, S. C. Schindler, J. P. Smol, A. J. Uutala, J. R. White & R. J. Wise, 1987. Paleolimnological evidence for recent acidification of Big Moose Lake, Adirondack Mountains, N.Y. (USA). Biogeochemistry 3: 267–296.

Cholnoky, B. J., 1968. Die Okologie der Diatomeen in Binnengewasser. J. Cramer, Lehre, Germany.

Christie, C. E. & J. P. Smol, 1986. Recent and long-term acidification of Upper Wallface Pond (N.Y.) as indicated by mallomonadacean microfossils. Hydrobiologia 143: 355–360.

van Cleve, K., L. A. Viereck & R. L. Schlentner, 1971. Accumulation of nitrogen in alder (*Alnus*) ecosystems near Fairbanks, Alaska. Arct. Alp. Res. 3: 101–114.

Connally, G. G. & L. A. Sirkin, 1971. Luzerne readvance near Glens Falls, New York. Geol. Soc. Am. Bull. 82: 89–1008.

Connally, G. G. & L. A. Sirkin, 1973. Wisconsinan history of the Hudson-Champlain lobe. Geol. Soc. Am. Memoir 136: 47–69.

Craft, J. L., 1979. Evidence of local glaciation, Adirondack Mountains, N.Y. 42nd Annual Reunion, Eastern Friends of the Pleistocene, 75 pp.

Crisman, T. L. & D. R. Whitehead, 1978. Paleolimnological studies of small New England (USA) ponds II: Cladoceran community responses to trophic oscillations. Pol. Arch. Hydrobiol. 25: 75–86.

Crocker, R. L. & J. Major, 1955. Soil development in relation to vegetation and surface age at Glacier Bay, Alaska. J. Ecol. 43: 427–448.

Cronan, C. S., 1980. Solution Chemistry of a New Hampshire subalpine ecosystem: a biogeochemical analysis. Oikos 34: 272–281.

Cronan, C. S., 1984. Biogeochemical responses of forest canopies to acid precipitation. In R. A. Linthurst (ed.). Direct and Indirect Effects of Acidic Deposition on Vegetation. Butterworth Publishers, 65–79.

Cronan, C. S. & G. R. Aiken, 1985. Chemistry and transport of soluble humic substances in forested watersheds of the Adirondack Park, New York. Geochim. Cosmochim. Acta 49: 697–1705.

Cronan, C. S., W. A. Reiners, R. C. Reynolds, Jr. & G. E. Lang, 1978. Forest floor leaching: contributions from mineral, organic, and carbonic acids in New Hampshire subalpine forests. Science 200: 309–311.

Cushing, E. J. & H. E. Wright, Jr., 1965. Hand-operated piston corers for lake sediments. Ecology 46: 380–384.

Davis, M. B., 1981a. Outbreaks of forest pathogens in Quaternary history. Proc. IV. Int. Paynol. Conf. Lucknow (1976–77) 3: 216–227.

Davis, M. B., 1981b. Quaternary history and the stability of forest communities. In D. C. West, H. H. Shugart & D. B. Botkin (eds). Forest Succession: Concepts and Applications. Springer-Verlag, New York, N.Y., 132–153.

Davis, M. B., 1983. Holocene vegetational history of eastern United States. In H. E. Wright, Jr. (ed.). Late Quaternary Environments of the United States. Volume 2. The Holocene. University of Minnesota, Minneapolis, 166–181.

Davis, M. B., R. W. Spear & L. C. K. Shane, 1980. Holocene climate of New England. Quat. Res. 14: 240–250.

Davis, R. B. & D. S. Anderson, 1985. Methods of pH calibration of sedimentary diatom remains for reconstructing history of pH in lakes. Hydrobiologia 120: 69–87.

Davis, R. B., D. S. Anderson & F. Berge, 1985. Paleolimnological evidence that lake acidification is accompanied by loss of organic matter. Nature 316: 436–438.

Davis, R. B. & G. L. Jacobson, 1985. Late-glacial and early Holocene landscapes in northern New England and adjacent areas of Canada. Quaternary Research 23: 341–368.

189

Denny, C. S., 1974. Pleistocene geology of the northeast Adirondack region, New York. U.S. Geol. Surv. Prof. Paper 786, 50 pp.

Dethier, B. E., 1966. Precipitation in New York State. Cornell Univ. Agric. Exp. Station Bull. 1009, 78 pp.

DiNunzio, M. G., 1972. A vegetational survey of the alpine zone of the Adirondack Mountains, New York. Master's thesis. State Univ. Coll. Forest. at Syracuse Univ., Syracuse, N.Y., 109 pp.

Eaton, J. S., G. E. Likens & F. H. Bormann, 1973. Throughfall and stemflow chemistry in a northern hardwood forest. J. Ecol. 61: 495–508.

Engstrom, D. R., 1987. The influence of vegetation and hydrology on the humus budgets of Labrador Lakes. Can. J. Fish. aquat. Sci. (in press).

Engstrom, D. R. & B. C. S. Hansen, 1985. Postglacial vegetational change and soil development in southeastern Labrador as inferred from pollen and chemical stratigraphy. Can. J. Bot. 63: 543–561.

Engstrom, D. R. & H. E. Wright, Jr., 1984. Chemical stratigraphy of lake sediments as a record of environmental change. In E. Y. Haworth & J. W. G. Lund (eds.). Lake Sediments and Environmental History. University of Minnesota Press, Minneapolis, Minn., 11–67.

Engstrom, D. R., E. B. Swain & J. C. Kingston, 1985. A paleolimnological record of human disturbance from Harvey's Lake, Vermont: geochemistry, pigments and diatoms. Freshwat. Biol. 15: 61–288.

Ford, M. S. (J.), 1984. The influence of lithology on ecosystem development in New England: a comparative ecological study. Ph.D. Thesis, University of Minnesota, Minneapolis.

Gaudreau, D. C. & T. Webb, III, 1985. Late-Quaternary pollen stratigraphy and isochronic maps for the northeastern United States. In V. M. Bryant, Jr. & P. G. Holloway (eds.). Pollen records of late Quaternary North American sediments. American Association of Stratigraphic Palynologists Foundation, Dallas, 247–280.

Heimberger, C. C., 1934. Forest-type studies in the Adirondack region. Cornell Univ. Agric. Exp. Station Memoir 165, 122 pp.

Holroyd, III, E. W., 1970. Prevailing winds on Whiteface Mountain as indicated by flag trees. Forest Sci. 16: 222–229.

Holway, J. G., J. T. Scott & S. A. Nicholson, 1969. Vegetation of the Whiteface Mountain region of the Adirondacks. In Atmospheric Sciences Research Center Report 92. State Univ. of New York, Albany, N.Y., 1–49.

Hustedt, F., 1939. Systematische und Ökologische Untersuchungen über die Diatomeen-Flora von Java, Bali, und Sumatra nach dem Material der Deutschen Limnologischen Sunda-Expedition III. Die ökologischen Factorin und ihr Einfluss auf die Diatomeenflora. Arch. Hydrobiol. supplement 16: 274–394.

Jackson, S. T., 1983. Late-glacial and Holocene vegetational changes in the Adirondack Mountains (New York): a macrofossil study. Doctoral dissertation. Indiana University, Bloomington, Indiana, 182 pp.

Jackson, S. T., 1988. Postglacial changes along an elevational gradient in the Adirondack Mountains (New York). New York State Museum and Science Service Bulletin (in press).

Jacobson, Jr., G. L. & H. J. B. Birks, 1980. Soil development on recent end moraines of the Klutlan Glacier, Yukon Territory, Canada. Quat. Res. 14: 87–100.

Likens, G. E., F. H. Bormann, R. S. Pierce, J. S. Eaton & N. M. Johnson, 1977. Biogeochemistry of a Forested Ecosystem. Springer-Verlag, New York, N.Y., 146 pp.

Lovett, G. M., W. A. Reiners & R. K. Olson, 1982. Cloud droplet deposition in subalpine balsam fir forests: hydrological and chemical inputs. Science 218: 1303–1304.

Lowe, R. L., 1974. Environmental requirements of pollution tolerance of freshwater diatoms. EPA-670/4–74–005. U.S. Envir. Protection Agency, Cincinnati, Ohio, USA.

Mackereth, F. J. H., 1966. Some chemical observations on post-glacial lake sediments. Philos. Trans. R. Soc. London, Ser. B, 250: 165–213.

Meriläinen, J., 1967. The diatom flora and the hydrogen-ion concentration of the water. Ann. bot. fenn. 4: 1–58.

Mordoff, R. A., 1949. The climate of New York State. N.Y. State Coll. Agric. Bull. 704, Cornell Univ., Ithaca, N.Y., 72 pp.

Nihlgård, B., 1979. Precipitation, its chemical composition and effect on soil water in beech and a spruce forest in south Sweden. Oikos 21: 208–217.

Nilsson, S. E., H. G. Miller & J. D. Miller, 1982. Forest growth as a possible cause of soil and water acidification: an examination of the concepts. Oikos 39: 40–49.

Nygaard, G., 1956. Ancient and recent flora of diatoms and Chrysophyceae in Lake Gribsø. In K. Berg & I. C. Peterson (eds.). Studies in Humic, Acid Lake Gribsø. Folia limnol. scand. 8: 32–94.

Olson, R. K., W. A. Reiners, C. S. Cronan & G. E. Lang, 1981. The chemistry and flux of throughfall and stemflow in subalpine balsam fir forests. Holarct. Ecol. 4: 291–300.

Overpeck, J. T., 1985. A pollen study of a late Quaternary peat bog, south-central Adirondack Mountains, New York. Geological Society of America Bulletin 96: 145–154.

Pennington, W., E. Y. Haworth, A. P. Bonny & J. P. Lishman, 1972. Lake sediments in northern Scotland. Philos. Trans. R. Soc. London, Ser. B, 264: 191–294.

Quennerstedt, N., 1955. Diatomeerna i Langans sjovegetation [Diatoms in the lake vegetation of the Langan drainage area, Jamtland, Sweden]. Acta Phytogeographica Suecica 36: 1–208.

Rapp, G., Jr., J. D. Allert, B. W. Liukkonen, J. A. Ilse, O. L. Loucks & G. E. Glass, 1985. Acid deposition and watershed characteristics in relation to lake chemistry in northeastern Minnesota. Environ. Int. 11: 425–440.

Raynal, D. J., F. S. Raleigh & A. V. Mollitor, 1983. Characterization of atmospheric deposition and ionic input at

Huntington Forest, Adirondack Mountains, N.Y. State Univ. N.Y. College of Envir. Sci. and Forest. (ESF-83–003), Syracuse, N.Y., 85 pp.

Reed, S., 1982. The late-glacial and postglacial diatoms of Heart Lake and Upper Wallface Pond in the Adirondack Mountains, New York. Masters thesis, San Francisco State University, San Francisco, Calif.

Reiners, W. A. & G. E. Lang, 1979. Vegetational patterns and processes in the balsam fir zone, White Mountains, New Hampshire. Ecology 60: 403–417.

Renberg, I., 1976. Paleolimnological investigations in Lake Prastsjon. Early Norrland 9: 113–159.

Renberg, I. & T. Hellberg, 1982. The pH history of lakes in southwestern Sweden, as calculated from the subfossil diatom flora of the sediments. Ambio 11: 30–33.

Sanger, J. E. & E. Gorham, 1972. Stratigraphy of fossil pigments as a guide to the post-glacial history of Kirchner Marsh, Minnesota. Limnol. Oceanogr. 17: 840–854.

Schindler, D. W. & T. A. Jackson, 1975. The biogeochemistry of phosphorus in an experimental lake environment: evidence for the formation of humic-metal phosphate complexes. Verh. int. Ver. Limnol. 19: 11–221.

Schlesinger, W. H. & W. A. Reiners, 1974. Deposition of water and cations on artificial foliar collectors in fir Krummholz of New England Mountains. Ecology 55: 378–386.

Scott, J. T. & J. G. Holway, 1969. Comparison of topographic and vegetation gradients in forest of Whiteface Mountain, New York. In Atmospheric Sciences Research Center Report 92. State Univ. of N.Y., Albany, N.Y., 44–87.

Smol, J. P., 1982. Postglacial changes in fossil algal assemblages from three Canadian lakes. Doctoral dissertation. Queen's Univ., Kingston, Ontario, Canada.

Smol, J. P., 1986. Chrysophycean microfossils as indicators of lakewater pH. In J. P. Smol, R. W. Battarbee, R. B. Davis & J. Meriläinen (eds.). Diatoms and Lake Acidity. W. Junk, Dordrecht.

Smol, J. P. & M. M. Boucherle, 1985. Postglacial changes in algal and cladoceran assemblages in Little Round Lake, Ontario. Arch. Hydrobiol. 13: 25–49.

Spear, R. W., 1981. History of alpine and subalpine vegetation in the White Mountains, New Hampshire. Doctoral dissertation. Univ. of Minnesota, Minneapolis, Minn.

Sprugel, D. G., 1976. Dynamic structure of wave-generated *Abies balsamea* forests in the northeastern United States. J. Ecol. 64: 889–911.

Swain, E. B., 1985. Measurement and interpretation of sedimentary pigments. Freshwat. Biol. 15: 53–75.

Tipping, E. & M. Ohnstad, 1984. Aggregation of aquatic humic substances, Chemical Geology 44: 349–357.

Vitousek, P. M., 1977. The regulation of element concentrations in mountain streams in northeastern United States. Ecol. Monogr. 47: 65–87.

Webb, III, T., 1981. The past 11 000 years of vegetational change in eastern North America. BioScience 31: 501–506.

Webb, III, T., 1988. Eastern North America. In B. Huntley & T. Webb, III (eds.). Vegetation History. Kluwer Academic Publishers.

Webb, III, T., E. J. Cushing & H. E. Wright, Jr., 1983. Holocene changes in the vegetation of the Midwest. In H. E. Wright, Jr. (ed.). Late-Quaternary Environments of the United States. Volume 2. The Holocene. Univ. of Minnesota Press, Minneapolis, Minn., 42–165.

Whitehead, D. R. & T. L. Crisman, 1978. Paleolimnological studies of small New England (USA) ponds I: late glacial and postglacial trophic oscillations. Pol. Arch. Hydrobiol. 25: 471–481.

Whitehead, D. R. & S. T. Jackson, 1988. The vegetational history of the High Peaks region of the Adirondack Mountains, N.Y.: inferences from palynology. New York State Museum and Science Service Bulletin submitted).

Whitehead, D. R., D. F. Charles, S. T. Jackson, S. E. Reed & M. C. Sheehan, 1986. Late-glacial and Holocene acidity changes in Adirondack (N.Y.) lakes. In J. P. Smol, R. W. Battarbee, R. B. Davis & J. Mariläinen (eds.). Diatoms and Lake Acidity, W. Junk, Dordrecht, 251–274.

Whitehead, D. R., H. Rochester, Jr., S. W. Rissing, D. B. Douglass & M. C. Sheenan, 1973. Late glacial and postglacial productivity changes in a New England pond. Science 181: 744–747.

Witty, J. E., 1968. Classification and distribution of soils on Whiteface Mountain, Essex County, New York. Doctoral dissertation. Cornell Univ., Ithaca, N.Y., 291 pp.

Young, V. A., 1934. Plant distribution as influenced by soil heterogeneity in Cranberry Lake region of the Adirondack Mountains. Ecology 15: 154–196.

Role of carotenoids in lake sediments for reconstructing trophic history during the late Quaternary

Hans Züllig

Züllig Ltd., CH-9424 Rheineck, Switzerland

Key words: lake history, plankton carotenoids, photosynthetic carotenoids, carotenoid stratigraphy, natural and anthropogenic eutrophication

Abstract

Cultural eutrophication of lakes occurring over the last 100 years is well known. Less well known is the eutrophication of lakes in earlier, late Quaternary time due to human and other causes. The recent and earlier trophic changes are documented in the sedimentary record by several groups of parameters. Among the most revealing of these are the diverse carotenoid pigments that originate from phytoplankton, photosynthetic bacteria, and other biota. The interpretation of the carotenoids in ancient sediments is facilitated by the study of carotenoids in recent sediments from lakes with relevant limnological and historical information. I support these contentions with evidence from several Swiss lakes, with emphasis on the late Quaternary development of Pfaffikersee and Soppensee.

Introduction

The greatly accelerated input of nutrients to Swiss lakes in this century has led to well known symptoms of eutrophication. Even certain high elevation lakes in mountain valleys, e.g., in the lake chain in the Upper Engadine (Grisons), Switzerland, at 1900 m, have been affected by eutrophication. One lake in that chain, the St. Moritzersee, had barely been influenced by civilization up to the end of the 19th century (Fig. 1). Since then, with the development of

St. Moritz as a holiday resort (Fig. 2), the lake has changed from extreme oligotrophy to eutrophy. The present eutrophic condition is reflected by high diatom production. The increase in productivity since 1900 is evidenced by increased diatom (frustule) accumulation rates in the sediment, shifts in Cladocera remains, determined by M. Boucherle (Boucherle & Züllig, 1983), and the initiation of a type of varve reflecting anaerobic bottom water (Züllig, 1982a and b) (Fig. 3).

Usually, discussions of lake eutrophication center around recent cultural phenomena, as at St. Moritz, but anthropogenic changes in the trophic state of lakes also occurred earlier in the late Quaternary due to activities such as the initial settlement of the land (deforestation and associated production of ash and erosional inwash), and artificial drainage and associated water level fluctuations. Probable trophic effects of these ac-

Originally published in
Journal of Paleolimnology **2**: 23–40.

Fig. 1. St. Moritz before 1800. Engraving, author unknown (Züllig, 1982 a, b).

Fig. 2. St. Moritz 1980, Photo Duschletta (Züllig, 1982 a, b).

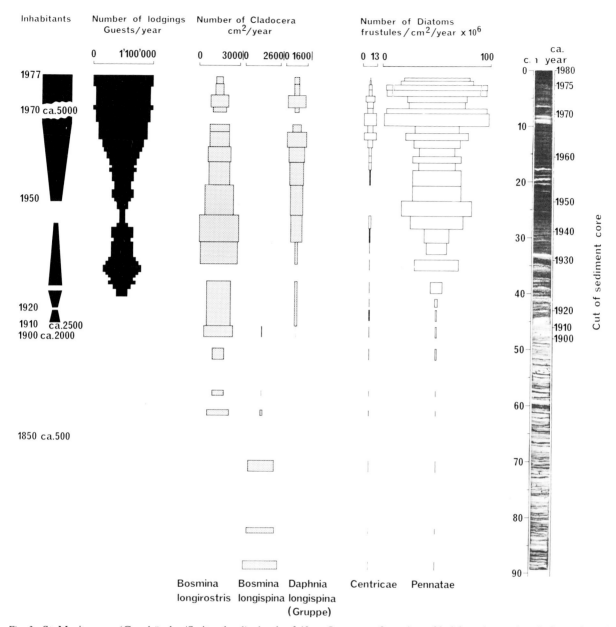

Fig. 3. St. Moritzersee (Graubünden/Switzerland), depth of 43 m. Increase of number of lodgings (guests/year), formation of anoxic, annual laminated sediments since 1920 and increase of number of diatom frustules since 1900 (Züllig, 1982 a, b), and changes of cladoceran remains.

tivities have been inferred from analysis of sediments deposited during the Stone and Bronze Age and the period of Roman colonization (Ammann, 1986). Changing non-anthropogenic conditions including physical (e.g., climate; developmental changes in lake morphology), chemical, and biological factors also have affected lake tro-

phic state during the late Quaternary (Ammann, 1985; Kelts, 1978).

The past trophic condition of lakes can be inferred from stratigraphic study of a wide range of sedimentary parameters. For example, in a core taken at 60 m water depth in the Baldeggersee (Sturm, pers. commun.) we found in Atlantic

194

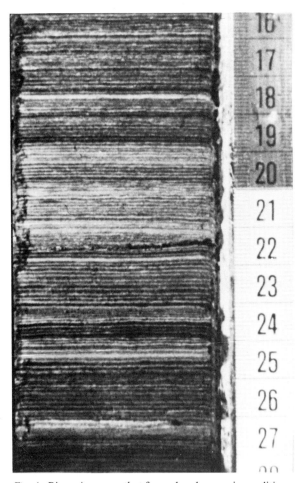

Fig. 4. Biogenic varves that formed under anoxic conditions, deposited in the Atlantic period (core from Baldeggersee, taken by M. Sturm).

Period sediments biogenic varves that could only have been formed under anoxic conditions (Fig. 4). Depending on limnological circumstances, anoxia may be interpreted as indicative of eutrophic lake conditions. The biological groups Cladocera and diatoms, already used as examples (Fig. 3), are only two of the many different groups of biota whose sedimentary remains are useful for paleolimnological inference of trophic state (Frey, 1974; Boucherle & Züllig, 1983).

Sedimentary pigments also yield information relevant to past trophic state. In lake sediment cores Brown (1968), Brown *et al.*, (1984), and Züllig (1981, 1982a, 1985a and b) found large fluctuations in carotenoid pigments originating from phytoplankton, photosynthetic bacteria,

and other sources (Sanger, 1988). To the extent that different groups of lacustrine microbiota bearing specific carotenoid pigments are known to be associated with particular trophic conditions, the trophic history of lakes can be read from the sediments as one would read a colored script. The objective of this paper is to illustrate the usefulness of sedimentary carotenoid pigment analysis for reconstructing lake trophic history.

Carotenoid pigments and their preservation in the sediments

The interpretation of carotenoid stratigraphy necessitates consideration of the preservation of the pigments. Paleolimnological research has shown that certain carotenoids are extraordinarily stable. For example, Vallentyne (1960) found carotenoids in 20 000 year old sediment from Searles Lake. Watts *et al.* (1977) described the occurrence of carotenoids in 5000 to 340 000 year old marine sediments. Sediment samples as old as 14 000 yr B.P. from Lobsigensee (Züllig, 1985a and b) and Soppensee (see later) provide evidence of long term preservation of most known planktonic and bacterial carotenoids (e.g., oscillaxanthin, myxoxanthophyll, okenone, spheroidenone, etc.). Nevertheless, it is known that certain carotenoids are unstable, for example fucoxanthin (Repeta & Gagosian, 1982) and peridinin (Kjosen *et al.*, 1976). While it may be reasonable to expect the formation of epoxy- and furanoxy carotenoids by oxidation of zeaxanthin during settling through an oxygen rich water column, and by acid reactions in the sediments, such oxygenated forms have not been clearly demonstrated in sediments (Züllig, 1982a). A review of research on the sedimentary preservation of carotenoids and critical discussion of the subject have been given by Swain (1985) and Sanger (1988). Sanger (1988) described the factors that promote pigment preservation, and he stated that 'No evidence has been presented that carotenoids decompose significantly once buried in the sediment, and if slow decompositions does occur it is similar to that or organic matter as a whole.'

Carotenoid pigments and their extraction

Important pigments for distinguishing between plant groups are:
1. myxoxanthophyll and its derivatives for Cyanophyta, oscillaxanthin for *Oscillatoria rubescens* and (but not observed in Swiss lakes) *O. aghardii,*
2. lutein for Chlorophyta and higher green plants,
3. fucoxanthin for Chrysophyta,
4. alloxanthin for Cryptophyta, and
5. peridinin for Pyrrhophyta.

A useful summary of the distribution of carotenoids in photosynthetic bacteria and algae has been given by Goodwin (1976).

For the study of lake trophic development, carotenoid pigments (carotenes and their derivatives) can easily be extracted from wet sediment with a mixture of acetone and alcohol. The determination of total carotenes and other carotenoids, the so-called crude carotenoids (Züllig, 1981), gives an approximate index of total bioproduction. Crude carotenoids have been measured by the optical density (1 cm light path) at 450 and 600 nm of acidified acetone/alcohol extracts (Züllig, 1985a) (Fig. 5).

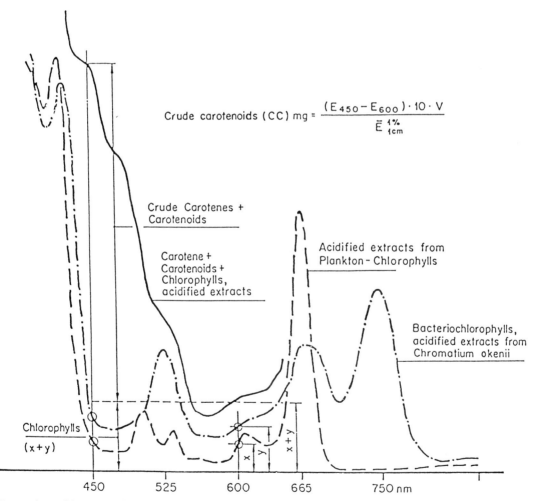

$$\text{Crude carotenoids (CC) mg} = \frac{(E_{450} - E_{600}) \cdot 10 \cdot V}{\bar{E}\,^{1\%}_{1cm}}$$

Fig. 5. Comparison of the absorption curves for chlorophylls and bacteriochlorophyll, both contained in acidified azetone/alcohol extracts 1:1, with the absorption curves for those chlorophylls as well as carotenes and carotenoids in sediment extracts (Züllig, 1985 a).

The individual carotenoid pigments have been separated with thin-layer chromatography (Kieselgel-Plate 60 Merck) while plankton carotenoids have been developed in several steps in mixtures of hexane/acetone/isopropanol (Züllig, 1981, 1982a). Chromatograms of extracts from photosynthetic bacteria are developed with benzine (Cp 110/140 °C) and acetone (80 : 20). The main pigments occurring in photosynthetic bacteria have been given by Züllig (1985a, 1986). Spheroidene and spheroidenone from *Rhodopseudomonas sphaeroides* and okenone from *Chromatium okenii*, as well as isorenieratene from *Phaeobium*, occur geographically widely in lake sediments and have been reported by Brown (1968). *Chromatium okenii* can form a visible plate at the sediment surface. Such thin layers were found at lago Cadagno (see later). The occurrence of certain bacterial pigments provide evidence for permanent or transitory anoxic conditions at the sediment surface (McIntosh, 1983; Züllig, 1985a).

Expression of carotenoid data

To use sedimentary carotenoid pigments for reconstruction of changes in trophic state of a lake, or for comparing trophic states of different lakes, it is important to express carotenoid contents in comparable units. Usually, carotenoid contents are expressed as weight percent of dried (80 °C) sediment. (At temperatures higher than 80 °C, clay minerals start to dewater and organic chemicals begin to volatilize; Iberg, 1954). However, percentage data can be misleading because data are expressed relative to each other. Because of the variable (from year to year) contributions of clay, carbonates, and other materials, the concentrations of carotenoids will also vary, unrelated to past biotic changes. To circumvent this effect, I have tried to use the true accumulation rate of carotenoids, according to the following equations (Züllig, 1981).

Specific annual quantity of sediments = SAQS

$$SAQS = h \times 1\ cm^2 \times D_f \times s \times 1000\ mg$$

where h is thickness (cm) of the annual layer of sediment, D_f the dry weight factor, and s the specific weight (conveniently 2.8).

Specific annual accumulation rate of a carotenoid = SAAC

$$SAAC = \frac{\%\ carotenoid \times SAQS \times mg}{100}$$

Some Swiss lakes, especially Zurichsee, have clear annual laminae. In the absence of such laminae, the accumulation rate of total sediment can be determined by recent chronostratigraphic markers (e.g., ^{137}Cs; pollen, as in Lotter, 1983, 1987) or radiometric techniques including ^{210}Pb and ^{14}C (e.g., Dominik *et al.*, 1981). These matters were discussed by Engstrom *et al.* (1985), Swain (1985), and Sanger (1988). For the stratigraphies discussed in this paper, accumulation rates could be calculated only for Zurichsee. For Walensee, Lago Cadagno, and Soppensee there was either inadequate dating control or the dry weight factor was missing.

Study of recent sediments, plankton, and trophic state

To facilitate interpretation of the early history of lakes from carotenoid stratigraphy, it is helpful to first study the carotenoid contents of recent sediments in lakes with available limnological and historical data.

Walensee

Walensee (a_o = 24.1 km^2, z_m = 145 m, elev. = 419 m) is one of the few prealpine Swiss lakes still in an oligotrophic condition. From the phytoplankton biomass data collected by the Wasserversorgung Zürich in 1985 and 1986, viz. the mean values of 4 mg l^{-1} (1985) and 1.2 mg l^{-1} (1986), I would expect low values for carotenoids in sediments. This is the case. The maximum value for crude carotenoids in surface sediment at 140 m water depth is 6.5×10^{-3} percent of dry

Walensee (Switzerland) »Unterterzen« 140m

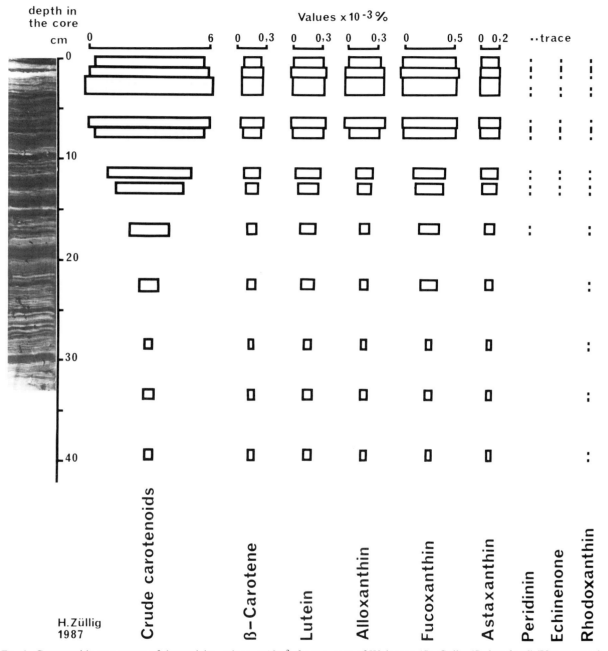

Fig. 6. Carotenoids as percent of dry weight, values × 10⁻³, from a core of Walensee (St. Gallen/Switzerland) 'Unterterzen', water depth of 140 m. The sediment from 33–40 cm was disturbed during preparation and therefore could not photographed.

weight (Fig. 6), whereas I found in the eutrophic Zurichsee (Züllig, 1982a) more than 30 × 10⁻³ percent. The stratigraphy does show that Walensee, while still oligotrophic, has undergone a considerable percentage increase in sedimentary pigment concentration, suggesting eutrophication. In most recent years I have observed a decrease of crude and single carotenoids in Wa-

lensee sediments, which indicates a slight restoration of the lake, and which is confirmed by
Bürgi *et al.* (1988) based on plankton analyses.

Zurichsee

Zurichsee (a_o = 90.1 km^2, z_m = 136 m, elev. =
406 m) is a prealpine Swiss lake that has become

more eutrophic than Walensee. Phytoplankton
biomass averaging 6 mg l^{-1} and with a maximum
value of 40 mg l^{-1} was measured in Zurichsee in
1985 and 1986 by Wasserversorgung Zurich.
Oscillatoria rubescens was prominent in the plankton between 1899 and the mid 1960s. The reasons
for the cultural eutrophication are associated with
greatly increased density of human population in
the past 100 years (Jaag, 1949). In a classic study,

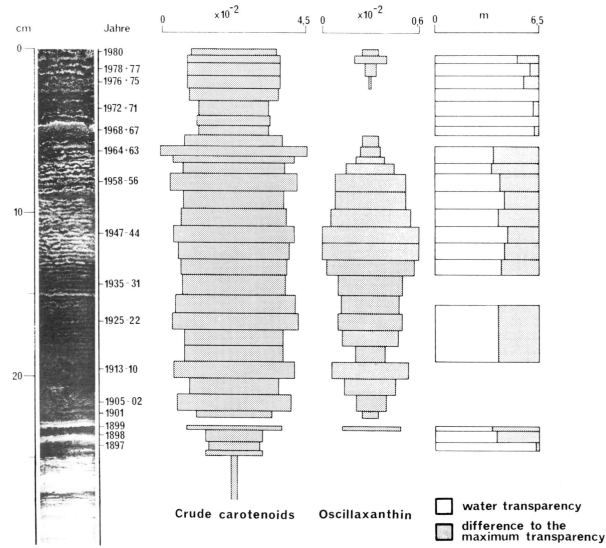

Fig. 7. Zurichsee, 'Meilen' (Switzerland), core from 125 m water depth, taken 4 April 1981. Relationships between crude
carotenoids, oscillaxanthin, and water transparency. Dates are based on varve counts (Züllig, 1982). Values mg/cm^2/year.

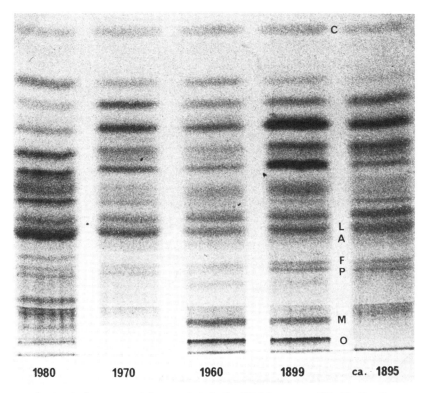

Fig. 8. Chromatograms of extracts from several time periods in the Zürichsee core (Fig. 7), C = β-carotene; A = alloxanthin; L = lutein; F = fucoxanthin; P = peridinin; M = myxoxanthophyll; and O = oscillaxanthin (Züllig, 1982 a).

Nipkow (1920) found that varves first appeared in the sediment in 1886. The recent trophic history is documented by pigment analysis of a core taken in 1981 at 125 m water depth (Züllig, 1982a) (Fig. 7). Accumulation rates of oscillaxanthin reflect the appearance, disappearance, and reappearance of *O. rubescens*. Evidence of these fluctuations consists of chromatograms of extracts of individual varves from several years, which show changes in concentrations of oscillaxanthin (the main pigment of *O. rubescens*) and other individual carotenoids (Fig. 8).

Oscillatoria rubescens has colored Swiss lakes red-brown during this century and has depleted oxygen from the deep layers of the lakes, especially Zurichsee. The appearance of this alga could be described as an epidemic. The fluctuations of *O. rubescens* (see later) are rather difficult to understand.

Lago Cadagno

An unusual biomass distribution occurs in meromictic Lago Cadagno. This lake ($a_o = 0.27$ km², $z_m = 21$ m, elev. = 1950 m) is located in gypsum and dolomite in the southern Swiss Alps. Above the anaerobic and H_2S-charged monimolimnion lies the much less saline and highly oxygenated mixolimnion. In a narrow redox transition zone (RTZ) in the chemocline, light and concentrations of oxygen and sulfide are optimal for photosynthetic sulfur cycling. Based on limited data, the depth of the RTZ varies with season between approximately 6 and approximately 13 m (Fig. 9). Sulfide oxidation in the light by purple sulfur bacteria takes place near the lower, oxygen-free boundary of the RTZ. High population densities of *Chromatium okenii* (and the pigment okenone) occur at that boundary (Hanselmann, 1985; Del Don *et al.*, 1985).

200

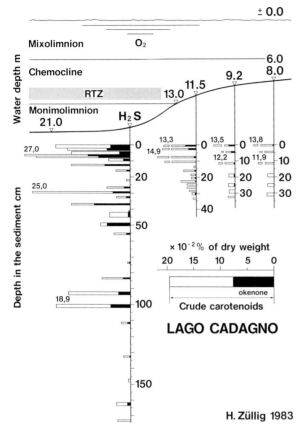

Fig. 9. Lago Cadagno (Ticino/Switzerland), max. water depth 21 m. The mixolimnion is rich in oxygen and slightly mineralized. The monimolimnion is free of oxygen and contains large quantities of sulfides. There is a chemocline between approx. 6 and 13 m with a narrow redox transition zone (RTZ) at a depth of approx. 13 m during the summer. Sulfide oxidation in the light by purple sulfur bacteria (Chromatium okenii) takes place near the lower oxygen-free boundary of the RTZ (Hanselmann, 1985, Del Don et al. 1985) and produces the pigment okenone which we can find in sediment cores at different water depths (Züllig, 1985a).

A series of cores was taken by Züllig (1985a) at water depths just above, in, and below the RTZ (Fig. 9). In the surface sediment of these cores the proportion of total carotenoids that consists of okenone is specific to the water depth of the sediment sample in relation to the RTZ, demonstrating that a depth series of sediment cores can be used to detect changes through time in RTZ depth.

One of the Lago Cadagno cores, 4 m long ('long core'), taken by A. Losher (Swiss Federal Insti-

tute of Geology, Zurich) at 19.5 m water depth (Fig. 10) contains irregular layers of sand, clay, and thin plates (probably annual layers) of fossilized sulfur bacteria – an uncommon sedimentary phenomenon. In such biogenic layers we found high quantities of okenone as well as pigments from *Rhodopseudomonas sphaeroides* (Figs. 11 and 12).

The carotenoids spheroidene and spheroidenone have a special significance. Under strictly anoxic conditions, *Rhodopseudomonas* produces yellow spheroidene only. Red spheroidenone is produced when oxygen is present, even very small amounts of oxygen (Fig. 11). If only spheroidene is present and spheroidenone is absent from sediment, it is inferred that the lake was meromictic (Züllig, 1985a).

Sediment core studies

Lago Cadagno

Pigments of Cyanophyta are absent in the long core from Lago Cadagno (Fig. 10). Apart from okenone and spheroidene, most consistently present in substantial concentrations are lutein, alloxanthin and beta-carotene. In single chromatograms from several depths in this core (Fig. 12), I find, besides okenone from *Chromatium*, only the yellow spheroidene from *Rhodopseudomonas*, and no red spheroidenone. This may be taken as evidence for meromixis characterized by absence of oxygen in the monimolimnion. Although detailed dating is not available for this core, it appears that the meromictic condition has existed for at least four thousand years, because the bottom of the core has a ^{14}C age of about 4000 years (Losher, pers. communication).

Soppensee

The eutrophic (Stadelmann and Sturm, pers. communication) Soppensee (a_o = 0.24 km², z_m = 27 m, elev. = 596 m), located in central Switzerland, has no stream inflows and therefore

201

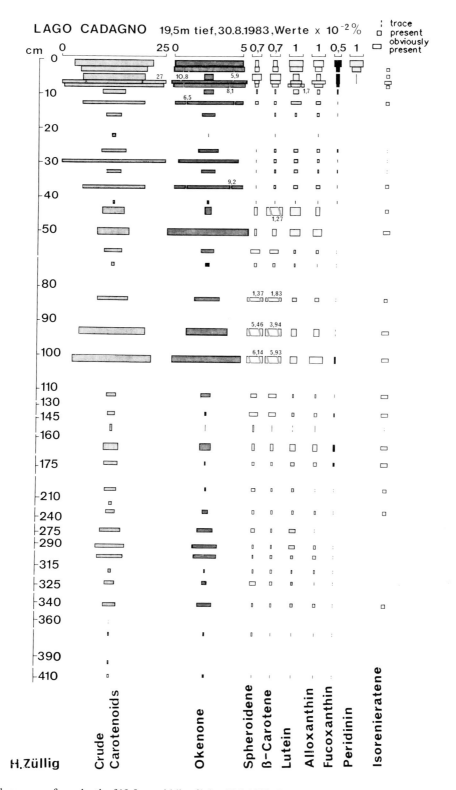

Fig. 10. Lago Cadagno, core from depth of 19,5 m, middle of lake, 30.8.1983. Carotenoids as percent of dry weight, values × 10⁻². Spheroidenone is strictly absent, indicating a meromictic state throughout the long period represented by the core.

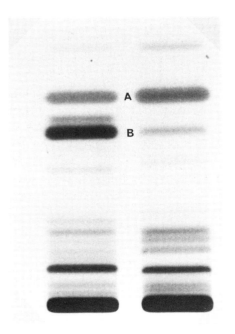

has minimal influences from the surface of the terrestrial watershed. A core taken at 27 m water depth by M. Sturm (Swiss Federal Institute for Water Resources and Water Pollution Control, Zurich) contains thousands of varves, providing a chronology for stratigraphies of pigments and other trophic indicators.

Sediment deposited during the early late-glacial (Oldest Dryas) has very low carotenoid concentrations (Fig. 13); the lake probably had low planktonic productivity. In this sediment, alloxanthin (from Cryptophyta) is dominant, and traces of lutein (from Chlorophyta) and fucoxanthin (from Chrysophyta) are present. Pigments of

◄ *Fig. 11.* Chromatograms of extracts of *Rhodopseudomonas sphaeroides* cultures. Left: not strictly anaerobic; Right: strictly anaerobic. Developed by Acetone-Benzine 20:80. A = yellow spheroidene; B = red spheroidenone.

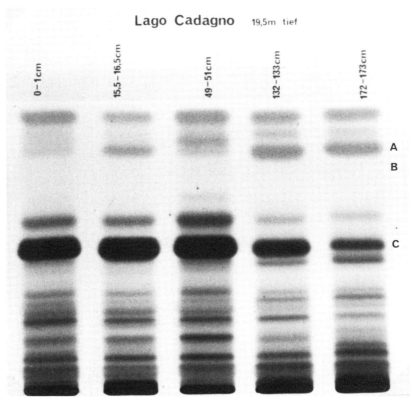

Fig. 12. Lago Cadagno, chromatographs of extracts of sediment from a series of depths (Fig. 10). Developed by Acetone-Benzine 20:80. A = yellow spheroidene; B = red spheroidenone is strictly absent; C = red okenone.

203

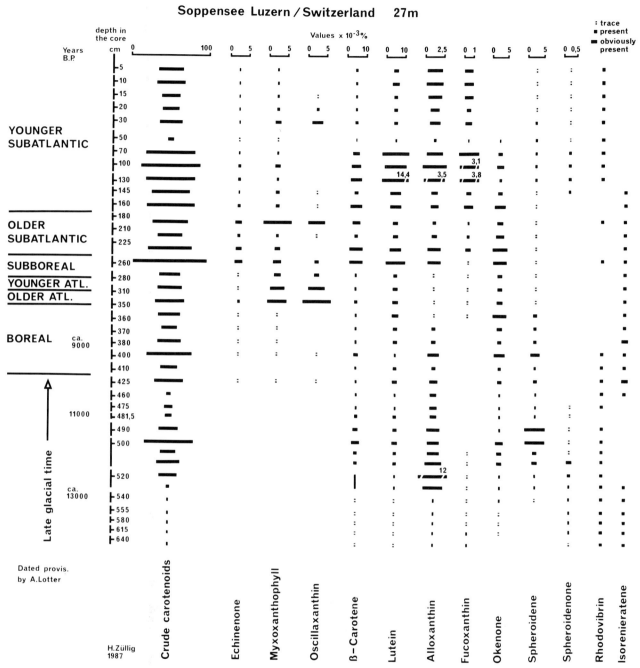

Fig. 13. Soppensee (Luzern/Switzerland), core from water depth of 27 m, middle of lake, showing distribution and increase/decrease of carotenoids (in percent of dry weight).

Cyanophyta are absent, but pigments of photosynthetic bacteria are present including okenone and the red spheroidenone. This means that the lake was holomictic, but that periods of deep water anoxia occurred (probably annually).

In sediment deposited at the end of the Oldest Dryas, dated by pollen analysis (Lotter, 1987), yellow spheroidene appears; later in the late-glacial this pigment increases, and spheroidenone disappears, indicating the establishment of me-

204

Pfaffikersee, 35m tief, Okt. 1984

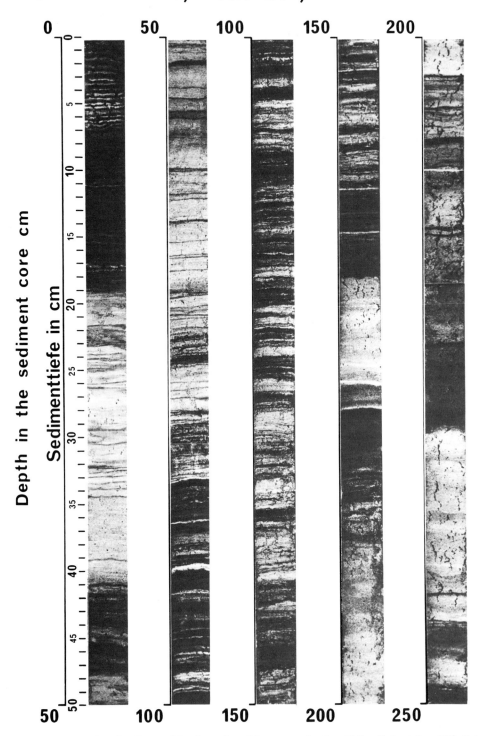

Fig. 14. Pfaffikersee (Zurich/Switzerland), core 2.5 m long, from 35 m water depth, middle of lake, taken 30th Oct. 1984. Recent sediments varved since ~ 1930 (~ 19 cm).

romixis. Pigments of Cyanophyta appear in the Boreal; their concentrations increase greatly in the Older Atlantic (e.g. oscillaxanthin from *Oscillatoria rubescens*) and remain ample during the mid-Holocene, indicating a probable change to a warmer climate.

The Chlorophyta, indicated by lutein, as well as Cryptophyta, indicated by alloxanthin, increased at the end of the Subboreal (Fig. 13). These pigments remained among the most abundant ones until the middle of the Younger Subatlantic. Oscillaxanthin was absent (*Oscillatoria rubescens* disappeared) in the early half of the Younger Subatlantic: a previously unknown temporal response of *O. rubescens*. In that period, the great increase in Chlorophyta and diatoms, the latter characterized by fucoxanthin, as well as alloxanthin, reflects a eutrophication that may have resulted from human settlement in Roman times, as shown by pollen spectra (Lotter, 1987). An alternative explanation for this eutrophication is suggested by the reappearance of spheroidenone. This reappearance indicates that oxygenated water reached the lake bottom and that the long period of meromixis came to an end. The circulation would have introduced nutrient-rich bottom water to the euphotic zone, and resulted in an increase in planktonic productivity. Due to a reduction of all pigments in the middle (70–50 cm) of the Younger Subatlantic, I propose that a reduction in trophic state occurred in that period.

By comparison, my first carotenoid stratigraphy from a Zurichsee core taken at 136 m water depth in 1952, showed high concentrations of crude carotenoids (not analyzed for single carotenoids) first appearing in the Boreal/Older Atlantic (Züllig, 1956). The earliest peaks in pigments definitely indicate that Cyanophyta and Chlorophyta appeared in the Boreal. This was also demonstrated in a core from the small (a_o = 2 ha), closed basin of Lobsigensee near Bern on the Swiss Plateau (Ammann, 1985; Züllig, 1985a and b).

Pfaffikersee

The important role and usefulness of carotenoid stratigraphy is further demonstrated by studies at Pfaffikersee, located 18 km east of Zurich and at 537 m elevation in a molasse formation. The lake was formed by glacial erosion into a 'Zungenbecken' and dammed by a moraine. Natural shallowing by accumulation of sediments, and by lowering of the outlet by erosion of the moraine, have caused a considerable reduction in the lake's original surface area (presently 3.25 km²) and depth (presently 35 m) (Schüepp *et al.*, 1977). Today, the surrounding area is dominated by agricultural and marsh areas. In the neighbouring region, there are small farm villages as well as the larger town of Pfaffikon. Since about 1940 it has been known (Züllig, unpublished) that in the sediments of the wind-exposed Pfaffikersee varves had been forming since 1930.

A series of sediment cores was taken in 1984 with a Ramming piston corer (Züllig, 1969). One of the cores, from 35 m water depth, is illustrated in Fig. 14. This and the other cores have not yet been dated (except for the top series of varves), but this core has several varved sections and the entire core has been provisionally dated by extrapolation of varve-based sediment accumulation rates to non-varved sections.

The sediment underlying the recent (to ca. 1930) varves consists of a monotone series of layers of light gray to light yellowish-brown chalky carbonate (Fig. 14). Below 0.8 m, there are irregular laminations of FeS, reflecting a reduced condition. It was surprising to find these laminae, for it was believed that Pfaffikersee had been oligotrophic and had oxygenated bottom water before 1930 (Thomas, 1942 and 1966).

According to the carotenoid stratigraphy (Fig. 15), the lake has been alternating between mesotrophic and eutrophic at irregular intervals, starting more than 2000 years ago. Reasons for the early eutrophication episodes are unclear. Possibly human impacts in the Bronze Age, afterwards during periods of Roman colonization, and later settlement by farmers (especially during the little climatic optimum 1200–800 yr B.P. (Lamb,

206

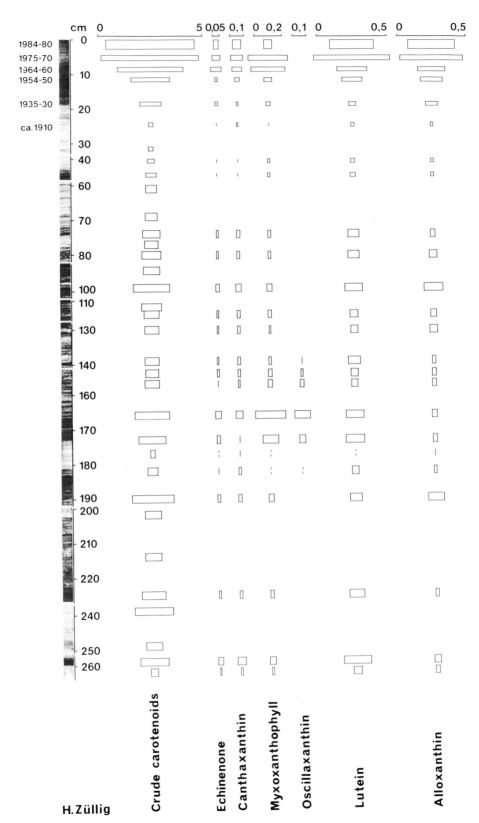

Fig. 15. Pfaffikersee, core from 35 m water depth, middle of lake. Carotenoid stratigraphy, carotenoids in percent of dry weight, values $\times 10^{-2}$, approx. 19 cm in the core corresponds to 1930 (Züllig, unpublished).

1977) involving deforestation by fire and inwash to the lake of ash and eroded soils caused these episodes.

Of particular interest is the temporary appearance of *Oscillatoria rubescens* following a major erosional episode. While *O. rubescens* did not appear later in Pfaffikersee, according to the carotenoid stratigraphy, the neighbouring Zurichsee has had such Cyanophyta for the past 90 years (forming dense blooms, at times). It is puzzling that *O. rubescens* has remained absent in Pfaffikersee in recent times, despite the lake's high nutrient status (Thomas, 1966). Other types of Cyanophyta have increased during the most recent eutrophication period in Pfaffikersee, as indicated by the myxoxanthophyll stratigraphy. Therefore, the absence of oscillaxanthin does not necessarily indicate the absence of eutrophic conditions.

Conclusions

1. New studies on pigments in ancient sediments of Swiss lakes, e.g. in late glacial sediments, show good preservation of most of the carotenoids that are known from recent sediments. Surprising is the occurrence of carotenoids of photosynthetic bacteria in sediments of the Oldest Dryas period when lakes are thought to have been oligotrophic.
2. The Swiss results coincide with those from Canada and USA and encourage the further use of 'fossil' carotenoids for study of the trophic development of lakes.
3. Exact between core and between-lake comparisons of trophic development are only possible if carotenoid values are expressed as annual accumulation rates. The calculation of such rates requires whole sediment data on annual accumulation rates expressed on a dry weight basis.
4. In cores from several small lakes, the carotenoid values in the warm part of the Boreal period as well as in the Older and Younger Atlantic, until the beginning of the Subboreal period, are surprisingly high. It remains to be determined whether such large carotenoid accumulation rates (and anoxic varve formation) also occurred in big Swiss lakes during the same periods. It is not easy to understand that such striking indications of natural eutrophication could have been produced solely as a result of the temperatures of the aforementioned periods.
5. The inferred early trophic development of the investigated lakes demonstrate the sensitive trophic responses of lakes to natural influences such as climate, and in small lakes to the activities of people during times of early settlement. We believe that such perceptions can be important for the choice of techniques (e.g. sewage treatment plant with high efficiency or only aeration) to be used for restoration of lakes. Unfortunately, so far there are no examples of the application of knowledge from carotenoid stratigraphic study to lake restoration strategy.

Acknowledgements

I thank Dr. M. Sturm and Dr. A. Losher for the samples from cores of Soppensee and Lago Cadagno, and W. Wüger for the organization and help in taking cores from Lago Cadagno and Pfaffikersee. I also appreciate very much the assistance of Prof. Ronald B. Davis and Dr. John P. Smol in the revision of my manuscript.

References

Ammann, B., 1985. Lobsigensee. Late glacial and Holocene Environments of a lake on the Central Swiss Plateau, Introduction and Palynology. In: G. Lang (ed.), Swiss lake and mire environments during the last 15 000 years. Diss. Bot. 87: 127–134. J. Cramer, Vaduz.

Ammann, B., 1986. Litho- and palynostratigraphy at Lobsigensee: Evidences for trophic changes during the Holocene. Hydrobiol. 143: 301–307.

Boucherle, M. M. & H. Züllig, 1983. Cladoceran remains as evidence of change in trophic state in three Swiss lakes. Hydrobiol. 103: 141–146.

Brown, S. R., 1968. Bacterial carotenoids from freshwater sediments. Limnol. Oceanogr. 13: 233–241.

208

Brown, S. R., H. J. Mc Intosh & J. P. Smol, 1984. Recent paleolimnology of a meromictic lake. Fossil pigments of photosynthetic bacteria. Verh. int. Ver. Limnol. 22: 1357–1360.

Bürgi, H. R., H. Ambühl, H. Bürer & E. Szabo, 1988. Wie reagiert das Seenplankton auf die Phosphorentlastung. Mitt. EAWAG, Zürich, 24: 4–5.

Del Don, C., K. W. Hanselmann, R. Peduzzi & H. Züllig, 1985. Phototrophic bacteria in the redox transition zone of Lago Cadagno, a meromictic, alpine lake. Experientia 41. Birkhäuser Verlag, CH-4010 Basel.

Dominik, J., A. Mangini & G. Müller, 1981. Determination of recent deposition rates in lake Constance with radioisotopic methods. Sedimentol. 28: 653–677.

Engstrom, D. R., E. B. Swain & J. C. Kingston, 1985. A paleolimnological record of human disturbance from Harvey's Lake, Vermont: Geochemistry, oscillaxanthin and diatoms. Freshwater Biol., 15: 261–288.

Frey, D. G., 1974. Paleolimnology, pp. 95–123. In W. Rodhe (ed.), Jubilee Symposium: 50 Years of Limnological Research. Mitt. int. Ver. Limnol. 20, 402 pp.

Goodwin, T. W., 1976. Chemistry and biochemistry of plant pigments. Academic Press London, New York, San Francisco.

Hanselmann, K. W., 1985. Lago Cadagno, ein 'Reagensglas' für Umweltforschung in der Natur. Neue Zürcher Zeitung 121, 83.

Iberg, R., 1954. Beitrag zur Kenntnis von Tonmineralien einiger Schweizer Böden. Diss. ETH Zürich, No. 2296.

Jaag, O., 1949. Die neuere Entwicklung und der heutige Zustand der Schweizer Seen. Verh. int. Ver. Limnol. 10: 192–209.

Kelts, K., 1978. Geological and sedimentary evolution of Lake Zürich and Zug. Diss. ETH 6146.

Kjøsen, H., S. Nordgard, S. Liaaen-Jensen, W. A. Svec, H. H. Straim, P. Wegfahot, H. Rapoport & F. T. Haxo, 1976. Algal carotenoids XV. Structural studies on peridinin. Acta chem. Scand. (B) 30: (2) 2–120.

Lamb, H. H., 1977. Climate. Present, Past and Future. Vol. 2. Climatic History and the Future. London Methuen.

Lotter, A., 1983. Pollenanalytische und sedimentologische Untersuchungen am Amsoldingersee bei Thun. Diplomarbeit Univ. Bern, 66 pp. (unpublished).

Lotter, A., 1987. Palynologische Untersuchung des Profils Soppensee 5086–14. Univ. Bern, System. Geobot. Inst., (unpublished).

Mc Intosh, H. J., 1983. A paleolimnological investigation of the bacterial carotenoids of Sunfish Lake. Thesis. Queen's Univ., Kingston, Ont.

Nipkow, F., 1920. Vorläufige Mitteilungen über Untersuchungen des Schlammabsatzes im Zurichsee. Schweiz. Z. Hydrobiol. 1: 10–122.

Repeta, J. R. & R. G. Gagosian, 1982. Carotenoid transformation in coastal marine waters. Nature: 295, No. 5844, 51–54.

Sanger, J. E., 1988. Fossil pigments in paleoecology and paleolimnology. Palaeogeogr., Palaeoclim., Palaeoecol., 62: 343–359.

Schüepp, M., M. Bider, M. Bouët, CH. Urfer, 1977. Regionale Klimabeschreibungen, 1. Teil Klimatologie der Schweiz, Band II; Beiheft zu Annalen der MZA, Zürich.

Swain, E. B., 1985. Measurement and interpretation of sedimentary pigments. Freshwat. Biol., 15: 53–75.

Thomas, E. A., 1942. Untersuchungen am Greifensee und am Pfaffikersee. Schweiz. Fisch Ztg. 2.

Thomas, E. A., 1966. Der Pfaffikersee vor, während und nach künstlicher Durchmischung. Verh. int. Ver. Limnol. 16, 144–152.

Vallentyne, J. R., 1954. Biochemical limnology. Science, 119: 605–606.

Vallentyne J. R., 1956. Epiphasic carotenoids in postglacial lake sediments. Limnol. Oceanogr. 1: 252–262.

Vallentyne, J. R., 1960. Fossil pigments. In: M. B. Allen, Comparative biochemistry of photoreactive systems. Academic Press, New York, N.Y., 83–105.

Watts, D. C. & J. R. Maxwell, 1977. Carotenoid diagenesis in marine sediment. Geochim. cosmochim. Acta 41, 493–497.

Züllig, H., 1956. Sedimente als Ausdruck des Zustandes eines Gewässers. Schweiz. Z. Hydrobiol. 18: 5–143.

Züllig, H., 1969. Ein Klein-Rammkolbenlot zur Gewinnung ungestörter Sedimentprofile. Schweiz. Z. Hydrol. 31: 128–131.

Züllig, H., 1981. On the use of carotenoid stratigraphy in lake sediments for detecting past developments of Phytoplankton. Limnol. Oceanogr. 26: 970–976.

Züllig, H., 1982a. Untersuchungen über die Stratigraphie von Carotinoiden im geschichteten Sediment von 10 Schweizerseen zur Erkundung früherer Plankton-Entfaltungen. Schweiz. Z. Hydrobiol. 44: 1–98.

Züllig, H., 1982b. Die Entwicklung von St. Moritz zum Weltkurort im Spiegel der Sedimente des St. Moritzersees. Zeitschr. 'Wasser, Energie, Luft'. 74: 7/8. Baden/Switzerland.

Züllig, H., 1985a. Pigmente phototropher Bakterien in Seesedimenten und ihre Bedeutung für die Seenforschung. Schweiz. Z. Hydrobiol. 47/2: 87–126.

Züllig, H., 1985b. Carotenoids from plankton and phototrophic bacteria in sediments as indicators of trophic changes: evidence from the Late-glacial and the early Holocene of Lobsigensee. In: G. Lang (ed.), Swiss lake and mire environments during the last 15 000 years. Diss. Bot. 87: 127–134. J. Cramer, Vaduz.

Zullig, H., 1986. Carotenoids from plankton and photosynthetic bacteria in sediments as indicators of trophic changes in Lake Lobsigen during the last 14 000 years. Hydrobiol. 143: 315–319.

Classification of lake basins and lacustrine deposits of Estonia*

Leili Saarse
Institute of Geology, Academy of Sciences of the Estonian SSR, 200101 Tallinn, 7 Estonia Avenue, Estonian SSR, USSR

Key words: classification, lake basins, lacustrine deposits Estonia

Abstract

Based on extensive data from a long-term investigation, a new genetic classification of lake basins is proposed for Estonia. Eight lake groups are distinguished, tectonic-denudation, glacial, chemical, fluvial, coastal (neotectonic), telmatogenic, cosmogenic and artificial, containing 13 subgroups and 19 basin types. Also proposed is a new lithological classification of Estonia's organic and calcareous lake sediments, based on analyses of more than 2000 sediment samples from 90 contemporary and 50 late-glacial (extinct) lakes. Of the ca. 1 150 Estonian lake basins that formed on mineral substrate, the two largest basins are of preglacial, tectonic-denudation origin, later modified by glaciers. Eight hundred lakes are of glacial origin, and 300 of other origins in the Holocene. In addition, ca 20 000 bog pools formed on peat in the Holocene. Only minerogenous sedimentation occurred in the lakes in the late-glacial period. After that, organic (gyttjas) and/or calcareous sediments have formed. Azonal factors have been largely responsible for the wide variation in Estonia's lacustrine deposits.

Introduction

Early genetic classifications of lake basins were worked out by Penck (1894) and Woldstedt (1926). Both became well known in Estonia. Since then many classifications have been published, among them the widely used one by Hutchinson (1957) which covers lake types on a world-wide basis. Håkanson & Jansson (1983) summarized Hutchinson's genetic classification, and reviewed concepts of classification of lakes based on trophic state and thermal stratification.

In Estonia, Kask's (1979) classification is the most complete, though it doesn't cover all varieties of lake basins. The investigation of Estonian lakes during the last 15 years as part of two projects, 1) IGCP Project 158 and 2) a more local project 'Formation of lake and mire deposits and displacement of ancient shorelines in the north Baltic region' (Saarse, 1985, 1987), has enabled us to collect new data on the litho- and biostratigraphy of lacustrine deposits, and has provided opportunity to study the morphology, genesis and development of lake basins. These studies have led to the development of a new genetic classification of Estonian lakes (Table 1). Our investigations indicate that the following agents are most important in the formation of

Originally published in
Journal of Paleolimnology 3: 1–12.

210

Table 1. Genetic classification of Estonian lake basins.

Genetic group	Genetic subgroup	Genetic type of lake basin
I Tectonic-denudation	Tectonic-denudation	Tectonic-denudation lakes deepened by ice
II Glacial	Ice-barrier and near-ice	Ice-walled lakes Proglacial (ice-marginal) lakes Near-ice lakes in hollows in stagnant ice terrain
	Dammed by glacial deposits	Dammed valley lakes
	Glaciokarst	Kettle hole valley lakes Kame field lakes Esker lakes Kettle hole moraine lakes
	Accumulation	Glacial accumulative lakes Residual accumulative lakes
	Exaration-accumulation	Drumlin field lakes
III Chemical	Dissolution	Karstic lakes
IV Coastal (neotectonic)	Coastal-residual	Relict lakes in coastal area
	Coastal-accumulation	Lakes obstructed by spits and bars
V Fluvial	Flood-plain	Oxbow lakes
VI Telmatogenic	Organic	Bog pools
VII Cosmogenic	Meteor	Crater lakes
VIII Artificial	Man-made	Storage lakes

Estonian lake basins: tectonic, glacial, fluvial, neotectonic, telmatogenic, chemical, cosmic and anthropogenic.

Among the ca. 1,150 small (< 14.4 km^2) lakes of Estonia, more than 70 percent are of glacial origin. The first lakes to appear ca. 13 500 yr B.P. were small water bodies on glacial ice which developed into ice-walled lakes. During deglaciation, proglacial lakes and lakes in depressions removed from the ice were formed (Saarse, 1979). Due to various processes 12 000 to 8 000 yr B.P. (glacial deposition and abrasion, erosion, glaciokarst, etc.), a diversity of basin types were formed. In our region glaciokarst processes continued into the early Holocene (to ca. 8 000 yr B.P.). Since then karstic, oxbow, bog secondary pools, coastal, cosmic, and anthropogenic water bodies have been formed in Estonia.

This paper offers: 1) a new classification of Estonian lake basins with brief characterizations of accumulated sediment, and 2) a lithological classification of Estonian organic and calcareous lacustrine deposits.

Part I. Genetic classification of Estonia's lake basins

Tectonic-denudation lake group

Two large Estonian lakes, Peipsi-Pihkva and Võrtsjärv belong to the tectonic-denudation lake group (Fig. 1, Table 1). Both are located in tectonic-denudated depressions of preglacial origin which were modified by glaciers, partly by glacial erosion, partly by filling with glacial deposits (Raukas & Rähni, 1969). Peipsi-Pihkva, with an area of 3558 km², and maximum and average water depth of 15.3 and 7.1 m, respectively, is located in a basin that had been occupied by a glacial lobe. The contemporary Peipsi-Pihkva contains glacial and glacioaquatic deposits (tills, varved clays, fluvioglacial sand and gravel) overlain by Holocene minerogenous lacustrine deposits. Thin layers of gyttja, with high organic content (up to 83%) occur in deep, calm bays (Raukas, 1981).

Vōrtsjärv has an area of 270 km², and maximum and average depths of 6.0 and 2.8 m, respectively. The basin is filled with till and sandy deposits that are covered by varved clays formed in the Glacial Vōrtsjärv stage, ca. 13000–12000 yr B.P. Upwards are lacustrine clays, silts and sands accumulated in Primeval Vōrtsjärv during Allerød and Younger Dryas times. Lacustrine lime and lake marl (see Part II on lithological classification) were deposited during the Preboreal and Boreal (Orviku, 1973). Starting in the Atlantic period, sediments with greater organic content have been formed including weakly calcareous or siliceous/minerogenous gyttjas (Veber, 1973).

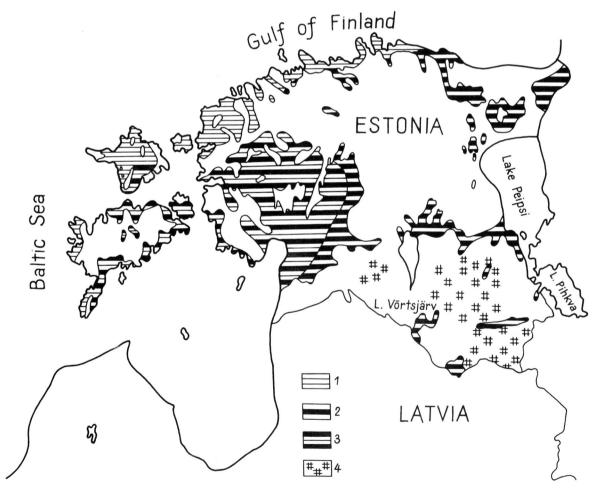

Fig. 1. Distribution of glaciolacustrine deposits in Estonia. 1 – deposits of Baltic Ice Lake, 2 – local proglacial lakes overlain by Baltic Ice Lake deposits, 3 – local proglacial lake deposits, 4 – ice-walled lake deposits.

Glacial lake group

This group has been divided into five subgroups (Table 1). *Ice-barrier* and *near-ice* lakes appeared in Estonia during the recession of the Weichselian (Valdaian) ice sheet. There are three different types: ice-walled, proglacial or ice-marginal, and near-ice in hollows. The first glacial lakes formed in the southeastern part of Estonia. They originated on top of glacial ice. With melting of underlying ice, they became ice-walled lakes (Fig. 1). These lakes developed ca. 13 300 to ca. 12 600 yr B.P. (Saarse, 1979). Lacustrine clays, with thicknesses from 1 to 10 m cover the tops of morainic hills and large plateaus. The uppermost part of the clay unit is massive; the lower part stratified with individual sandy laminae marking periods of intensive influx of material in meltwater. The stratification is so irregular that identification and enumeration of varves is virtually impossible. The absence of clear varves in these small basins was caused by the erratic melting of the ice walls, from different sides, and the insufficient distance for proper sorting of the sediment load. On the average, these stratified sediments are highly silty (64.4% 0.002–0.05 mm); moderately clayey (24.2% < 0.002 mm); and slightly sandy (11.4% 0.05–2.00 mm).

In the ice-barrier and near-ice lake group, most widespread were the local ice-marginal or proglacial lakes, which formed at different times ca. 12 700 to ca. 10 800 yr B.P. (Kessel & Raukas, 1979). The deposits of local proglacial lakes are scattered over large areas of Upper Estonia (i.e. in areas above the highest limit of the Baltic Ice Lake) where the deposits are mostly visible on the surface. In the areas of Lower Estonia that were inundated by the Baltic Ice Lake or the Baltic Sea local proglacial lake deposits are buried (Figs. 1 & 2). The glaciolacustrine deposits in these short-lived lakes are well laminated, with distinct varves and clearly developed facies. In individual summer layers, 20–80 diurnal laminae have been identified. The precondition for the formation of diurnal laminae was an elongated basin of sedimentation with unidirectional inflow. The current-ripple marks, cross-bedded summer layers, and erosional traces on the surface of winter layers give evidence of shallowness of these lakes. Their unstable water level is evidenced by coexistence of drainage varves and classical varve couplets in the same sequences (Pirrus, 1968). Average grain-size composition of local proglacial lake deposits is similar to that of ice-walled lakes: silty fraction 59.5%; clayey and sandy fractions 28.0 and 12.4%, respectively.

The Baltic Ice Lake also belongs to the proglacial lake type, but it was relatively long-lived. The actual duration has been subject to debate: about 1000 yr (e.g. Pärna, 1960; Synge, 1980; Donner, 1982) or about 2000 yr (e.g. Kvasov, 1979; Kessel & Raukas, 1979). Deposits of the Baltic Ice Lake cover large areas of the west-Estonian lowland and archipelago (Fig. 1). A special trait of these clayey sediments is the similarity in grain-size of summer and winter layers, the winter layer being only a bit more clayey. Boundaries between varves are transitional, which is explained by the high dispersivity of initial material and large area of the basin of sedimentation. In complete sequences proximal, central and sometimes also distal facies are distinguishable. Internal stratification in summer layers reflects intensive melting periods or stormy days, when coarse material drifted over large areas. Current-ripple marks, erosional features, cross-bedded summer layers, and contemporaneous disturbances of bedding are rare (Pirrus & Saarse, 1979). The thickness of these laminated clays is remarkable, up to 28 m and with hundreds of varves and only one or two drainage laminae. Grain-size composition is distinguished by high clay content, 40–60 percent.

In southern Estonia several of the hollows in dead-ice hummocky terrain have been studied and found to contain weakly laminated lacustrine silts and clays, which formed at the final stage of regional deglaciation. These lakes were near but not in contact with the stagnant ice. The hollows were fed by glacial meltwater that carried a substantial sediment load. As in the case for most other small glaciolacustrine basins, the hollows dried up in the late-glacial period. The laminated deposits of the hollows, remaining below the ground-water table, haven't consolidated. They

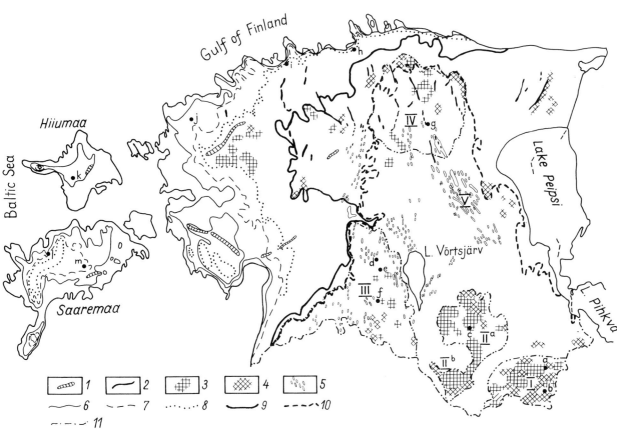

Fig. 2. Distribution of major elevatioral, glacial and late-glacial features of Estonia. 1 – esker, 2 – end moraine, 3 – moraine (till) relief, 4 – kame field, 5 – drumlin field, 6 – limit of Limnea Sea, 7 – limit of Littorina Sea, 8 – limit of Ancylus Lake, 9 – stage III limit of Baltic Ice Lake, 10 – stage I limit of Baltic Ice Lake, 11 – boundary of heights and uplands. Locations of lakes mentioned in paper: a – Kirikumäe, b – Pulli, c – Pühajärv, d – Võistre, e – Viljandi, F – Õisu, g – Äntu, h – Kahala, i – Ülemiste, j – Tänavjärv, k – Tihu, l – Ohtja, m – Kaali. I – Haanja Heights, II^a – Otepää Heights, II^b – Karula Heights, III – Sakala Uplands, IV – Pandivere Uplands, V – Saadjärv drumlin field.

remain liquid, with high porosity and compressibility. They have a grey or blue-grey color. Their average grain-size composition is 50.9% silt, 36.8% clay, and 12.3% sand. Minerological composition differs from the ice-walled and proglacial lake deposits in that much more chlorite (23%) and mixed-layer mineral (7%) content is present.

Lakes dammed by glacial deposits formed in river valleys (Valgejõe, Loobu, Kunda, etc.) by the damming of the water flow by esker ridges and kames. The basal, late-glacial unit of the lacustrine deposit consists of silts and sands bearing disseminated organic matter. That unit is conformably overlain by lake marl and lacustrine lime, with gyttja beds in the deepest parts of basins. The marl/lime/gyttja deposits started to

accumulate after the beginning of the Holocene (Männil, 1961). Due to the rise in water level in the second half of the Atlantic period, the rivers eroded channels through the barriers and these dammed valley lakes dried up. Brightly colored lacustrine deposits now form exposures on the river banks.

Glaciokarstic water bodies are the most widespread glacial lake type in Estonia. They occur in the ancient valleys and hummocky moraines of South Estonia, and in esker ridges and kame fields in North Estonia (Fig. 2). Kettle hole valley lakes are characteristic of the Sakala Uplands (Fig. 2), where undulating morainic relief is intersected by deep ancient valleys. Lake basins here are elongated, their floors are uneven, and sedi-

214

ment composition is heterogenous. In the Holocene parts of the lacustrine deposits, calcareous
and siliceous/minerogenous gyttjas dominate.
Deep kettle holes with steep underwater slopes
and negligible populations of aquatic plants are
poor in lacustrine deposits. In deep, steep-sided
Viljandi, for example, during the entire Holocene
only one meter of sediment (siliceous/clayey
gyttja) has accumulated, whereas in the shallow
lakes Õisu and Võistre 8–10 m has accumulated.
According to pollen chronostratigraphy, minerogenous lacustrine sediments formed since the
Allerød (Lõokene & Pirrus, 1979), to be succeeded in the Preboreal by organic and calcareous
deposits.

Glaciokarstic lakes in kame fields (Fig. 2) are
characteristic of North Estonia. The deposits of
these lakes consist of gyttja with low amounts of
siliceous/minerogenous and calcareous material.
The absence of basal glaciolacustrine and
lacustrine minerogenous clay-silt deposits, and
the presence of basal peat layers overlain directly
by gyttja testify to the delay in the formation of
these lakes – not until the end of the late-glacial
or during the Holocene. The total depths and
bedding conditions of the deposits are variable
due to differences in bottom topography, water
chemistry, trophic state and biological productivity.

The glaciokarstic lakes of esker ridges (Table 1)
in North Estonia are of two subtypes: 1) typical
kettle holes between eskers or near them, and
2) residual lakes dammed by esker ridges. The
morphology of the first subtype is similar to
glaciokarstic lakes of kame fields; the second subtype is similar to the accumulative glacial lakes of
South Estonia. In the second subtype, siliceous or
organic gyttjas are dominant. Where carbonate
bedrock and calcareous beds comprise the catchment, calcareous sediments have accumulated in
the lakes as well. Lacustrine sediments deposited
in the Younger Dryas are minerogenous and silty
and sandy. Minerogenous deposits continue to
the Preboreal, and in some of these lakes to the
Boreal as well, indicating that glaciokarst
processes were rather long-lasting in this region.

Kettle holes in areas of hummocky moraine are
small in area, but with water depths as great as
38 m. The depths and characteristics of the
lacustrine deposits depend on sediment supply,
water chemistry, and primary production of the
lakes. Glaciolacustrine varved clays are commonly absent; lacustrine sand-silt beds are thin or
absent. Basal telmatic peats occur in some basins.
The lacustrine sediment usually consists of lime
gyttja or calcareous gyttja (see Part II on lithological classification), interbeded with layers of
lacustrine lime and lake marl. Carbon-14 dates of
bottom sediments (e.g. 10 770 ± 130 yr B.P.,
TA-216, Sarv, 1983; 8820 ± 60 yr B.P., Tln-650,
Punning et al., 1985) indicate that these glaciokarst hollows were formed at various times in the
late-glacial and early postglacial; in some areas
the frequency of basin formation decreased in the
Boreal period.

Glacial accumulative lakes, a subgroup of glacial accumulation lakes (Table 1), occur in Upper
Estonia, mostly on the heights and uplands
(Fig. 2). These basins were formed as depressions
in glacial drift. For a short period during ice recession the basins were occupied by small proglacial
lakes. After ice disappearance and lowering of the
water level, lakes remained in the deepest parts of
the basins. These changes are recorded in the
bottom deposits: till or glaciofluvial sand and
gravel are covered with glaciolacustrine varved
clays and/or glaciolacustrine laminated sands
and silts. These early deposits are covered by
organic gyttja (e.g. Kirikumäe and Pulli in Fig. 2)
or calcareous gyttja and organic gyttja (Pühajärv
in Fig. 2). This complete succession of late- and
postglacial deposits is of importance for stratigraphic study.

Residual accumulative lakes (Table 1) are
located in Lower Estonia, on the terraces of the
Baltic Ice Lake (Fig. 2). They occupy depressions
in gently rolling glacial relief or in bedrock
hollows. These lakes are relatively large in area
(compared, for example, to most glaciokarstic
lakes), but shallow and similar to the typical
accumulative basins of Upper Estonia. Their
independent development began at the end of the
Younger Dryas, immediately after the drainage of
the Baltic Ice Lake 10 400–10 200 years ago

(Björck & Digerfeldt, 1984). The basal glacial and glaciolacustrine deposits are commonly covered by clayey lacustrine deposits, lacustrine lime and gyttjas. Minerogenous sediments continue into the Preboreal, though such late minerogenous deposits are thin and in some sequences they contain alternating minerogenous and organogenous layers.

The *exaration-accumulation* lakes (Table 1) in the Saadjärv drumlin field (Fig. 2) are a special type, drumlin field lakes, resulting from a complicated combination of erosion and deposition by the glacier (Pirrus *et al.*, 1987). These interdrumlin basins contain a complete stratigraphic succession of glacial, late-glacial and Holocene deposits. Older Dryas sediments consist of yellowish-brown varved clays with a thickness of 1–3 m. Allerød units are dark grey and greenish-grey silts and clays with dispersed organic matter. Younger Dryas laminated silts are coarser, frequently with fine-grained sand and moss interbeds (Pirrus & Saarse, 1978). The boundary between late-glacial and Holocene deposits is sharp; the minerogenous deposits are replaced by gyttjas or lacustrine lime. In shallow and medium depth lakes, during the Holocene two carbonate rich units were formed, with organic gyttja between them (Saarse & Kärson, 1982). The lake basins themselves were formed during Gothiglacial, between the ice recession of the Otepää and Pandivere stages about 12 400–12 200 years ago (Raukas, 1986), at the same time as the formation of the main drumlin field. In the earliest period (duration 100–200 years) they were submerged by local proglacial lakes, at the bottom of which clay was deposited as varves. After the lower northern thresholds were freed of ice, the water level in the proglacial lakes dropped and independent development of the inter-drumlin lakes started, about 12 000 years ago.

Chemical lake group

The chemical lake group is represented by the karstic lakes of the Pandivere Uplands (Fig. 2), where Ordovician and Silurian limestones crop out or lie under a thin Quaternary mantle. The basins were formed by calcite dissolution in cracks and fissures. Fed by springs, these lakes are very clear, 13.5 m transparency in Lake Äntu Sinijärv which is the clearest lake in Estonia. The bottoms of all three Äntu lakes are rich in calcareous deposits, but most other karstic lakes are temporary water bodies in sink holes that lack lacustrine sediment.

Coastal (neotectonic) lake group

Coastal (neotectonic) lakes are located on the rising terraces of the Baltic Sea (Fig. 2). As the formation of coastal lakes has been occurring since the recession of the Yoldia Sea, they are metachronous. *Coastal-residual lakes* (coastal relict lakes; Table 1) occupy former marine depressions; *coastal-accumulation lakes* (Table 1) were formed by obstruction by spits and bars. Marine lagoon, lagoon-lake, and lake phases can be distinguished in the sediment of lakes in both subgroups.

On the Yoldia Sea terrace there is only one relict lake (Kahala), which was isolated from the sea in Preboreal time. The basin is filled with glacial, glaciolacustrine, marine, and lacustrine deposits. The basal sediment of the littoral zone consists of sand. The deepest sediment recovered from the profundal zone is laminated clay that was formed in the Baltic Ice Lake. The clay is overlain by Yoldia silt. The uppermost part of the silt has abundant plant remains accumulated in the Preboreal in a Yoldia Sea lagoon and lagoon-like. Gyttjas, as much as 6.5 m thick are distributed over the bottom of the contemporary lake, except for a narrow belt in the eastern littoral zone. The gyttjas have been deposited in the lake since the Early Boreal.

On the Ancylus Lake terrace there are several lakes which were isolated during Boreal time, among them Ülemiste in the vicinity of the city of Tallinn. Ülemiste belongs to the coastal accumulation lake type, dammed by a bar which formed during the Ancylus transgression. Carbonate bedrock with overlying till, and sand and silt crop out at the eastern part of the basin. Elsewhere in the basin, these units are overlain by lacustrine lime, and calcareous, silicic and organic gyttjas. Geological, palynological, ^{14}C and diatom data indicate that the lacustrine unit, with thicknesses of 4–8 m, started to accumulate in the Early Boreal. The isolation contact of organic gyttja (just above sand) was dated 8300 ± 90 yr B.P. (TA-691; Ilves, 1980).

There are few lakes left on the terrace of the Littorina Sea. Among the most throughly studied ones are Tänavjärv, Ohtja and Tihu (Fig. 2). These are all shallow lakes, their littoral zone coverd by floating mats; the basins filled with gyttjas. Tänavjärv differs from the others by its oligotrophic state and its distinctive aquatic plant communities that include *Lobelia dortmanna* which is at its southeastern geographic limit. Stratigraphy of bottom deposits is rather simple: marine sand overlain by silt rich in dispersed organic matter and containing a brackish water diatom assemblage, all deposited during the Late Atlantic period. These marine and brackish water deposits are mantled by siliceous gyttja formed since the Late Atlantic period. The isolation contact in these lakes is lithologically well marked. In the overgrown part of L. Tänavjärv it is dated 4350 ± 60 yr B.P. (Tln-1100). This date is probably too young. According to the chronostratigraphy of the Littorina Sea, the isolation of Tänavjärv should have occurred about 5300 years ago.

Residual and accumulation lakes on the Limnea Sea terraces are metachronous, starting their separate development since the Early Subboreal, about 4000 years ago. Lithology of bottom sediment is similar to the coastal lakes of the Littorina terrace, also with clear isolation contacts confirmed by diatom analyses. The very young lakes of the alvar area of Saaremaa Island lack lacustrine deposits. (An alvar area is an almost tree- and shrub-free area which has a very thin soil cover or no soil at all upon carbonaceous bedrock (Königsson, 1968). It got his name from the Great Alvar on the island of Öland in the Baltic).

Fluvial lake group

Fluvial lakes are small in number compared to coastal lakes. Oxbow lakes (Table 1) are most common on the flood plain of the Emajõgi River, but they also occur along the middle courses of North and West Estonian rivers. Oxbow lake sediments consist mostly of silts with laminae containing dispersed organic matter, sometimes with peaty interbeds. These units are underlain by channel facies whose composition varies from place to place depending on the bedrock and Quaternary sediments eroded by the river systems (Hang & Miidel, 1987).

Telmatogenic lake group

There is a total of ca. 20 000 bog pools on Estonian raised bogs (Kask, 1979). The pools were formed at local sites of peat degradation on bog surfaces (therefore, the pools are secondary water bodies), first appearing along with development of convexity of the bogs (Masing, 1968). Parallel pool complexes formed on marginal slopes of convex bogs, and central pool complexes formed on the raised central parts of convex bogs. At some bogs, bog hollows coalesced to form bog pools which have existed for hundreds of years (Masing, 1968). Residual lakes (primary water bodies) are commonly located at the outskirts of raised bogs. The bottom deposits of these lakes consist of peat in the littoral zone, coarse highly humic sediments in the sublittoral with finer organic, highly humic sediments at the lake center.

Cosmogenic lake group

In Estonia only one lake, Kaali on Saaremaa Island is cosmogenic. It occupies a meteor crater formed ca 3500 years ago (Kessel, 1981). On the bottom, dolomite debris is coverd by silty sand, silicic gyttja, woody peat and organic gyttja, with a total thickness of lacustrine deposits of 5.8 m.

Part II. Lithological classification of Estonian lake deposits

Nomenclature is needed for the description of lake sediments. Only a few lithological classification systems for lake deposits have been worked out. Unquestionably the classification system that has been in most widespread use in Europe is von Post's (1924). In the Soviet Union, a similar system had been developed by Sukhachov *et al.* (1943) in which fine detritus, coarse detritus, clayey, sandy, diatomitic, calcareous, and algal sapropels, and dopplerite were distinguished. The limitation of these classification systems is the absence of numerical values for the constituent components of the lake sediment. The classification on unconsolidated sediment by Troels-Smith (1955) is free of this shortage, but it is too complicated for everyday usage.

Håkanson & Jansson (1983) reviewed genetic and descriptive classification systems for lake sediments, indicating that the wide diversity of the systems reflects the differing principles on which they are based, and concluding that '....no generally accepted classification system is available'. Those authors focused their attention on determining the minimum number of parameters that need to be considered for studying and describing limnic sediments. They proposed an approach to sediment description based on the organic and nitrogen contents of the sediments. They demonstrated the link between these constituents and lake trophic state, but did not develop the approach into a practical system for general classification of lake sediments.

Lithological studies of bottom deposits from more than 140 lakes in Estonia (> 2000 samples) offered the possibility to revise the lithological classifications in use, and to produce a new system for organic and calcareous lacustrine deposits. This new system does not apply to lacustrine deposits lacking (< 1%) organic matter. Three main lacustrine sediment groups are distinguished: 1) calcareous sediments, 2) gyttjas, and 3) highly minerogenous gyttjas.

The calcareous sediment group contains three subgroups, all with more than 50 percent calcareous matter and less than 50 percent organic matter. The first two contain less than 25 percent (of dry wt) non-calcareous allochthonous matter: lacustrine chalk (75% calcareous and 25% organic matter) and lacustrine lime (50–75% calcareous and < 50% organic matter) (Table 2). The third subgroup, lake marl, has less than 15

Table 2. Lithological classification of the organic and calcareous lacustrine deposits of Estonia (% dry wt).

Nomenclature of deposits	Organic component	Calcareous component	Allochthonous mineral and organic component
lacustrine lime	< 50	50–75	< 25
lacustrine chalk	< 25	> 75	< 25
lake marl	< 15	50–75	> 25
lime gyttja	> 50	25–50	< 25
lime silicic gyttja	15–50	25–50	< 75
calcareous gyttja	> 50	10–25	< 50
calcareous silicic gyttja	15–50	10–25	< 75
organic gyttja	> 50	< 10	< 50
silicic gyttja	15–50	< 10	< 85
highly minerogenous limy clay-silt gyttja (limy mud)	< 15	25–50	> 50
highly minerogenous calcareous clay-silt gyttja (calcareous mud)	< 15	10–25	> 75
highly minerogenous clay-silt gyttja (mud)	< 15	< 10	> 75

218

percent organic matter, more than 25 percent allochthonous non-calcareous matter, and 50 to 75 percent calcareous matter.

Gyttjas are largely organogenous lacustrine deposits consisting of more than 15 percent organic matter, less than 50 percent calcareous matter, and less than 85 percent non-calcareous allochthonous matter. Due to the varying allochthonous, calcareous and organic contents of gyttjas, six subgroups are distinguished: lime gyttja, lime-silicic gyttja, calcareous gyttja, calcareous-silicic gyttja, organic gyttja and silicic gyttja, as detailed in Table 2.

The third group, highly minerogenous gyttjas contain less than 15 percent organic matter, less than 50 percent calcareous matter, and more than 50 percent allochthonous non-calcareous matter. Highly minerogenous limy silt-clay gyttjas (limy muds) contain 25–50% calcareous matter. Highly

minerogenous calcareous clay-silt gyttjas (calcareous muds) contain 10–25% calcareous matter, highly minerogenous clay-silt gyttjas (muds) contain less than 10 percent calcareous matter.

For mineral lacustrine deposits lacking organic matter, the widely used lithological classifications of glaciogenic deposits are recommended (e.g. Wentworth, 1922; Karlsson *et al.*, 1981).

Discussion and conclusions

Estonia's diverse assemblage of lakes provides ample opportunity for study of the morphology and origin of lakes basins, and the litho- and biostratigraphy of late-glacial and postglacial lake sediments. There were at least eight major factors responsible for the formation of Estonian lake

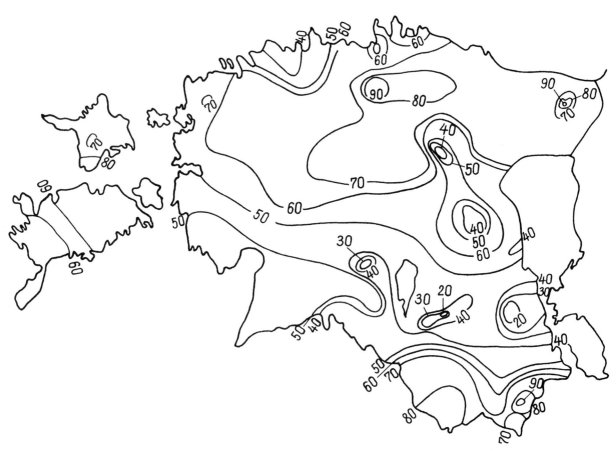

Fig. 3. Distribution of organic matter (% dry wt) in Estonian lacustrine deposits.

basins: tectonic, neotectonic, glacial, chemical, fluvial, telmatic, cosmic, and human. The formation of lake basins has occurred over a long period. Two large basins are preglacial, more than 800 water bodies are of glacial origin, and over 300 of postglacial origin (not counting some 20 000 bog pools). Stratigraphically important sequences occur in glacial accumulative and exaration-accumulative basins, which contain complete and undisturbed lacustrine sedimentary sequences starting in the Older Dryas. The glacio-karstic lakes offer good opportunities to study accumulation rates of lacustrine deposits, as the catchment areas of these lakes weren't cultivated. In inter-drumlin lakes the alternation of shallow- and deep-water deposits with indicative diatoms provide evidence of water level fluctuations. Oxbow lakes enable us to reconstruct fluvial events, and coastal lakes – the history of the Baltic Sea.

During the late-glacial period, only minerogenous sedimentation took place. In some lakes, minerogenous sedimentation lasted up to the Boreal period. After that, organogenous or calcareous sedimentation took place. But in large tectonic-denudation lakes minerogenous sedimentation has continued to the present day, except in deep calm bays.

Geographic patterns of organic content in lake surficial sediment are mapped in Fig. 3. Comparison with the map in Fig. 2 indicates that highly organic sediments are characteristic of lakes in the kame fields of North Estonia where average organic contents are more than 70 percent. Next come the lakes of Hiiumaa Island, Haanja and Karula Heights (Fig. 2), with organic percentages of more than 60 percent. All these lakes occupy closed basins, i.e. seepage lakes fed by precipitation and groundwater. Their catchments are covered by sand, gravel or peat. On the other hand, the sediment of the drainage lakes of the Saadjärv drumlin field, Sakala Uplands and Central Estonia differ in their relatively low organic content, caused by greater terrigenous mineral input carried by streams and surface wash. These observations, and many of those presented earlier indicate that azonal factors are largely responsible for the variation in character of Estonia's lacustrine deposits.

Acknowledgements

The pollen analyses were carried out by A. Sarv (Tallinn), diatom analyses by E. Vishnevskaya (Leningrad), and ^{14}C dating by E. Ilves (Tartu). The diagrams and maps were drawn by A. Ronk (Tallinn). My sincere thanks are due to all these persons. I am indebted to R. B. Davis (Orono, U.S.A.), M. L. Schwartz (Bellingham, WA, U.S.A.) and H. Kukk (Tallinn) for their reviews of the manuscript.

References

Björck, S. & G. Digerfeldt, 1984. Climatic changes at Pleistocene/Hocene boundary in the Middle Swedish endmoraine zone, mainly inferred from stratigraphic indications. In N.A. Mörner & W. Karlén (eds.), Climatic Changes on a Yearly to Millennial Basis. D. Riedel Publ. Comp.: 37–56.

Donner, J. J., 1982. Fluctuations in water level of the Baltic Ice Lake. Ann. Acad. Sci. Fenn. Ser. A. III. Geologica-Geographica 134: 13–28.

Håkanson, L. & M. Jansson, 1983. Principles of Lake Sedimentology. Springer-Verlag, Berlin: 316 pp.

Hang, T. & A. Miidel, 1987. Geology and geomorphology of Estonian river valleys as a basis for palaeohydrological research. In A. Raukas & L. Saarse (eds.), Palaeohydrology of the temperate zone I. Rivers and Lakes. Valgus, Tallinn: 84–98.

Hutchinson, G. E. A treatise on limnology, 1. J. Wiley & Sons, N.Y. 1015 pp.

Ilves, E., 1980. Tartu radiocarbon dates X. Radiocarbon 22: 1084–1089.

Karlsson, R., S. Hansbo & Swedish Geotechnical Society, 1981. Soil classification and identification. Swedish Council for Building Research, St Göransg. 66, S-112 33 Stockholm, Sweden. 49 pp.

Kask, I., 1979. Eesti järvede arengust ja järvenõgude klassifikatsioonist. In A. Raukas (ed.), Eesti NSV saarkõrgustike ja järvenõgude kujunemine. Valgus, Tallinn: 88–103.

Kessel, H., 1981. Kui vanad on Kaali järviku põhjasetted. Eesti Loodus 4: 231–235.

Kessel, H. & A. Raukas, 1979. The Quaternary history of the Baltic, Estonia. In V. Gudelis & L.-K. Königsson (eds.), The Quaternary history of the Baltic. Acta Univ. Upsaliensis, Symposia Universitatis Upsaliensis, Annum Quingentesimum Celebrantis 1: 127–146.

220

Kvasov, D. D., 1979. The Late-Quaternary history of large lakes and inland seas of eastern Europe. Ann. Acad. Sci. Fenn. Ser. A. III. Geologica-Geographica 127: 71 pp.

Königsson, L.-K., 1968. The Holocene History of the Great Alvar of Öland, Acta phytogeogr. Suecica 55. Uppsala: 172 pp.

Lõokene, E. & R. Pirrus, 1979. Õisu-Mõõnaste ja Päidre järve geoloogiast. In A. Raukas (ed.), Eesti NSV saarkõrgustike ja järvenõgude kujunemine. Valgus, Tallinn: 170–181.

Masing, V., 1968. Rabadest, nende arengust ja uurimisest. Eesti Loodus 8: 451–457.

Männil, R., 1961. Pandivere kõrgustiku piirkonnas esinevaist holotseensetest järvesetetest. ENSV TA Geoloogia Instituudi uurimused VIII. Tallinn: 115–133.

Orviku, K., 1973. Võrtsjärve geoloogilisest arengust. In T. Timm (ed.), Võrtsjärv. Valgus, Tallinn: 26–32.

Pirrus, E. A., 1968. Varved clays of Estonia. Tallinn. 143 pp. (in Russian).

Pirrus, E. & L. Saarse, 1979. Tekstuurid jääpaisjärvesavide tekketingimuste peegeldajana. In A. Raukas (ed.), Eesti NSV saarkõrgustike ja järvenõgude kujunemine. Valgus, Tallinn: 182–192.

Pirrus, R., A.-M. Rõuk & A. Liiva, 1987. Geology and stratigraphy of the reference site of Lake Raigastvere in Saadjärv drumlin field. In A. Raukas & L. Saarse (eds.), Palaeohydrology of the temperate zone II. Lakes. Valgus, Tallinn: 101–122.

Pirrus, R. & L. Saarse, 1978. Late-Glacial lake sediments in Estonia. Pol. Arch. Hydrobiol. 1/2 25: 333–336.

Punning, J.-M., M. Ilomets, T. Koff, R. Rajamäe, I. Petersen & T. Tiits, 1985. Stratigraphic and palaeogeographic investigations of lacustrine and bog sediments from Vällamäe kettle hole (SE Estonia). Acad. Sci. E.S.S.R. Preprint. Tallinn: 58 pp. (in Russian).

Pärna, K. K., 1960. Geology of the Baltic Ice Lake and large local proglacial lakes of Estonia. Eesti NSV TA Geoloogia Instituudi uurimused V. Tallinn: 268–275 (in Russian).

Penck, A., 1894. Morphologie der Erdoberflähce. 2. Teil X. Stuttgart. 696 S.

von Post, L., 1924. Das genetische System der Organogenen Bildungen Schwedens. Comité internat. d. Pédologie IV, communication 22.

Raukas, A., 1981. On the composition of the bottom deposits of L. Pihkva-Peipsi. In A. Raukas (ed.), Bottom deposits of L. Pihkva-Peipsi. Ühiselu, Tallinn: 23–41 (in Russian).

Raukas, A., 1986. Deglaciation of the Gulf of Finland and adjoining areas. Bullet. Geol. Soc. Finland 58. Part 2: 21–33.

Raukas, A. & E. Rähni, 1969. On the geological development of the Peipsi-Pihkva depression and the basins distributed in that region. Eesti NSV Tead. Akad. Toimet. 18: 114–127. (in Russian).

Saarse, L., 1979. Distribution and bedding conditions of the South Estonian limnoglacial clayey deposits. Eesti NSV Tead. Akad. Toimet. 28: 145–151 (in Russian).

Saarse, L., 1985. Peculiarities of sedimentation in the small Estonian lakes. In INQUA Eurosiberian Subcommission of the Holocene & IGCP Project 158 Palaeohydrology of the temperature zone in the last 15000 years. Symp. in Switzerland. Abstr. of papers and posters. Bern: 39.

Saarse, L., 1987. The Holocene palaeoecological changes in Estonia. In M.-J. Gaillard (ed.), IGCP 158 Palaeohydrological changes in the temperate zone in the last 15000 years. Symp. at Höör, Sweden, Abstr. of lectures and posters. Lund: 121–124.

Saarse, L. & J. Kärson, 1982. Sedimentation peculiarities in L. Elistvere, Prossa and Pikkjärv. Eesti NSV Tead. Akad. Toimet. 30: 12–19 (in Russian).

Sarv, A., 1983. Stratigraphical subdivision of the Holocene bog and lacustrine deposits of the Remmeski and Senno sections. In T. Bartosh (ed.), Palynologic researches in geologic studies of the Baltic Region and the Baltic Sea. Zinatne, Riga: 77–83 (in Russian).

Sukhachev, I. A., I. A. Baryshnikova & T. P. Borodina, 1943. Sapropel and its importance in agriculture. Acad. Sci. U.S.S.R., Moscow-Leningrad: 53 pp. (in Russian).

Synge, F. M., 1980. A morphometric comparison of raised shorelines in Fennoscandia, Scotland and Ireland. Geol. Fören. i Stockholm Förh. 102: 235–249.

Veber, K., 1973. Põhjasetete geoloogiast ja levikust. In T. Timm (ed.), Võrtsjärv. Valgus, Tallinn: 33–36.

Troels-Smith, J., 1955. Characterization of unconsolidated sediments. Danm. Geol. Unders. IV, 3. 73 pp.

Wentworth, C. K., 1922. A scale of grade and class terms for clastic sediments. J. Geol. 30: 377–392.

Woldstedt, P., 1926. Probleme der Seenbildung in Norddeutschland. Z. Ges. Erdk. Berl. 103–124.

Factors affecting the interpretation of caddisfly assemblages from Quaternary sediments

Nancy E. Williams
Division of Life Sciences, University of Toronto, Scarborough Campus, 1265 Military Trail, Scarborough, Ont. Canada M1C 1A4

Key words: paleoecology, Quaternary, Trichoptera

Abstract

The use of assemblages of caddisflies (Trichoptera) in paleoecology has been explored over the past ca. 12 years. During this time, sites in North America, Great Britain and Europe have been studied and progress has been made in both the mechanics of identification and in the understanding of factors relevant to the interpretation of assemblages of this wholly aquatic group. Quaternary caddisfly fossils are abundant and usually well-preserved in waterlaid sediments. Individual larval sclerites can be identified by reference to shapes, textures, colour patterns muscle scar patterns and setal distributions. The flat frontoclypeus is particularly easily identified.

Study of the biological and distributional data relevant to a caddisfly assemblage yields information at two levels. First the probable local habitat and second the climate can be described. This information is derived from both modern collections and the literature as the morphology of species, and hence it is assumed their environmental requirements, have not changed during the Quaternary. Comparisons with fossil assemblages of other plant and animal groups suggest that there are important factors to be considered in interpreting caddisfly assemblages, particularly those from sediments deposited during cold periods: for example, caddisflies may be slower to migrate than certain terrestrial insects, in particular the beetles (Coleoptera), and some glacial assemblages may therefore be dominated by 'hangers on'. Lotic and lentic species also may migrate at different rates.

Although caddisflies have not yet been fully exploited as paleoecological indicators, they have already contributed to our understanding of past environments and warrant much greater use.

Introduction

Most paleoentomologists have long been aware of the existence of fossils of aquatic caddisly larvae (Trichoptera). However, it is only during the past ca. 12 years that the potential of caddisfly remains as indicators of past environments has been explored in North America and Europe. This means that caddisfly paleoecologists have not yet achieved the levels of sophistication in interpretation reached by some workers involved with other groups of organisms. On the other hand,

Originally published in
Journal of Paleolimnology **1**: 241–248.

considerable progress has been made. Already it has been established (Williams & Morgan, 1977; Williams *et al.*, 1981; Wilkinson, 1981; 1984, 1987 & Williams, 1987) that Quaternary larval caddisfly fossils are abundant and identifiable, and that their assemblages yield considerable information about past habitats and climates.

The caddisflies (about 40 families worldwide) are close relatives of the moths and butterflies, although the caterpillar-like larvae are virtually all aquatic. Their ability to spin and use silk has led to great diversity in habits and habitats among the 10 000 or more species (Wiggins, 1977). Three general categories of life style are related to silk use. The first of these comprises the free-living, mostly predaceous caddis larvae which use silk strands to aid in maintaining station in running water. The second group, the netspinning caddisflies construct silken shelters and nets of various types with which they capture plant, animal and detrital food. The third group the casebuilding caddisflies construct diverse portable cases of silk, and plant and mineral particles. Species in this group occupy a wide variety of trophic categories including shredder, collector-gatherer, predator, scraper and piercer (Wiggins, 1978). Members of the latter two groups are found in all types of lotic and lentic waters. In North America, all caddisfly families found have representatives in running waters while 61% have representatives in lakes, marshes, and ponds (Wiggin, 1977). The Trichoptera then, have some inherent qualities which facilitate their use as paleoecological indicators.

Caddisfly paleoecology has progressed from foundations laid by coleopterists, but differences in life stages used, skeletal structure and insect habitats must be considered in comparative interpretation of the two groups. Since, unlike beetles, caddis larvae cannot fly into sites, their remains are likely to represent fairly local assemblages even though lotic depositional sites may include specimens transported from various distances upstream. However, caddisfly chitin is somewhat more fragile than that of beetles and it seems likely (although untested) that many caddisfly sclerites would be degraded if transported over long dis-

tances and/or redeposited from older sediments. In addition, members of the two orders probably differ in their ability to respond rapidly to climate changes, and this will be discussed more fully in the interpretation section.

Identification

Like the caterpillars, larval caddisflies have chitinous exoskeletons mainly covering the head and thorax, and it is mostly the disarticulated sclerites of these body segments which occur as fossils.

Caddisfly chitin provides a variety of characters useful in separating genera and species. Many of these have been used little in conventional keys for identification of larvae perhaps because skeletal details are obscured in whole larvae by the presence of other body tissues. For example, the frontoclypeal sclerite (dorsal covering of the head) illustrated in Fig. 1a can be placed in the family Limnephilidae on the basis of its shape. It can be further identified to species [*Grensia praeterita* (Walker)] by looking at the proportions of its anterior and posterior parts, its colour, its texture, the distribution and structure of spines and the distribution of setal alveoli (points of attachment) (see Williams & Wiggins, 1981). Similarly, the shape of the frontoclypeal sclerites illustrated in Fig. 1c and 1d place them in the genus *Chimarra* of the family Philopotamidae, while the two species, *obscura* (Walker), and *socia* Hagen, are separated on the basis of details of the shape of the anterior margin. While some members of this genus are unknown as larvae, it seems reasonable to assume that any undescribed larvae whose adults are separable may have sclerites similar but not identical to sclerites of these species. Colour patterns and distribution and colour of muscle scars (points of muscle attachment) are also useful diagnostic characters.

Comparison with modern reference material is essential and therefore access to identified sclerites of modern specimens is important. These are most useful when kept on microscope slides but this unfortunately is not the custom in most

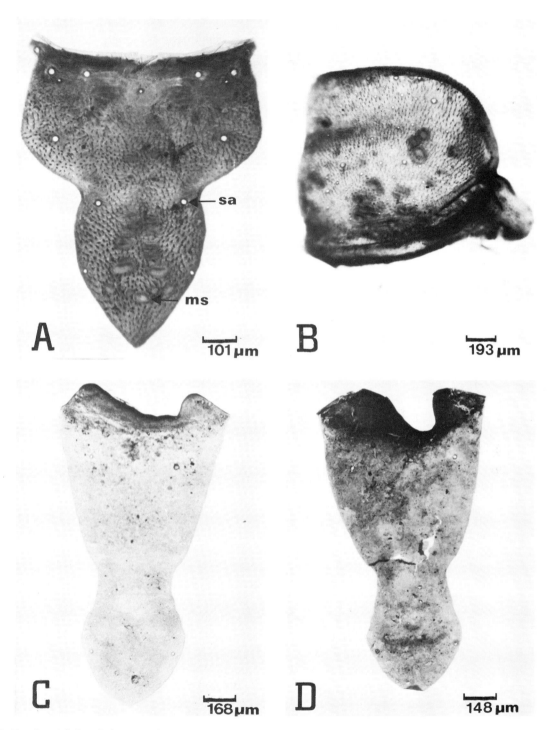

Fig. 1. Fossil caddisfly sclerites. (A). frontoclypeus of *Grensia praeterita* (Walker) from Kuskokwim River, Alaska (Boutellier Interstadial). sa – setal alveolus (point of muscle attachment of a seta. ms – muscle scar (point of attachment of a muscle). (B). right pronotal sclerite of *Grensia praeterita* from Kuskokwim River, Alaska. (C) *Chimarra socia* Hagen from Au Sable River, Michigan (approximate age 4000 B.P.). (D). *Chimarra obscura* (Walker) from the Don Valley, Toronto, Ontario (? Sangamonian). All specimens are part of the author's collections.

museum collections. In addition, there are still many species whose larvae are not found in any collection.

In genera where good specific differences have not yet been found on whole larvae or fossils, the fossil identification is of necessity left at the genus level. Perhaps 25% of North American caddisfly genera fall into this category, while larvae of some species in other genera are unknown. The net result is that the proportion of North American fossils identified to genus or less precise taxonomic category in studies so far, ranges from 32 to 73%. However, this should not be considered as evidence that identification is a major problem. While no one would deny that 100% specific identifications would be ideal, the level of identification achieved by specialists in various other groups varies considerably. A brief survey of some recent North American paleoecology papers shows that the number of taxa identified only to genus or higher taxonomic level ranged from 40 to 72% for beetles, 63 to 100% for chironomids, 75 to 100% for pollen, 41 to 57% for plant macrofossils, 8 to 17% for diatoms and 8 to 28% for molluscs. Since each of these groups has contributed a great deal to our understanding of past environments, it is clear that the information content of a fossil assemblage is not entirely determined by percentage of identifications at the species level. In any case caddisfly larvae would appear to be about as identifiable as the beetles.

Photographs have proven to be the easiest means of comparing modern and fossil specimens, since they can be viewed simultaneously, lend themselves to measurement and enhance textural details on the chitin surface. As my own collections of North American and European Trichoptera grow, they are supplemented by a series of photographs comprising an atlas of head and thoracic sclerites (Williams in prep.). Figure 2 illustrates the photographic comparison of modern and fossil specimens of the frontoclypeus of two *Hydropsyche* species (Hydropsychidae). Both are members of the 'checkerboard' group of species in which this sclerite is characterized by a dark background with six or seven light patches. However the two species [*morosa* Hagen and

alternans (Walker)] have differences in sclerite shape, length/width ratio, muscle scar colour and distribution of setal alveoli. These features can be seen even on well-worn fossil specimens.

The caddisfly frontoclypeus has proven to be particularly easy to work with primarily because it is flat. This means that characters are readily seen, measured and photographed. Other sclerites such as the labrum, parietals, mandibles, pronotum (Fig. 1b), mesonotum, metanotum, leg segments and anal hooks are also abundant but are somewhat less easily identified. Our present needs include hastening the publication of identification aids and training other workers in sclerite identification.

Interpretation

In developing ground rules for interpretation of caddisfly assemblages, the previous experience of other paleoentomologists has been an advantage. However, the wholly aquatic habitat of caddis larvae must be considered as well as other factors specific to caddisflies.

It is assumed, as for Coleoptera (Coope, 1977), that morphological stability implies physiological stability and therefore similarity in ecological requirements. So far, Quaternary caddisfly assemblages have supported the assumption of ecological stability, since in most cases species composition and associations, and relative abundance of taxa have been similar to conditions of modern communities and these assemblages have led to deductions about climate and environment which are in agreement with conclusions reached independently from other animal groups (e.g. Williams & Morgan, 1977; Williams *et al.*, 1981; Elias & Wilkinson, 1983; Wilkinson, 1987; Elias & Williams, in prep.; Williams & Morgan, in prep.)

The first paleoenvironmental interpretation based primarily upon caddisfly fossils was a sample from the Don beds at Toronto, Ontario (Williams & Morgan, 1977). Fortuitously, this was an interglacial assemblage of well-preserved and abundant fossil material (densities of up to

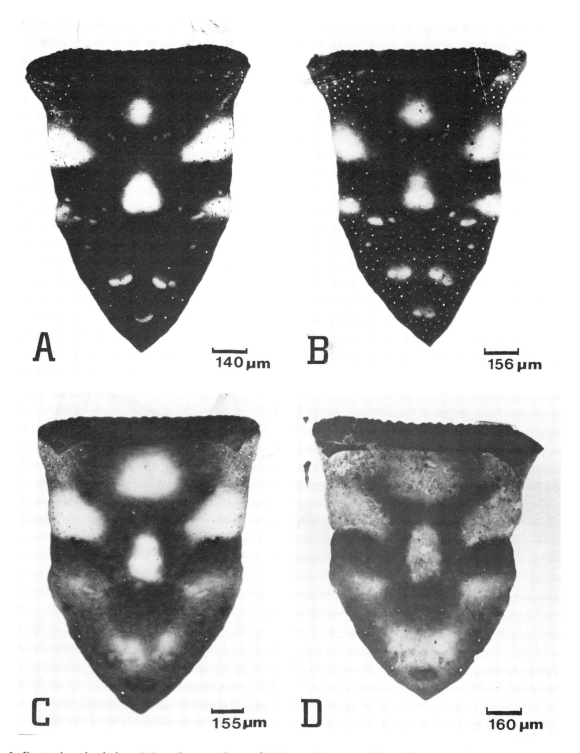

Fig. 2. Frontoclypeal sclerites. (A). modern specimen of *Hydropsyche morosa* Hagen from Uxbridge, Ontario. (B). fossil of *H. morosa* from Au Sable River, Michigan (approximate age 4000 B.P.). (C). modern specimen of *H. alternans* (Walker) from L. Winnipeg, Manitoba. (D). fossil of *H. alternans* from Scarborough Bluffs, Toronto, Ontario (? Wisconsinan).

226

1000 caddisfly sclerites/l of sediment). This and other early attempts (Williams *et al.*, 1981; Wilkinson, 1981) showed that it is possible to acquire information at two levels: local habitat and climate. First, a probable local habitat can be described, and with the wholly aquatic caddis larvae, this means that information such as type of water body, water depth, current speed, substrate, aquatic vegetation and surrounding terrestrial vegetation can be deduced. Species, genera and even families of caddisflies are indicative of particular local conditions. At the least, lotic and lentic habitats can be separated, as the presence of any members of the families Hydropsychidae, Philopotamidae, Rhyacophilidae, Brachycentridae, Beraeidae, Calamoceratidae and Odontoceridae and certain genera of the Polycentropodidae, Psychomyiidae, Glossosomatidae and Hydroptilidae almost always indicates running water, while some genera of the Phryganeidae, Limnephilidae and Leptoceridae are indicative of lentic water.

The Hydropsychidae are particularly useful in determining, more specifically, the nature of the local habitat. For example, the North American species *Hydropsyche confusa* (Walker) is found on the rocky substrate of wave-swept shores of very large lakes, eg. Lake Erie, where the turbulent water brings a supply of plant and animal food into its silken capture net (Barton & Hynes, 1978). In contrast, the related, European net-spinning species *H. bulgaromanorum* Malicky lives mostly in the deeper parts of the lower reaches of large rivers (Malicky, 1984). Many other caddis larvae have somewhat restricted substrate and depth requirements which can be used to indicate substrates and depths present in past water bodies. Elias & Wilkinson (1983) for example, found trichopteran indicators of stony and sandy lake shores in a lateglacial assemblage from Switzerland, while various riverine caddisflies found in the Scarborough Formation, Toronto, Ontario (Williams *et al.*, 1981) indicated the presence of gravel, sand and silt substrates associated with riffles and deeper water.

Feeding habits of caddisflies give clues as to the size and abundance of suspended organic matter,

presence of aquatic algae, sponges and macrophytes, and presence and type of streamside vegetation (Williams, 1988) although species are not usually dependent upon specific host plants or animals. Many leptocerids are among the caddisflies known to be particularly sensitive to siltation and organic and inorganic pollution (Resh & Unzicker, 1975), but water temperature, pH and dissolved oxygen tolerances as yet are unknown for many caddisfly species. In addition it should be noted here that only a few caddisflies (some molannids, phryganeids and leptocerids) occupy the deeper parts of lakes (20–100 m) (Wiggins, 1977) so that although deep lake water may sometimes be indicated by their presence (e.g. Wilkinson, 1987), their absence from lake sediments may or may not indicate shallower or anoxic water.

Attempts to improve the accessibility of information are underway. Habitat information derived from both collections and the literature is now becoming easier to access as it is entered into a microcomputer data bank at the University of Toronto, Scarborough Campus. At present categories we have included in the database are location, collection date, species, water temperature, current, depth, pH, substrate particle size, water body width or diameter, aquatic vegetation and terrestrial vegetation. The database also facilitates searching for modern analogues and distribution data.

When the distribution data are added to the local habitat information, second level, or climatic deductions, are possible. Since caddisfly fossils up to at least 500 000 years old can be identified as extant species (Williams, unpublished data), and many of these species occur now in areas which have been recently glaciated, it seems reasonable to assume that during the Quaternary, caddisflies have responded to climate change by migrating rather than by evolving rapidly or suffering extinction. Caddisflies are not usually dependent on particular plant species for food and are less able than terrestrial insects to take advantage of microclimate.

Climate has direct and indirect effects upon caddisflies. While air temperature, precipitation and winds directly affect the ability of adults to move about, mate and oviposit, the development of the aquatic larval and pupal stages is directly influenced by water temperature as well as indirectly through the results of air temperature and precipitation upon terrestrial and aquatic vegetation and soils, which in turn affect water chemistry, and supply food for larval caddisflies. Because small lakes, rivers and larger streams have proportionally large contact areas with shoreline vegetation and soils compared with larger lakes, and are less influenced by groundwater than small streams, caddisflies within the former habitats are probably most influenced by macroclimate.

There are many species with restricted modern distributions. For example, some stream and river caddisflies are restricted to areas of the boreal coniferous forest, montane coniferous forest or temperate deciduous forest e.g. the deciduous forest river species *Potamyia flava* (Hagen), *Hydropsyche cuanis* Ross, *H. aerata* Ross and *H. phalerata* Hagen (Hydropsychidae) (Ross, 1963), while a number of lake and pond species are restricted to far northern areas e.g. the limnephilids *Limnephilus pallens* Banks and *Grensia praeterita* (Walker) (Lehmkuhl & Kerst, 1979). Further search of the literature and museum records reveals a great variety of present-day distribution patterns ranging from holarctic e.g. the coldwater species *Apatania zonella* (Zett.) (Limnephilidae) and *Glossosoma intermedium* (Klapalek) (Glossosomatidae) (Wiggins, 1977), to restricted to a single water body e.g. *Hydropsyche aenigma* Schefter, Wiggins & Unzicker (Hydropsychidae), from the Beaverkill River, New York.

We are just beginning to find evidence of the distances caddisflies can migrate and the rates at which they have done so in the past. For example, the tundra pool species *Grensia praeterita* (Fig. 1a and b) occurred several times in the past 100 000 years, in parts of southern Alaska which are now forested (Elias & Williams, in prep.). Distinctive head and thoracic sclerites of another limnephilid species *Clistoronia* sp., were deposited in Michigan (Williams, unpublished data) about 14 000 years B.P., even though this genus is now restricted to western parts of the United States and Canada (Arizona is the most easterly record). However, we need to study caddisfly fossils from a great many more sites before the extent of caddisfly distributional shifts is known.

So far, in both North America and Europe, most climatic inferences from caddisfly assemblages have been supported by conclusions reached independently from other groups. For example, Williams *et al.* (1981) surmised from Scarborough Formation chironomid, beetle and caddisfly assemblages that the climate of southern Ontario was slightly cooler than at present during presumed Early Wisconsinan times. Williams & Morgan (1977) concluded from a caddisfly assemblage that the southern Ontario climate was at least as warm as at present during the last interglacial, agreeing with conclusions reached from pollen (Terasmae, 1960) and diatoms (Duthie & Mannada Rani, 1967). Elias & Wilkinson (1983) inferred from both beetle and caddisfly assemblages, that a cold tundra environment existed at Lobsigensee, Switzerland, about 13 500 B.P. but was replaced by a more temperate climate after 13 000 B.P.

There is now evidence to suggest that the rate of response to climate change may not be the same for all caddisflies under all conditions. Special considerations may be needed in interpreting glacial assemblages in particular. For example, Briggs *et al.* (1985) described an assemblage of Arctic and Alpine beetles from a British site at Queensford deposited about 39 000 B.P. However, the caddisflies consisted of six northern and central European taxa, all of which still occur in Britain today. Williams (1987) proposed that the beetles (known to disperse rapidly) were probably true indicators of a rapidly cooling macroclimate, while the caddisflies were 'hangers on' in a shallow river which was relatively warm in summer and rich in algal food due to the unshaded treeless environment. This explanation depends on the hypothesis that some caddisflies disperse more slowly than beetles and indeed there is biogeographical evidence of differences in migration rates within the caddisflies, since lotic (running water) limnephilid caddisflies have dispersed more slowly, post-glacially across Scandinavia than lentic (still water forms) (Otto, 1982). Similarly, comparisons of the lotic and lentic limnephilids found in Britain and northern and central Europe today show that Britain and continental Europe have far more lentic than lotic species in common. Therefore, future interpretations involving caddisfly fossils must consider the possibility of a time lag in arrival of some species and departure of others, particularly where rapid climatic deterioration is a possibility.

Conclusions

Although few people are as yet making use of them, Quaternary caddisfly fossils have shown great potential as paleoecological tools. They are abundant, diverse and identifiable, have already contributed to our understanding of past environments and will continue to do so, particularly when used in combination with other groups of organisms.

Acknowledgements

I thank Susan Marrone for preparing the figures and typing the manuscript and D. Dudley Williams for a critical reading of the manuscript.

References

Barton, D. R. & H. B. N. Hynes, 1978. The wave-zone macrobenthos of the exposed Canadian shores of the St. Lawrence Great Lakes. J. Great Lakes Res. 4: 27–45.

Briggs, D. J., G. R. Coope & D. D. Gilbertson, 1985. The chronology and environmental framework of early man in the Upper Thames Valley. Brit. Archaeol. Rep. 15: 1–176.

Coope, G. R., 1977. Fossil coleopteran assemblages as sensitive indicators of climatic changes during the Devensian (last) cold stage. Phil. Trans. r. Soc. Lond. B, 280: 313–340.

Duthie, H. C. & R. G. Mannada Rani, 1967. Diatom assemblages from Pleistocene interglacial beds at Toronto, Ontario. Can. J. Bot. 45: 2249–2261.

Elias, S. A. & B. Wilkinson, 1983. Lateglacial insect fossil assemblages from Lobsigensee, Swiss Plateau. Studies in the Late Quaternary of Lobsigensee 3. Rev. Paleobiol. 2: 89–204.

Elias, S. A. & N. E. Williams, in prep. Quaternary insect fossils from southern Alaska.

Lehmkuhl, D. M. & C. D. Kerst, 1979. Zoogeographical affinities and identification of central arctic caddisflies (Trichoptera). Musk Ox 25: 12–28.

Malicky, H., 1984. The distribution of *Hydropsyche guttata* Pictet and *H. bulgaromanorum* Malicky (Trichoptera Hydropsychidae), with notes on their bionomics. Entomologist's Gaz. 35: 257–264.

Otto, C., 1982. Habitat, size and distribution of Scandinavian limnephilid caddisflies. Oikos 38: 355–360.

Resh, V. H. & J. D. Unzicker, 1975. Water quality monitoring and aquatic organisms: the importance of species identification. J. Wat. Pollut. Cont. Fed. 47: 9–19.

Ross, H. H., 1963. Stream communities and terrestrial biomes. Arch. Hydrobiol. 59: 235–242.

Schefter, P. W. & G. B. Wiggins, 1986. A systematic study of the Nearctic larvae of the *Hydropsyche morosa* group (Trichoptera: Hydropsychidae). Life Sci. Misc. Publ., Roy. Ont. Mus.: 1–94.

Terasmae, J., 1960. A palynological study of Pleistocene interglacial beds at Toronto, Ontario. Geol. Surv. Canada Bull. 56: 23–40.

Wiggins, G. B., 1977. Larvae of the North American Caddisfly Genera (Trichoptera). Univ. Toronto Press, Toronto. 1–401.

Wiggins, G. B., 1978. Trichoptera. In: R. W. Merritt & K. W. Cummins, (eds.), An Introduction to the Aquatic Insects of North America. Kendall/Hunt Publ. Co, Dubuque, Iowa: 147–185.

Wilkinson, B., 1981. Quaternary sub-fossil Trichoptera larvae from a site in the English Lake District. In G. P. Moretti (ed.), Proc. 3rd Int. Symp. Trichoptera, Junk, The Hague: 409–419.

Wilkinson, B., 1984. Interpretation of past environments from sub-fossil caddis larvae. In J. C. Morse (ed.), Proc. 4th Int. Symp. Trichoptera, Junk, The Hague: 447–452.

Wilkinson, B., 1987. Trichoptera sub-fossils from temperate running water sediments. in M. Bournaud & H. Tachet (eds.), Proc. 5th Int. Symp. Trichoptera, Junk, The Hague: 61–66.

Williams, N. E., 1987. Caddisflies and quaternary palaeoecology – what have we learned so far? In M. Bournaud & H. Tachet (eds.), Proc. 5th Int. Symp. Trichoptera, Junk, The Hague: 57–60.

Williams, N. E., 1988. The use of caddisflies (Trichoptera) in palaeoecology. Palaeogeogr., Palaeoclimatol. Palaeoecol. 62: 493–500.

Williams, N. E. & A. V. Morgan, 1977. Fossil caddisflies (Insecta: Trichoptera) from the Don Formation, Toronto, Ontario, and their use in paleoecology. Can. J. Zool. 55: 519–527.

Williams, N. E. & A. V. Morgan, in prep. Late Pleistocene and Holocene caddisfly assemblages (Insecta:Trichoptera) from eastern North America.

Williams, N. E., J. A. Westgate, D. D. Williams, A. Morgan & A. V. Morgan, 1981. Invertebrate fossils (Insecta: Trichoptera, Diptera, Coleoptera) from the Pleistocene Scarborough formation at Toronto, Ontario, and their palaeoenvironmental significance. Quat. Res. 16: 146–166.

Williams, N. E. & G. B. Wiggins, 1981. A proposed setal nomenclature and homology for larval Trichoptera. In G. P. Moretti (ed.), Proc. 3rd Int. Symp. Trichoptera, Junk, The Hague: 421–429.

Sequence slotting for stratigraphic correlation between cores: theory and practice

R. Thompson[1] & R. M. Clark[2]

[1]*Dept. Geophysics, Univ. Edinburgh, Mayfield Rd. Scotland EH9 3JZ;* [2]*R. M. Clark, Dept. Mathematics, Univ. Monash, Australia*

Key words: sequence slotting, magnetic susceptibility, isotopes, palaeomagnetism

Abstract

Sequence slotting is an objective numerical method which allows stratigraphic records to be compared and matched. Quantitative core correlation can be easily performed using sequence slotting on many types of paleolimnological and geological data. Dynamic programming algorithms have greatly enhanced the speed with which sequence slotting can be carried out.

We have further modified the sequence slotting method to limit or even to prevent the formation of long blocks in the slotted sequences. Such blocking or clumping has previously restricted the application of sequence slotting in many practical situations. We have applied the modified dynamic sequence slotting technique, based on common path length summation of Euclidean distances, to magnetic susceptibility data, to isotopic measurements and to palaeomagnetic directions on the sphere.

Introduction

The sequence slotting method permits matching and comparison of records which lack precise dating information but which are internally ordered. Such ordered records, in which the data sequence is known unambiguously, are obtained from many stratigraphic successions such as marine, lake and cave sediments.

When geological marker horizons, for example ash beds, are available, stratigraphic correlations can be made between successions without undue difficulty. In most circumstances, however, such clear geological information is scarce or unavaila-ble. Stratigraphic comparison between records then tends to be made either (i) 'by eye' by looking for features which appear similar – the technique of 'bump matching', or else (ii) by resorting to time series analysis, such as the calculation of lagged cross correlation coefficients, even though the data on which the calculations are performed are rarely independently dated or equally spaced, or (iii) by erecting a simple model of the depositional relationship between the sequences e.g. a linear stretching. In this paper we investigate a range of stratigraphic situations in which a fourth approach, sequence slotting, may be of value.

A convenient method for visualizing these four

Originally published in
Journal of Paleolimnology **2**: 173–184.

230

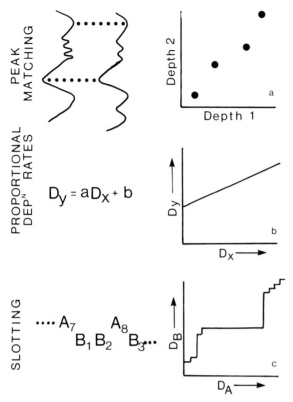

Fig. 1. Core correlation methods. Depth-depth plots at right hand side

a. Correlation by eye – 'bump matching'.

b. Simple linear deposition model e.g. time series cross correlation.

c. Sequence slotting.

approaches is in terms of depth-depth plots (Fig. 1). Visual matching of maxima and minima fixes isolated points on a depth-depth plot (Fig. 1a), whereas the linear relationship of a simple depositional model appears in Figure 1b. Cross correlation using time series analysis also leads to a linear relationship as depicted in Fig. 1b. Sequence slotting produces a step-like plot for the depositional relationship (Fig. 1c).

All four approaches involve a 'search' for a 'good' fit. Such searches necessarily yield appealing fits or high correlation coefficients, because poor fits and low coefficients are rejected in the search procedures. High correlation coefficients discovered in such analyses may be of little significance on account of the excessive searching and data manipulation that may have gone into producing the optimal fits.

We have investigated a variety of approaches to the method of sequence slotting and considered ways in which earlier versions of the technique may be improved or modified for them to handle geological and geophysical data. The most important of these modifications concerns removing the characteristic disjoint-blocking pattern produced by sequence slotting. This annoying effect has previously hindered practical application of the technique. We have applied our modified slotting method to three types of data – isotopic, susceptibility and palaeomagnetic directions – and then discussed the results of these analyses. Finally we examined the quality of match of stratigraphic records caused by search procedures. In particular, we examined the quality of fit of palaeomagnetic records by making use of a slotting statistic and a simulation study.

Sequence slotting

The sequence slotting procedure involves combining two sequences to form a single joint sequence such that similar objects are placed together from the two sequences, while the original ordering within each sequence is preserved (Gordon, 1973). Dynamic programming methods allow all possible slottings to be assessed in an efficient manner, so that the dissimilarity between the sequences is minimized subject, of course, to the constraints of the original orderings (Delgoigne & Hansen, 1975).

Consider two sequences A and B, of length m and n respectively, of ordered observations or objects at stratigraphic horizons denoted by A_1, $A_2 \ldots A_m$ and $B_1, B_2 \ldots B_n$. Each observation can consist of measurements on p variables X_1, $X_2 \ldots X_p$. A dissimilarity measure is then set up to assess the resemblance of the observations to one another. Call the dissimilarity between the jth object in sequence A and the kth object in sequence B $d(A_j, B_k)$.

One dissimilarity measure (Gordon, 1973) is then

$$d(A_j, B_k) = \sum_{i=1}^{p} w_i \, | \, X_{ij}(A) - X_{ik}(B) \, |$$

where w_i are weights on each variable. For data on the unit sphere the arc length between directions can be used as a dissimilarity measure (Embleton et al., 1983) which yields a formule of

$$d(A_j, B_k) = \sum_{i=1}^{p} w_i(X_{ij}(A) \cdot X_{ik}(B))$$

An alternative measure for points in p dimensional space is the weighted Euclidean distance between them (Clark, 1985). In this case

$$d(A_j, B_k) = \left(\sum_{i=1}^{p} w_i(X_{ij}(A) - X_{ik}(B))^2 \right)^{1/2}$$

These different dissimilarity measures may themselves be combined in a variety of ways to estimate the concordance between the two sequences. One assessment of total concordance is to use the sum of local dissimilarity measures (Fig. 2b), namely the dissimilarities between a given point and the immediately preceding and following points in the other sequence (Gordon, 1973; Gordon & Reyment, 1979). Another calculation yields a total concordance measure equivalent to the combined path length (CPL) of Figure 2a. Holmquist (1989) has suggested that there are advantages in summing the products of the local dissimilarities about a given point when forming the total dissimilarity measure, particularly when the correlation between adjacent objects within the sequences is low. Gordon (1980) has also taken the minima of the local dis-

$$A_1 \qquad A_2 - A_3$$
$$\diagdown \qquad \diagup$$
$$B_1 - B_2$$

(Clark 1985)

$$A_1 \qquad A_2 \quad A_3$$
$$B_1 \quad B_2$$

(Gordon 1973)

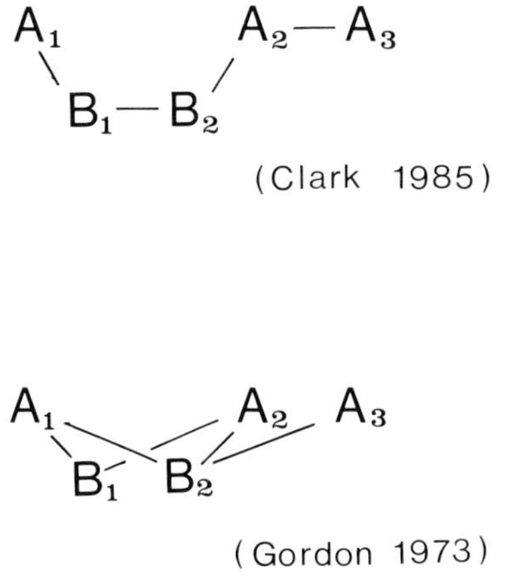

Fig. 2. a. Common path along joint slotting. b. Local dissimilarities between two sequences.

similarities. We have followed Embleton et al., (1983) and Clark (1985), and minimized the total combined path length (CPL). Our path lengths were then based on the Euclidean distances between adjacent horizons in the combined sequence.

The introduction to the sequence slotting method of the dynamic programming technique, by making use of optimality principles, has made sequence slotting extremely efficient. Optimal slotting of two sets of several hundred objects requires only a few seconds on a mainframe computer whichever dissimilarity or concordance measure is adopted. The key to the dynamic programming approach is that the total dissimilarity minimization problem can be broken down into a series of extremely simple subproblems that can be solved in a well structured cascade. The dynamic programming method thus provides a fast, exact algorithm to solve the total discordance minimization problem.

Standardization

Data standardization is one extremely important practical aspect of using any mathematical approach to match sequences. In almost every practical application of sequence slotting, some form of standardization or data scaling needs to be carried out. Scaling methods developed for multivariate analyses can be directly applied to sequence slotting data. Rummel (1970, p 289–296) for example has reviewed the conventional approaches to scaling in factor analyses. Zhou et al. (1983) summarize some of the effects of scaling in factor analysis using both artificial and real data and point out its crucial importance in many practical situations.

We have found no universally satisfactory standardization procedure. Nevertheless, one approach we find relatively reliable for many multivariate data sets is that of (i) transforming the data spread to roughly that of a normal distribution e.g. by using a logarithmic or square root transformation, (ii) subtracting the mean of each sequence individually, (iii) individually setting the variance of each sequence to unity, and finally (iv) giving each variable equal weight. Steps (ii) to (iv) of this approach correspond to Gordon's (1980) second standardization procedure. On the sphere we standardize by rotating the data to make the mean directions equal. Compositional data need especially careful scaling (see for example, Clark *et al.* (1986)) before sequence slotting is applied.

One must be careful to guard against excessive manipulations of the data during standardization as extreme ordination attempts could lead to spurious data matchings. Nevertheless an iterative approach to standardization can prove useful in some circumstances. For example, it can be used to standardize only over the range of overlap of the slotted sequences. Alternatively we have used iterative scaling to produce significantly better concordance measures through the use of search techniques based on linear programming such as simplex optimization.

Additional approaches to data manipulation that we have experimented with, prior to slotting, include robust smoothing and the calculation of first and second differences or gradients. Unfortunately neither of these manipulations was able to help appreciably with a major difficulty in sequence slotting – namely the problem of clumping or blocking. This blocking problem forms the theme of the next section.

Unblocking

The heart of the practical problem of adapting available sequence slotting programmes to cope with 'messy geological data' lies with the necessity of removing unrealistically long blocks or groups of horizons from the optimal solutions produced by sequence-slotting. This recurring problem of unnaturally long blocks can arise in two distinct situations. One of these situations is when the parameters being slotted vary little down the core. A slight discrepancy between cores in these circumstances can have a profound influence on the slotting. It can produce long blocks and have the effect of making it appear that the sequences grossly mismatch, when in fact they hardly differ. The other situation in which geologically un-

justifiable blocks are produced is when one variable differs markedly, over a short section, between cores while the other variables exhibit similar fluctuations in the cores. Optimal, unconstrained sequence slotting invariably creates long blocks in this type of situation.

An example of the first of these two mismatch situations, i.e. that associated with low variability within the core, arises in connection with the magnetic susceptibility data considered in the section on multicore slotting. A good example of the second type of blocking mismatch is found in the study of Anderson (1986). He describes how some unusually high concentrations of diatoms in one of his cores have an exaggerated effect on their slot sequencing.

Some form of unblocking technique is needed to get around these two common difficulties in slotting data. Ideally if data standardization could be carried out perfectly, there would be no need to include such unblocking procedures in the sequence slotting algorithms. Without such exact scaling methods, we have to present sequence slotting with more flexibility, by including a parameter which allows the optimal slotting to vary between that of a smooth monotonic depth/depth plot as in Fig. 1b and the unconstrained slotting of Fig. 1c. We have used two approaches to achieve this modification to optimal slotting. The first is to set explicitly the maximum number of consecutive horizons permitted from either sequence. The second is to encourage horizons from the two sequences to interleave with one another by downweighting the between sequence dissimilarities.

An example of the effect of the first explicit block constraint on sequence slotting is shown in Fig. 3. Standardized oxygen isotope ratios are plotted against the position in the joint slotted sequence for two oceanic sediment cores. The uppermost plot shows the unconstrained joint slotting with no restrictions on block length, while the lowermost plot shows the joint slotting with a maximum length for interior blocks, in either core, of three.

Such oxygen isotope records back to around stage 23 are generally regarded to be of high

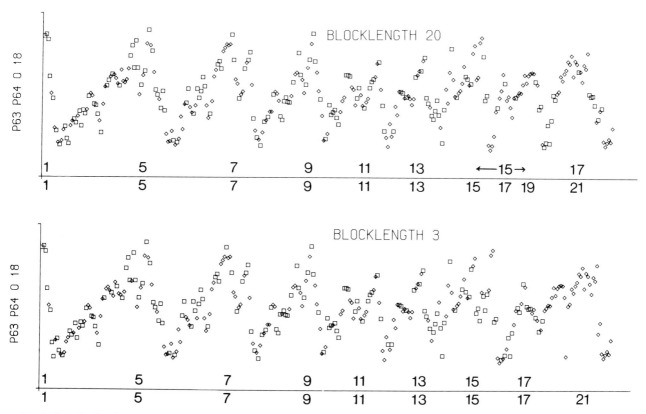

Fig. 3. Standardized oxygen isotope slotting of deep sea cores P6304–9 (squares) and P6408–9 (diamonds). (Data from Emiliani 1966 and 1978).

a. Optimal unconstrainted slotting

b. Slotting constrained by maximum internal block length of three in either core. Isotopic [18]O maxima labelled along horizontal, common path position axes. Isotopic features match correctly in unblocked analyses of b. See text for discussion of goodness of match statistic delta.

quality (Lowe & Walker, 1984), relatively easy to match by eye and to contain several distinctive features. Twenty-one of these features span approximately the last one million years. Nine successive interglacial maxima are labelled by odd numbers (Fig. 3). The remaining (even numbered) glacial isotopic minima make up the rest of the stage features.

Unconstrained sequence slotting of cores P6304-9 and P6408–9 manages to match the isotopic features correctly between stages 1 and 13 but then to mismatch the features 15 to 21 (Fig. 3a). We can observe in the region of mismatch how long blocks have been created by the slotting algorithm in order for it to obtain a good mathematical fit. Explicitly setting the maximum

internal block length to four creates a very slightly poorer fit, but yields the geologically correct result.

A depth-depth plot (Fig. 4) summarizes how the two isotope records fit together. Maximum internal block lengths of two, three or four all produce closely similar fits compared with the incorrect unconstrained (optimal) slotting (Fig. 4). We have yet to devise a routine method of determining the most appropriate maximum length for the blocks. However, for many earth science data sets, we have found maximum internal block lengths of around three or four give good results.

An example of the use of unblocking through downweighting of between-core dissimilarities,

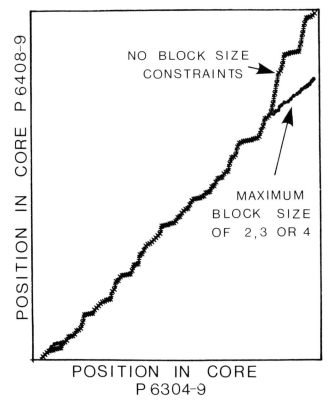

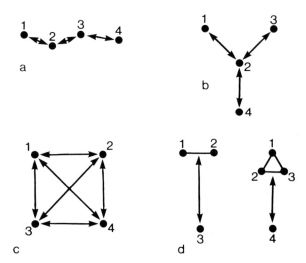

Fig. 5. Multicore slotting strategies
a. Transect
b. Master core
c. True multicore
d. Optimized iterative

Fig. 4. Depth-depth plot for slottings of cores P6304-9 and P6408-9 as presented in Fig. 3. The slottings with maximum internal block lengths of 2,3 and 4 are all very similar.

compared with within-core dissimilarities, is included in the next section which deals with multicore slotting.

Multicore slotting

The problem of matching several sequences, rather than just a pair of sequences, can be approached in several ways. The most common approach is to reduce the multicore problem into one involving matching only pairs of cores at a time. This is a natural approach to take with cores collected along a transect e.g. from shallow to deep water. The first sequence is matched with the second sequence (Fig. 5). Then the second sequence is matched to the third, without reference to the first slotting, and so on for all cores along the whole transect (Fig. 5a).

A similar approach is that of selecting a master core (Fig. 5b). Cores are matched individually to the master core. Once again only pairs of cores are dealt with at any one time. A similar drawback to that of core correlation along a transect is that the information recorded by some cores, for example cores 3 and 4 in Fig. 5b, is not utilized in the correlation of the remaining cores (e.g. 1 and 2 in Fig. 5b).

Ideally in multicore slotting, all cores should be matched simultaneously (see Fig. 5c). While such an approach is theoretically possible it has not so far proved tractable in the sequence slotting method. This intractability is because in the FORTRAN computer coding developed to date, the main array sizes increase according to the power of the number of cores to be slotted. Consequently storage space, even on a large mainframe computer, is totally inadequate for practical situations involving four or more cores.

An intermediate approach that we have adopted involves an approximate iterative optimization. As an example we have investigated three cores from Lake Kinneret (Fig. 6). Our multicore slotting method begins by taking two cores, standardizing the data and slotting the cores together by minimizing their combined path

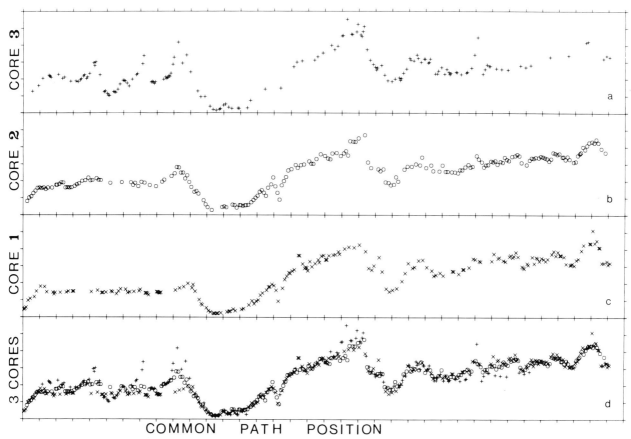

Fig. 6. Standardized Lake Kinneret magnetic susceptibility data from three coring sites. The susceptibility data are plotted against common path position of the multicore slot.

a. b. and c. show the data from individual cores

d. shows the three core slot. Between core dissimilarities have been down-weighted by a factor of 0.5 compared to within core dissimilarities as part of the unblocking procedure.

length exactly as in our sequence slotting procedure for two cores. This matched pair is then treated as one sequence and a third, standardized core slotted to it to produce a combined, joint sequence of three cores. We then begin the iterative optimization procedure by removing the first core but then immediately reslotting it back into the combined sequence of core two plus core three. During this iteration we retain the ordering of core two with core three which remained after removal of the core one horizons from the joint slotting of the three cores. Now, because of the order in which the cores were combined, a different joint slotting will almost certainly have been produced. We iterate a second time. In this step we remove core two from the combined sequence

and as before immediately reslot it. This second iteration step is followed by the same removal and replacement procedure but now using core three. We continue performing these three steps iteratively until the joint slotting of the three cores stabilizes. A fourth standardized core can now be added, and a further set of removal and replacement iterations performed. Our programme is written in such a way that the coding is able to handle any number of cores.

The iterative approach is relatively easy to program. The main modifications needed to the two core program concern (i) keeping track of which horizons in the joint slotting are from which core and (ii) increasing the array sizes. The key point here is that since slotting is always per-

236

formed pairwise, the array sizes are only marginally larger than that needed for slotting just two sequences.

Although slightly different solutions are arrived at according to the order in which the cores are slotted, the differences in the final multicore slot are rather small. Furthermore a perturbation analysis of the Kinneret susceptibility data (Fig. 6), performed by A. Gordon (Pers. Comm.), using a simulated annealing algorithm, could find no better slotting solution than the best obtained by our iterative procedure. These two results give us confidence that our approximate iterative method produces slottings that are close to the optimal slotting. Indeed in practical terms, especially in multicore work, such approximately optimal slottings are quite satisfactory.

The Kinneret susceptibility data of Fig. 6 are plotted against position in the best combined slotting. The upper three plots in Fig. 6 show susceptibility for each of the three cores individually, while the lowermost plot shows the combined slotting. Unblocking was again found necessary to achieve a realistic slotting. In this example, unblocking was obtained by down-weighting the between-core dissimilarities by a factor of one half compared to the within core dissimilarities. Without any unblocking, clear mismatches could be seen between the cores, particularly around common path position 400. This mismatching, in the unconstrained slotting, was a consequence of relatively low variability within the cores compared with between-core differences in this part of the sediment succession. However, with the inclusion of downweighting we see how all the major features in the three cores align well (Fig. 6).

Table 1. Quality of match statistic for ordered sequences on the unit sphere

Type of sequence	Data source	Delta	R
Identical data		0.00	1.00
Earth's rotation axis	VLBI/LASER	0.29	0.99
Palaeomagnetic demag. of sediment core	NRM/15mT	0.30	0.90
Geomagnetic Observatory Annual means	Eskdalemuir/Lerwick	0.51	0.98
Etna volcano secular variation	Historical/Lava Flows	0.56	0.92
Palaeomagnetic reversals	Bessasta/Borgarfj	0.60	0.86
Within lake secular var.	Lomond 1/Lomond 3	0.63	0.81
Paleomagnetic secular variation within lake	Windermere 1/ Windermere 3	0.65	0.79
ARIMA	101/100	0.72	0.67
Between lake sec. var.	Windermere/Lomond	0.73	0.70
Sec. var. between lake and stalactite	Akiyoshi/Kinki	0.77	0.65
No overlapping sections		>1.00	0.00

Quality of fit

One measure of the quality of fit is the Psi statistic of Gordon (1973) and Gordon and Birks (1974) which uses a standardized measure of the total local dissimilarities. Gordon (1982) suggests a statistic, delta, for use with the total path length approach to sequence slotting. The statistic, delta, is based on the difference between the length of the common path (CPL) and the lengths of the individual paths of core A (APL) and core B (BPL). If Euclidean distance is used as the dissimilarity measure, then delta measures the adequacy of piecewise linear interpolation within blocks in the combined sequence. More specifically,

$$\text{delta} = (2(\text{CPL})/(\text{APL} + \text{BPL})) - 1$$

As an example of the use of such quality of fit statistics, we have compared slottings of data sets on the sphere. Our particular interest was to assess the quality of match of palaeomagnetic secular variation records between lakes (Table 1). Based on this measure, we find poor agreement of palaeomagnetic records between lakes compared with the fits obtained for other types of geophysical data. Furthermore we find no better agreement of palaeomagnetic records between nearby lakes compared with lakes thousands of kilometres apart, as judged using the statistic delta. Such a result would not be anticipated geophysically if lake sediments are good recorders of both geomagnetic declination and inclination. We also investigated a simulation procedure of assessing quality of fit. After some trials we found that we could generate *random* autoregressive or integrated moving average (ARIMA) series which successfully mimicked the main characteristics of lake sediment palaeomagnetic secular variation data sets (see Fig. 7). An indication of the success of the simulation is that in informal questioning a number of experienced investigators have found some difficulty in distinguishing such ARIMA data from their own 'real' palaeomagnetic measurements. On slotting our ARIMA series with each other, we obtained an average delta

value of 0.67, which is the same delta value that we obtained for our between lake slottings. Our conclusion, once again, is that the palaeomagnetic data from different lakes are of poor quality, indeed of poorer quality than generally judged subjectively.

A further method of investigating quality of match is through an interpolated correlation coefficient, calculated following slotting. The right hand column of Table 1, tabulates such correlation coefficients for the data on the sphere that we have investigated. Low correlation coefficients, below about 0.7 (as found for randomly generated ARIMA data), reflect poor matches while the closer the coefficient to 1.0 the better the overall fit.

The above analyses do not make use of any dating information, only the stratigraphic ordering. This points to the importance of accurate, independent dating controls in geophysically assessing palaeomagnetic data. The analyses demonstrate how difficult it is to match lake sediment records using only the shapes of the palaeomagnetic secular variation curves. With such stratigraphic data, the human eye appears to be extremely good at 'picking out' correlations when they really exist, but to be rather poor at determining when data sets actually differ. The statistic delta may thus be particularly useful in quantifying quality of match of 'poor' data sets.

Discussion and comparison of alignment methods

Any method for cross-matching two sequences or series will produce an allegedly 'optimal' matching, no matter what data are used. A difficulty common to all such alignment methods is how to assess the practical significance of the final 'optimized correlation' or match.

In the case of sequence slotting the randomization test given by Clark (1988) provides a simple statistical test for assessing how well any given pair of sequences are matched, based on the value of delta. In this test, each sequence is split *at random* into two sub-sequences, each of which is then slotted against every other subsequence.

238

101 0.91 −0.41   100 0.8

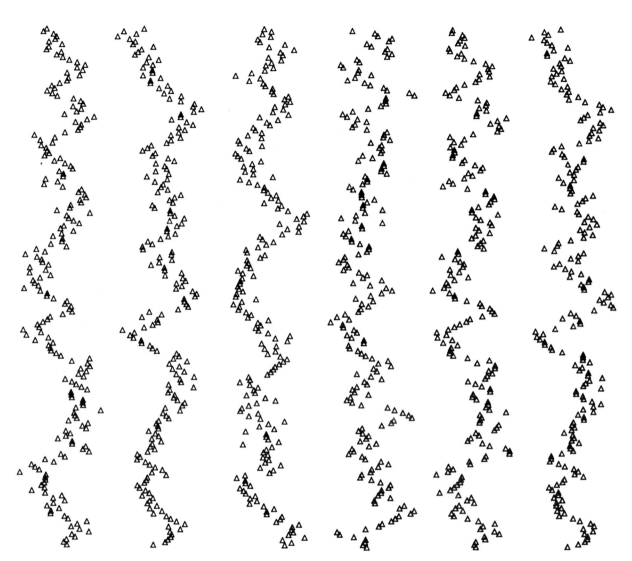

Fig. 7. ARIMA series. Data sets generated to mimic palaeomagnetic secular variation time series on the sphere. The three left hand examples shown were generated using a combined autoregressive moving average process with coefficients of 0.91 and − 0.41 respectively, while the three right hand examples were produced from a 100 process with an AR coefficient of 0.8. The choice of coefficients is not especially critical for generating series which resemble down core palaeomagnetic records.

This procedure is then repeated, say 40 or 50 times. The test then measures how well the original sequences and the combined sequence are reconstructed from these random subsequences. If the combined sequence can *not* be reconstructed as well as the original sequences individually, this indicates that the sequences do *not* arise from a common signal, i.e., the matching is spurious.

In principle this test leads to *exact* significance levels. It is easy to implement, does not involve any assumptions regarding the form of the fitted signal or the stretching function, or any simulation of data according to some assumed model. The examples in Clark (1988) show that the test works well and can be remarkably sensitive.

A variety of techniques have been developed in connection with the alignment or correlation of time series that can be applied to stratigraphic data. We finally briefly discuss the differences between three such techniques and note some of their advantages and disadvantages.

1. Clark & Thompson (1978) – linear stretching with cubic spline smoothing of the observed data
2. Martinson *et al.* (1982) – Fourier series estimation of the mapping function with coherence to be maximised
3. Present Paper – Sequence-slotting with un-blocking.

These three methods were developed to do three different things namely:

Method 1: to estimate a linear stretching function from noisy discrete data in which both 'signals' are distorted versions of some unknown reference signal;

Method 2: to estimate a general mapping function from two continuous signals with high signal-to-noise ratio, with one signal regarded as a known reference signal;

Method 3: to match two ordered sequences of observations (with multivariate responses), using only the relative *order* of the observations, and taking account of other stratigraphic constraints.

All three methods require, in their implementation, a number of more-or-less arbitrary decisions, e.g.,

Method 1: the number and location of the knots used to define the cubic splines fitted to the data.

Method 2: the number of Fourier coefficients, the constants Delta-a and k in equation 10, the degree of smoothing or filtering of the data prior to maximizing the coherence.

Method 3: the choice of distance measure and of 'local discordance', the preliminary standardization, the weights to be assigned to different variables.

Methods 1 and 2 require an iterative trial-and-error approach starting from an initial guess; in contrast method 3 uses a direct and efficient algorithm with no iteration (once the maximum block-length is specified).

Method 1 has the following advantages:
Being based on well-known and well-understood least-squares methods, its assumptions can be readily tested, and confidence limits for both the stretching function and the fitted signal can be found. The method treats the sequences symmetrically (unlike Method 2), in the sense that both sequences are treated as distorted versions of a common unknown signal. The least-squares criterion is more sensitive than the correlation coefficient, (see the examples and discussion in Clark & Thompson, 1978). In principe the method can test the adequacy of ANY given monotonic stretching function, but in practice this is difficult to do.

Its disadvantages are:
1. The choice of knots is arbitrary, although some guidelines are available to assist this choice.
2. A large number of parameters must be fitted.
3. A trial-and-error approach is needed to find the best values of alpha and beta (which define the stretching function).

Method 2 clearly works well in the situation for which it was designed. Also it is has been claimed to be superior to earlier versions of sequence slotting for matching palaeomagnetic data. However, it is not clear how well it would work with discrete data with a low signal-to-noise ratio or how it should be modified in order to deal with multivariate data. It has the following disadvantages:
1. Like method 2, it requires several arbitrary decisions and an iterative approach starting from an initial guess.
2. It cannot handle (i) cases in which only the order of the observations is known, (ii) stratigraphic constraints, and (iii) simultaneous matching of several sequences.
3. There is no guarantee that the fitted mapping function will turn out to be monotonic.
4. No statistical test is provided for interpreting or assessing the final maximised coherence C.

Method 3 has the following advantages:
1. It directly produces the optimum 'path' as well as an estimate of the stretching or mapping function.

2. It uses a fast and efficient algorithm, which requires no iteration, no initial guess, and no arbitrary prior smoothing of the data.
3. It can handle multivariate signals or responses.
4. Additional stratigraphic information is easily incorporated.
5. Only the *order* of the points in each sequence need be known; the corresponding depths or times are not needed.
6. The randomization test provides a conceptually-simple and easily-implemented statistical test of how well the series are matched or slotted together.
7. In its extended version (see Gordon, Clark & Thompson, 1988) it is possible to assess which parts of the slotted sequence are well-determined (and which parts are not), and to investigate the influence on the results of deleting single possibly-aberrant observations.

Method 3 has the following disadvantages:
1. The choice of distance function, relative weighting for different variables (in the multivariate case), initial standardization and maximum block length are arbitrary.
2. Sometimes it is not easy to interpret the final delta-value.
3. The algorithm requires a lot of computer memory, and in its extended form would be impractical on a mainframe computer if the sequences had more than 1000 points each.
4. It does not produce confidence bands for the fitted path or the implied stretching function.
5. Continuous signals would have to be digitized before analysis – so introducing another unwanted arbitrary choice.

References

Anderson, N.J., 1986. Diatom biostratigraphy and comparative core correlation within a small lake basin. Hydrobiologia 143: 105–112.

Clark, R.M., 1985. A FORTRAN program for contrained sequence- slotting based on minimum combined path length, Computers & Geosciences 11, 605–617.

Clark, R.M., 1988. A randomisation test for the comparison of ordered sequences. Research Report 177. Dept. Mathematics, Monash Univ.

Clark, R.M. & R. Thompson, 1979. A new statistical approach to the alignment of time series. Geophys. J. R. Astron. Soc. 58: 593–607.

Clark, J.S., J.T. Overpeck, III, T. Webb & III, W.A. Patterson, 1986. Pollen stratigraphic correlation and dating of barrier beach peat sections, Rev. of Palaeobot. Palynol. 47: 145–168.

Delcoigne, A. & P. Hansen, 1975. Sequence comparison by dynamic programming. Biometrika 62: 661–664.

Embleton, B.J.J., N.I. Fisher & P.W. Schmidt, 1983. Analytic comparison of apparent polar wander paths. Earth Planet. Sci. Letters 64: 276–282.

Emiliani, C., 1966. Paleotemperature analysis of Caribbean cores P6304-8 and P6304-9 and a generalized temperature curve for the past 425 000 years. J. Geol. 74: 109–124.

Emiliani, C., 1978. The cause of the ice ages. Earth Planet. Sci. Lett. 37; 349–352.

Gordon, A.D., 1973. A sequence-comparison statistic and algorithm. Biometrika 60: 1, 197–200.

Gordon, A.D., 1982. An investigation of two sequence-comparison statistics. Aust. J. Statist. 24: 332–342.

Gordon, A.D. & H.J.B. Birks, 1974. Numerical methods in Quaternary palaeoecology. II: Comparison of pollen diagrams. New Phytologist 73: 221–249.

Gordon, A. D., R. M. Clark & R. Thompson, 1988. The use of constraints in sequence slotting. In Data analysis and informatics 5. (Eds. E. Diday *et al.*) North Holland (In Press).

Gordon, A.D. & R.A. Reyment, 1979. Slotting of borehole sequences. Mathematical Geology 11: 3, 309–327.

Holmquist, B., 1989. Sequence-comparison statistics and dissimilarity measures for slotting two sequences. (Unpub. Manuscript).

Lowe, J.J. & M.J.C. Walker, 1984. Reconstructing Quaternary Environments. London: Longman.

Martinson, D.G., W. Menke & P. Stoffa, 1982. An inverse approach to signal correlation. J. Geophys. Res. 87: 4807–4818.

Rummel, R.J., 1970. Applied Factor analysis: Northwestern Univ. Press, Evanston, III., 617 p.

Zhou, D., T. Chang & J.C. Davis, 1983. Dual extraction of R-Mode and Q-Mode factor solutions, Math. Geol. 15: 5, 581–606.

A review of the origins of endemic species in Lake Biwa with special reference to the goby fish, *Chaenogobius isaza*

Sachiko Takahashi

Department of Zoology, Faculty of Science, Kyoto University, Sakyo, Kyoto, 606, Japan

Key words: ancient lakes, Lake Biwa, endemic organisms, relict species, adaptation, goby fish

Abstract

The 1400 m deep drilling of Lake Biwa has revealed that the lake probably originated ca. $5 \cdot 10^6$ yr ago. After a geographic shift to its present position, and by more than 3.10^5 yr ago, it had become a large and deep lake. The considerable longevity, large size and high diversity of habitats of this lake are considered to have contributed to its high abundance of endemic taxa. These taxa fall into two categories: (1) relict species of the Asiatic continent, higher latitude or marine origin; (2) species differentiated in the lake from littoral-lacustrine species, and having adapted to habitats peculiar to Lake Biwa. I discuss some of these endemic organisms, briefly review recent investigations on fossil organisms of the lake, and more fully discuss the origin of a representative endemic species, the pelagic gobiid fish *Chaenogobius isaza* Tanaka. This species is regarded as having differentiated from some littoral *Chaenogobius* species, creating a novel niche in the open water area after Lake Biwa was established as a deep lake.

Introduction

Lake Biwa (Fig. 1) in Japan is one of the oldest lakes in the world and has been studied by many scientists. Recently, a 1400 m deep drilling of the lake has been carried out (Horie, 1984, 1987), and the results have thrown light on the history of this ancient lake. These studies have suggested that Lake Biwa originated in the Pliocene about five million years ago, 100 km south of its present position, and shifted to its present position in the

Originally published in
Journal of Paleolimnology **1**: 279–292.

late Pliocene or early Quaternary. The lake had become large and deep by 300 000 years ago, and has continued in this condition to the present day (Yokoyama, 1984; Nakajima, 1987; Itihara, 1982).

Lake Biwa is now a warm monomictic lake located in the center ($35°00'-35°30'$ N) of Honshu Island. Morphometric characteristics of the lake are given in Table 1. The fauna and flora of the lake are rich in genera and species compared to other Japanese inland waters. The lake is also marked by the presence of a large number of endemic species (Uéno, 1975, 1984; Kawanabe, 1978). Mori (Sy. and Miura, 1980) recorded 296 species of plants (including algae) and 478 of animals, including 5 and 40 endemic species, respectively. Additional species, and descriptions of new species including endemic ones, have sub-

242

Table 1. Morphometric characteristics of Lake Biwa (Horie, 1984).

Altitude	85 m
Length	68.0 km
Maximum breadth	22.6 km
Length of shoreline	188.0 km
Area	674.4 km^2
Shoreline development	2.04
Maximum depth	104.0 m
Mean depth	41.2 m
Volume	27.8 km^3

sequently been reported (Hosoya, 1982; Watanabe, 1984; Morino, 1985; Sasa & Kawai, 1987).

Ancient lakes, such as Lake Tanganyika and Lake Baikal, often show a high endemicity. Baikal, which was established at the end of the Paleogene and/or Early Neogene (about 25·10^6 yr ago), has many endemic families (11) and genera (87) (Kozhov, 1963; Academy of Science of the USSR, 1979). Lake Tanganyka may have originated in the Pliocene. The long period of isolation is reflected in the numbers of endemic genera (38) and species (176) of fishes (Beadle, 1981). Lake Biwa, on the other hand, has no endemic families and few endemic genera; a difference which may be related to the respective longevities of the lakes.

The considerable longevity, large size and high diversity of habitats of Lake Biwa, in comparison with other Japanese lakes, are considered to have contributed to its abundance of endemic taxa. The endemic taxa are divisible into two categories (Kawanabe, 1978): (1) relict taxa which either migrated from the Asiatic continent when sea level was lower, advanced southward from high latitudes in some glacial age, or invaded from

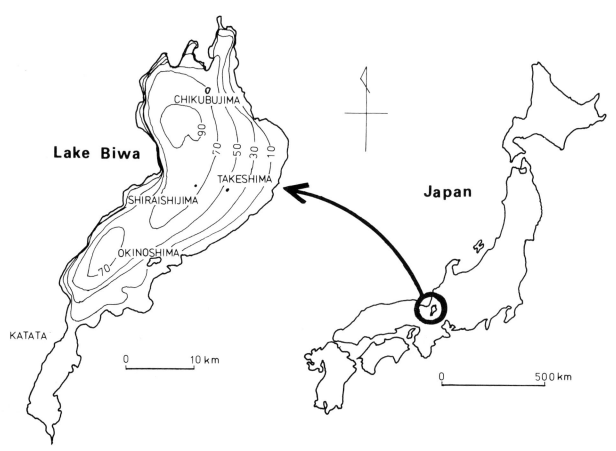

Fig. 1. Lake Biwa and its location in Japan. Isobaths in meters.

marine habitats but later disappeared from Japan except for Lake Biwa; (2) taxa which evolved in the lake from littoral species, adapting themselves to the particular habitats of Lake Biwa, especially to deep and open water or the rocky shore. I briefly discuss both types of endemics from Lake Biwa, review some recent investigations of fossil organisms of the lake, and consider the origin of an endemic goby fish *Chaenogobius isaza* Tanaka which I have been studying eco-physiologically.

Relict Taxa

Examples of the first category, as shown in Table 2, include cyprinid fishes such as *Opsariichthys uncirostris* (Fig. 2) and *Ischikauia steenackeri*; molluscs, such as *Hyriopsis schlegeli*, *Valvata piscinalis biwaensis*, *Pisidium kawamurai*, *Corbicula sandai*; a turbellarian, *Bdellocephala annandalei*; and a crustacean, *Kamaka biwae* (Fig. 3).

Opsariichthys uncirostris, a piscivorous fish distributed in Lake Biwa and Lake Mikata, which is near Lake Biwa, can be regarded as a semi-endemic species. This species has subspecies in the Korean Peninsula, the Amur drainage system and southern part of the Asiatic continent (Miyadi *et al.*, 1976). The herbivorous *Ischikauia steenackeri* has closely related genera widely distributed around East Asia, except Japan (Miyadi

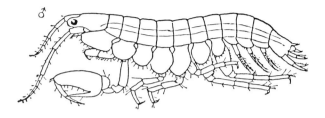
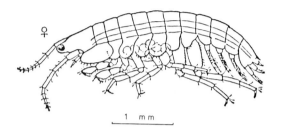

Fig. 3. Kamaka biwae Uéno (after Uéno, 1973).

et al., 1976; Research Group for Natural History of Lake Biwa, 1983). *Hyriopsis schlegeli*, which is utilized for culturing non-nucleated pearls, has its closest relative, *H. cumingii*, in China (Habe, 1973).

Kawanabe (1978) suggested that the ancestors of *Opsariichthys uncirostris* and *Ischikauia steenackeri* had come from the continent via a land bridge during the Pliocene or a more ancient period, and, for a time, were widely distributed throughout at least western Japan, and were then

Fig. 2. Opsariichthys uncirostris (Temminck et Schlegel). Photo credit: Biwakobunkakan (Aquarium of Lake Biwa).

Table 2. Examples of endemic taxa in Lake Biwa and their closely related taxa.

Endemic Taxa	Related Taxa

I. RELICT TAXA
Pisces

Opsariichthys uncirostris (Temmink et Schlegel)	{ *Opsariichthys uncirostris bidens* Günther[c] { *Opsariichthys uncirostris amurensis* Berg[c]
Ischikauia steenackeri (Sauvage)	{ Cultrinae around east Asia { fossils of *Ischikauia* spp. from Katata Formation

Mollusca

Hyriopsis schlegeli (v. Martens)[a]	*Hyriopsis cumingii* (Lea)[a]
Pisidium kawamurai Mori	*Pisidium* spp. in northern part of Japan
Corbicula sandai Reinhardt	{ *Corbicula japonica* Prime[a] { fossils of *Corbicula* sp. from Katata Formation
Valvata piscinalis biwaensis Preston[a]	*Valvata (Cincinna) piscinalis japonica* v. Martens[a]

Crustacea

Kamaka biwae Uéno	*Kamaka kuthae* Dershavin[e]

Turbellaria

Bdellocephala annandalei Ijima et Kaburaki	unidentified

II. ENDEMIC TAXA DIFFERENTIATED IN LAKE BIWA
Pisces

Gnathopogon caerulescens (Sauvage)	*Gnathopogon elongatus elongatus* (Temmink et Schlegel)[c]
Carassius auratus grandoculis (Temminck et Schlegel) }	{ *Carassius auratus gibelio* (Bloch)[d]
Carassius cuvieri Temminck et Schlegel[d]	{ *Carassius auratus* Linné Type Kinbuna[d]
Sarcocheilichthys biwaensis Hosoya[b]	{ *Sarcocheilichthys variegatus variegatus* (Temminck et Schlegel)[b] { *Sarcocheilichthys variegatus microoculus* Mori[b]
Parasilurus biwaensis Tomoda	
Parasilurus lithophilus Tomoda }	*Parasilurus asotus* Linnaeus[c]

Mollusca

Semisulcospira decipiens (Westerlund)[f]	
Semisulcospira multigranosa Bötiger[f]	
Semisulcospira reticulata Kajiyama et Habe[f]	
Semisulcospira nakasekoae Kuroda	fossils of *Semisulcospira* spp. from Katata Formation
Semisulcospira niponica (Smith)	
Semisulcospira morii Watanabe[f]	
Semisulcospira habei Burch[a]	

III. SOME ENDEMIC MOLLUSKS AND FOSSILS AND RELATED TAXA FROM THE KATATA FORMATION

Heterogen longispira (Smith)	fossils of *Heterogen longispira* (Smith)
Lanceolaria oxyrhyncha (v. Martens)	*Lanceolaria oxyrhyncha* (v. Martens)
Unio (Nodularia) biwae Kobelt	*Unio* sp.
Inversidens reiniana (Kobelt)	*Inversidens* spp.
Anodonta calipygos Kobelt	*Sinanodonta* spp.

Scientific names follow Mori (Sy.) & Miura (1980), except (a) Habe (1973), (b) Hosoya (1982), (c) Miyadi *et al.* (1976), (d) Tomoda (1984), (e) Uéno (1973), and (f) Watanabe (1984).

eliminated except for Lake Biwa. Yasuno (1983), however, considered that the genus *Ischikauia* differentiated in the Quaternary from the Cultrinae, a main component of the Miocene fish fauna of Japan, either directly or after it had spread to China and returned to Japan. *Ischikauia steenackeri* is, therefore, considered a Tertiary relict at subfamily level and, at the same time, can be considered as a fish which differentiated in Lake Biwa at the generic and species levels. *Opsariichthys uncirostris* is also considered to be a Tertiary relict (Nakajima, 1983). However, since the continental species are not differentiated so widely from their Japanese relatives, I believe that *O. uncirostris* probably came later from the continent, during the Quaternary via one of the several land bridges which must have occurred in that period (Lindberg, 1972).

Valvata piscinalis biwaensis, a small profundal bivalve, is regarded as a glacial relict (Annandale, 1922). *Valvata (Cincinna) piscinalis japonica*, a relative of *V. p. biwaensis*, is found in many lakes of high latitudes, in Hokkaido and Sakhalin. It is supposed that the common ancestral species might have lived in many lakes, even in the southwestern part of Japan, in some glacial ages, and was eliminated from shallow lakes during interglacial ages (Kawanabe, 1978). Some of the other inhabitants of the hypolimnion of Lake Biwa, such as *Pisidium kawamurai* and *Bdellocephala annandalei* are also regarded as being glacial relicts (Annandale, 1922).

Corbicula sandai was believed to have been derived in Lake Biwa from a marine species, *C. japonica*, because it is likewise dioecious and fertilizes eggs externally, unlike th fresh water species *C. leana* Prime (Mori (Sy.), 1984). Recently, however, a bivalve which is possibly *C. sandai* was found in China (Miyadi, 1984) and the speciation may therefore have occurred in China before the establishment of Lake Biwa. Further taxonomic study, such as an accurate classification based on the anatomy of the soft parts, is necessary to settle the complicated problem.

Kamaka biwae, a small crustacean of 4–5 mm body length (Fig. 3), was considered to be most likely a glacial marine relict (Uéno, 1984), because

other members of the family live in the marine habitat and the closest relative *K. kuthae* occurs in brackish or fresh waters of high latitudes along the coast of the Okhotsk Sea. However, *K. biwae* inhabits the littoral zone, where the water temperature rises in summer. Kawanabe (1978) stated that such animals could not possibly have invaded the lake directly from the sea during the glacial ages, when the sea level was much lower than in the interglacial and postglacial periods. That *K. biwae* is a glacial relict has been made more doubtful by the recent discoveries of *Kamaka* species at several shallow lakes in Honshu Island, Japan: Kasumigaura, Kitaura, Hinuma, Shinjiko and Nakaumi (Morino, personal communication). However, that all of these habitats are or were brackish lakes (Horie, 1964) argues that *K. biwae*, too, is a marine relict species.

Endemic Taxa that differentiated in Lake Biwa

Endemic taxa that differentiated from ancestral taxa in Lake Biwa (Table 2) are represented by the cyprinid fishes *Gnathopogon caerulescens*, *Carassius auratus grandoculis*, *C. cuvieri*, and *Sarcocheilichthys biwaensis* (Fig. 4); the catfishes *Parasilurus biwaensis* and *P. lithophilus* (Fig. 5); and diverse snail species in the genus *Semisulcospira*, such as *S. decipiens*, *S. multigranosa*, *S. reticulata*, *S. nakasekoae*, *S. niponica*, *S. morii* and *S. habei*.

Gnathopogon caerulescens, which lives in open water of the lake and feeds on zooplankters, is regarded (Tokui & Kawanabe, 1984) as having differentiated in Lake Biwa from a littoral inhabitant *G. elongatus elongatus*.

Carassius auratus grandoculis utilizes the deep and middle layers of the lake feeding on zooplankters and benthic animals. *C. cuvieri* is also found in open water but feeds on phytoplankters such as diatoms. Both species were believed to have developed as offshore types from the littoral *C. auratus*, Type Kinbuna (Tomoda, 1984). However, the recent discovery of fossil pharyngeal teeth intermediate between *C. cuvieri* and *C. auratus gibelio* from the Sakawa clays in Katata

Fig. 4. *Sarcocheilichthys biwaensis* Hosoya. Photo credit: Biwakobunkakan (Aquarium of Lake Biwa).

Fig. 5. *Parasilurus lithophilus* Tomoda. Photo credit: Biwakobunkakan (Aquarium of Lake Biwa).

Formation, the Paleo-Biwa stratum formed several hundred thousand years ago, suggests that *C. cuvieri* may have been derived from *C. a. gibelio* during the middle or late Quaternary (Research Group for Natural History of Lake Biwa, 1983). Whichever is correct, the ancestor must have been a benthic fish living in shallow water.

Parasilurus biwaensis is a large (to 1 m) piscivorous catfish that lives in the middle layer of open water of the northern part of Lake Biwa. *Parasilurus lithophilus*, on the other hand, is adapted to rocky shores. These catfishes are con-

sidered to have differentiated from *Parasilurus asotus*, a species inhabiting the muddy shore and the lower reaches of streams in Japan as well as in Lake Biwa (Tomoda, 1984). *Parasilurus lithophilus* is characterized by its comparatively protruded snout and eyes, this feature being regarded as an adaptation for foraging in rocky or stony areas (Tomoda, 1984). The protrusion of snout and eyes also occurs in one cyprinid fish, *Sarcocheilichthys biwaensis*, an inhabitant of rocky shores. That species never moves to the affluent or effluent rivers of Lake Biwa in any season, unlike the related species, *Sarcocheilichthys variegatus variegatus* and *S. variegatus microoculus* (Hosoya, 1982).

Ten species (or subspecies) of snails within the genus *Semisulcospira* occur in various habitats of Lake Biwa and its tributaries; seven of the species are endemic forms (Table 2). In the littoral zone, five species predominate: *S. niponica*, *S. decipiens*, *S. multigranosa*, *S. reticulata* and *S. habei*. *Semisulcospira niponica* exclusively occupies the rocky and stony bottom. The other four species live on the sandy or muddy bottom; their habitats differ in relation to depth and substratum (Davis, 1969; Watanabe, 1970, 1980).

Semisulcospira niponica shows morphological variation with habitat. A large and extremely nodulated form was found off three of four islands in Lake Biwa, Takeshima, Shiraishijima and Chikubujima Islands (Fig. 1), but not in other areas of the lake. The smaller the area of the island, the larger and the more intensely nodulated *S. niponica* is. *Semisulcospira morii* is isolated along two islands, Chikubujima and Takeshima. Three other types of *Semisulcospira* were found off Takeshima and Shiraishijima Islands (Watanabe, 1984). At this time it is not known if these types represent separate species. Further studies should reveal new *Semisulcospira* species. Nishino (1987) suggested that such diversification may be partly due to both low vagility and ovoviviparous reproductive systems.

Fossil evidence

Some fossil evidence exists on the origin of a few of these endemic species. The present day endemic species have not been developing through the five million year history of Lake Biwa. There have been many extinctions and species replacements. Fossils identical or closely related to certain extant species have been found from the uppermost layers of the sedimentary deposits of the Katata Formation, which is exposed in the Katata Hills on the west coast of Lake Biwa. These sediments were deposited 1 000 000–200 000 yr ago (Yokoyama, 1984).

Fossil fishes from Paleo-Lake Biwa are rare, in spite of the present day richness of the fish fauna of Lake Biwa. However, fossil pharyngeal teeth of cyprinids collected from Plio-Pleistocene sediments of Paleo-Lake Biwa and its river system have been examined recently (e.g., Nakajima, 1986; Kodera, 1985; Yasuno, 1983).

In the cyprinid fauna of Paleo-Lake Biwa of the early Pliocene, cyprinine and xenocypridine fishes were dominant, and cultrine and gobionine fishes follow them in abundance. This fauna is very different from that in present Lake Biwa, as well as present Japan, and is more akin to the Miocene Japanese fauna, and the Pliocene, and the present day fauna in China at the generic level. The Miocene cyprinid fauna of China consists of many extinct and some living genera, and differs from the present day Chinese fauna (Nakajima, 1986). From these fossil data, Nakajima (1986) concluded that the secondary radiation of the new cyprinid groups, the xenocypridine, cultrine fishes and so on, occurred in Japan in the early Miocene, and that the new fauna containing them spread from Japan at the margin of the Asian continent, to the interior of the continent during the middle of the Late Miocene. The xenocypridine and cultrine fishes, which were major groups in the Paleo-Lake Biwa of the early Pliocene, and the Miocene fauna in Japan, flourish in present day China, but with the exception of one cultrine fish, *Ischikauia steenackeri*, all are extinct in Japan.

The fossil pharyngeal teeth of *Ischikauia* spp. and the other genera of cultrine fishes are found from the upper layer, the Sakawa clays, and the middle layer, the Hiraen clays, of the Katata Formation, respectively (Research Group for Natural History of Lake Biwa, 1983; Yasuno, 1983). Xenocypridine fossils are also found in the Hiraen clays (Tomoda, 1984). Xenocypridine and cultrine fishes flourished until the mid-Quaternary.

Nakajima (1986) supposed that the extinction of these fish groups in Japan was caused by the disappearance of shallow and large lakes, and the appearance of torrential rivers during the Island Arc Movement of the Middle to Late Quaternary. The xenocypridine fishes also seem to have com-

248

peted for food with the osmerid 'ayu fish' *Plecoglossus altivelis*, Temminck et Schlegel. A decrease in production of algae during the glacial age, as postulated by Nakajima (1986), may therefore have accelerated their extinction.

Other important fish fossils found from the Katata Formation are the above-mentioned pharyngeal teeth of *Carassius* which are anatomically intermediate between *C. cuvieri*, an endemic species in Lake Biwa, and *C. auratus gibelio*, a common species in Japan.

Many molluscan fossils have been found in the Sakawa clays, the upper layer of the Katata Formation (Research Group for Natural History of Lake Biwa, 1983). They consist of Viviparidae, such as *Cipangopaludina japonica* (v. Martens) and *Heterogen longispira*; many Unionidae species, such as *Hyriopsis schlegeli*, *Inversidens japanensis* (Lea), *I. brandti* (Kobelt), *I.* spp., *Unio* sp., *Lanceolaria oxyrhyncha* and *Sinanodonta* spp.; Corbiculidae, such as *Corbicula leana* and *C.* sp.; and Pleuroceridae, such as *Semisulcospira* spp..

Cipangopaludina japonica, *Inversidens japanensis*, *I. brandti* and *Corbicula leana* are species found throughout Japan, whereas *Heterogen longispira*, *Hyriopsis schlegeli* and *Lanceolaria oxyrhyncha* are endemic species living in Lake Biwa. *Inversidens* spp., *Unio* sp., *Sinanodonta* spp., *Corbicula* sp. and *Semisulcospira* spp. may be closely related to the endemic species living in Lake Biwa (Table 2). Fossil mollusks which seem unrelated to the present day taxa are found from the lower part of the Paleo-Lake Biwa deposits.

Microfossils in core samples from the lake bottom were examined by Kadota (1987) and Mori (Sh.) (1984). Their data demonstrate that the number of taxa and abundance of organisms in Lake Biwa have gradually increased since the birth of the lake. In the deep clays and sand layers 582–732 m, ca. 2.0–2.4 · 10^6 yr B. P., fossil diatoms are scarce. Such fossils are also scarce in the alternating sands and silts with gravel between 250 and 582 m. These sediment particle-size frequencies indicate that the sedimentation occurred in shallow littoral areas. However, the upper core layer (several hundreds of thousands of years ago)

is characterized by the dominance of fossil diatoms. The existence of certain *Melosira* spp. and *Stephanodiscus* spp. are particularly interesting since their presence suggests that the deep oligotrophic character of Lake Biwa had appeared in this age and has continued up to the present day.

Origin of the Isaza, Chaenogobius isaza

An endemic gobiid fish, the Isaza, *Chaenogobius isaza* (Fig. 6), is famous for its remarkable diel vertical migration (Takahashi, 1981b). Though there is no fossil evidence, I will infer the origins of this taxon on the basis of my study (Takahashi, 1974, 1981a, 1981b, 1981c, 1982; Takahashi & Hidaka, 1984; Hidaka & Takahashi, 1987b) of its life history. There have been two hypotheses on the origin of the Isaza: (1) it is a glacial relict, since it is found in deep and cold water (Miura, 1970), and (2) it evolved in this lake as a pelagic species from some littoral gobiid ancestor, e.g. *C. annularis* Gill (Fig. 7) (Tokui & Kawanabe, 1984; Tomoda, 1984). Which hypothesis is more likely?

In addition to the Isaza, two littoral gobiid fish live in Lake Biwa and its inlet rivers: *Chaenogobius annularis*, which is scarce, and *Rhinogobius brunneus* (Temmink et Schlegel) (orange type) (Fig. 8), which is abundant all over the littoral zone. Both *C. annularis* and *R. brunneus* are predatory amphidromous fish distributed widely over East Asia. They eat eggs and larvae of other fishes. Those which inhabit Lake Biwa and its inlet rivers, however, are landlocked lacustrine types. Larvae of both gobies grow in the open water of Lake Biwa, and the young migrate to the lake shore or the inlet rivers. More than 300 000 yr ago, when Lake Biwa became a large and deep lake, ancestors of these two gobiid fishes probably occupied the lake shore and inlet rivers and became landlocked lacustrine types without serious modifications of their ontogeny. For the adults of the third gobiid fish, the Isaza, no space would have been left except for the offshore waters, which provided a new niche.

Life history of the Isaza in Lake Biwa. Adult

Fig. 6. Chaenogobius isaza Tanaka. Photo credit: Biwakobunkakan (Aquarium of Lake Biwa).

Fig. 7. Chaenogobius annularis Gill. Photo credit: Biwakobunkakan (Aquarium of Lake Biwa).

Isazas live in abundance in the open lake at around 50 m and greater depth. They remain in this cold bottom water in the daytime and migrate to the warm surface layer at night to feed on zooplankters (Takahashi, 1981b). However, they cannot reproduce offshore. Their demersal eggs, which are heavier than water, are deposited on the underside of stones near the shore. There are no stones in the profundal zone. Even if the fish were to deposit eggs on the profundal bottom, newly hatched larvae would not be able to swim the long way up to the water surface to take air into their gas bladders. This is because their fins are under-developed at this stage (Takahashi & Hidaka, 1984). The near-shore area is occupied throughout the year by *Rhinogobius brunneus*

Fig. 8. Rhinogobius brunneus (Temmink et Schlegel). Photo credit: Biwakobunkakan (Aquarium of Lake Biwa).

which also lays eggs under stones. *Chaenogobius annularis* also occupies the littoral zone, but it has never been found in abundance and it is rarely seen in the Isaza's spawning grounds. It is unlikely that this species significantly effects the reproduction of the Isaza.

Rhinogobius brunneus is inactive at the low temperatures of early spring but later becomes very active. The activity area includes the spawning sites of the Isaza. This activity starts in June when water temperatures rise to around 20 °C. With this rise of water temperature, the spawning areas of the Isaza become favorable feeding sites for many other predatory fishes. Experiments demonstrated (Hidaka & Takahashi, 1987b) that the Isaza is reproductively unsuccessful in the presence of active *R. brunneus*. At high temperature *R. brunneus* often peeped into and entered the nests of the Isaza. Even when an Isaza threatened and attempted to drive *R. brunneus* away, *R. brunneus* did not swim away immediately; instead, the disturbance attracted other nearby *R. brunneus*. The Isazas finally abandoned their nests. Even eggs which were spawned successfully would, if abandoned in the nest, be eaten by *R. brunneus*.

The Isaza does not appear to have evolved any effective defensive tactics and effective aggressive behaviour against *R. brunneus*.

Accordingly, it must have been a crucial problem for the Isaza to avoid competition with *Rhinogobius brunneus*. Isazas actually breed as early as late April to early May, when the water temperature is sufficiently low (around 13 °C) to inhibit the activity of *R. brunneus* and other predators of Isaza larvae (Hidaka & Takahashi, 1987b). But spawning that is too early would be disadvantageous, because spring propagation of zooplankton, on which Isaza larvae feed, begins in mid-April (Yamamoto, 1967, 1968). There is the further problem that mature Isaza cease to feed during reproduction. It may therefore be energetically unprofitable or impossible for the Isaza to stay and continue reproductive activity for a prolonged period. Thus, they spawn during the short period between the start of propagation of zooplankton and the onset of activity of *R. brunneus* and the appearance of other predatory fishes in their spawning area.

For reproduction to occur early in the spring and be highly concentrated over a very limited period, an early and synchronized gonadal maturation is essential. A seasonal survey of gonadal development in the population of Isaza (Takahashi, 1974, 1981a) showed a very synchronized onset of egg maturation at the end of November, that maturation proceeded steadily during winter, and that when fish arrived at the lake shore in March, all the females had almost mature eggs.

How is such a uniform onset of egg maturation timed? The physiological mechanism determining the timing of the limited breeding season of the Isaza is also closely related to the environment of Lake Biwa. The diel vertical migration noted earlier effectively creates a thermoperiod which is experienced by the Isaza and enables the Isaza to register the seasonal change in temperature of the surface layer. Experiments with various temperature regimes revealed that the decline of the temperature of the lake surface below a threshold around 17 °C triggers vitellogenesis of the Isaza starting in November (Takahashi, 1981c, 1982). The oocytes of the Isaza develop synchronously thereafter. In the early spring of the following year, the mature fish migrate to the lake shore and synchronously spawn in late April to early May when the water temperature of lake shore is around 13 °C. Experiments also indicated that the high temperatures (around 20 °C) of summer not only inhibit oogenesis but are also indispensable for normal and healthy development of ovarian eggs (Takahashi, 1982). This reproductive mechanism of the Isaza is in marked contrast to that of *R. brunneus* (Hidaka & Takahashi, 1987a). *Rhinogobius brunneus* reproduces asynchronously from early summer through autumn when water temperature is much higher, and the start of maturation is promoted by the combination of both long day and high temperature.

The success of the Isaza in Lake Biwa in utilizing the resources offshore would not be possible without physiological mechanisms that ensure the early and synchronous spawning at the lake shore.

Where did the Lake Biwa Isaza originate? The question is whether an ancestor of the Isaza which was preadapted for a pelagic life came into this lake, or whether some littoral gobiid fish subsequently differentiated to a pelagic fish. The Isaza, which is adapted to an open lake, is probably unable to reproduce in rivers because its larvae, which are intolerant of salt water, unlike *Chaenogobius annularis* (Kishi, personal communication), would drift down into the sea. Even if the larvae survived, they have no instinct for swimming up rivers. Once the Isaza had developed a life style adapted to an open lake, it would be very difficult for a fish to migrate through rivers. Artificial transplantation or culture has been successfully practiced in some of the endemic species, e.g., the cyprinid fishes *Gnathopogon caerulescens* and *Carassius cuvieri*. The Isaza, however, has never spread in any other region, though a great many individuals, mixed with other fish to be transplanted, must have been accidentally introduced into many rivers and lakes all over Japan. It seems very difficult for the Isaza to live and reproduce in other lakes. These factors favor the hypothesis that the Isaza differentiated in Lake Biwa from some littoral gobiid ancestor.

What, then, was the original littoral gobiid fish? Did the original littoral gobiid ancestor come from the northern district in the glacial period? The opinion that the Isaza is a glacial relict of northern origin is based on the fact that it inhabits deep cold water. But, as mentioned above, the effect of the high temperature (around 20 °C) at the surface layer of the lake before the onset of the vitellogenesis is indispensable for the Isaza, not only to synchronize its maturation but also for normal and healthy development of its oocytes. Therefore, I think that the low temperature of the hypolimnion, where Isazas stay in the daytime, does not provide a sound basis for the hypothesis of it being a glacial relict. Moreover, a gobiid species that may be closely related to the ancestor of the Isaza has never been reported from higher latitudes. It is therefore most unlikely that the Isaza is a glacial relict.

It is possible that *Chaenogobius annularis*, the other *Chaenogobius* species in Lake Biwa, is the ancestor of the Isaza. Miyadi (1984) hypothesized that the Isaza must have evolved from *C. annularis* neotenically. Tomoda (1984) also supposed *C. annularis* to be the ancestor of the Isaza. Takagi (1952), however, remarked that *C. annularis* has no more special morphological similarity to the Isaza than the other freshwater congeneric species in Japan, *C. castaneus* (O'Shaughnessy). The difference in the tolerance of salt water between the larvae of Isaza and *C. annularis* (Kishi, personal communication) mentioned above also contradicts a close relationship between the two species.

In addition to *Chaenogobius annularis*, there are some related species in and around the Japanese Islands. Takagi (1966a, 1966b) separated *Rhodontichthys laevis* (Steindachnen) from populations which had been classified into *C. castaneus*. Recently, a gobiid fish considered to be *C. taranetzi* Pinchuk, which had never been recorded in Japan, was found in Shinjiko (Hayashi, personal communication). There are also brackish water congeneric species, such as *C. heptachanthus* (Hilgendorf), *C. mororanus* (Jordan et Snyder), *C. macrognathus* (Bleeker), *C. uchidai* (Takagi), and *C. scrobiculatus* (Takagi). The phylogenetic relations of these species have not been made clear. The species closest to the Isaza cannot be determined precisely at present. This makes it difficult to hypothesize a mechanism for speciation of the Isaza.

I hypothesize that the Isaza differentiated into a pelagic fish from some littoral *Chaenogobius* species, creating a novel nich after Lake Biwa was established as a deep lake 300 000 years ago. *Chaenogobius* species have planktonic larvae and the adults are not as benthonic as most other gobiid species. Isaza larvae also carry out a diel vertical migration with other plankters in the open water of Lake Biwa (Nagoshi, 1982; Takahashi & Hidaka, 1984). Prolongation of the larval stage may have led to the pelagic life of the adult. The pelagic population would have become genetically isolated thereafter. Differentiation in time, or place, of spawning is considered to lead to intralacustrine genetic isolation (McKaye, 1980; Smith & Todd, 1984). When the Isaza differentiated, however, the spawning areas of pelagic population and remaining littoral population probably overlapped each other. Differentia-

tion of spawning season may have contributed to genetic isolation. Synchronous maturation, and early and concentrated spawning not only avoided competition with *Rhinogobius brunneus* and other predators but probably promoted genetic isolation ensuring a high frequency of encounters with intrapopulation mates.

The electrophoretic patterns of proteins and enzymes are useful genetic markers, and it is noticeable that sharp morphological differences are sometimes associated with small genetic distances (Humphries, 1984; Sage *et al.*, 1984). Though I cannot at present indicate fossil evidence or genetic data on speciation of the Isaza, I will present a more developed hypothesis with new data in the near future.

References

Academy of Sciences of the USSR, 1979. Lake Baikal. Guidebook. NAUKA, Moscow, 76 pp.

Annandale, N., 1922. The macroscopic fauna of Lake Biwa. Annot. Zool. Jap. 10: 127–153.

Beadle, L. C., 1981. The inland waters of tropical Africa. Longman, London and New York, 475 pp.

Davis, G. M., 1969. A taxonomic study of some species of *Semisulcospira* in Japan (Mesogastropoda: Pleuroceridae). Malacologia, 7: 211–294.

Habe, T., 1973. Nantaidobutsu, Mollusca. In M. Uéno (ed.) Tamiji Kawamura's freshwater biology of Japan. [Nippon tansui seibutsugaku.] Zukan no Hokuryukan, Tokyo: 309–341. (In Japanese).

Hidaka, T. & S. Takahashi, 1987a. Effects of temperature and day length on gonodal development of a goby *Rhinogobius brunneus* (orange type). Japan. J. Ichthyol. 34: 361–367.

Hidaka, T. & S. Takahashi, 1987b. Reproductive strategy and interspecific competition in the lake-living gobiid fish Isaza, *Chaenogobius isaza*. J. Ethol. 5: 185–196.

Horie, S., 1964. Nippon no mizu-umi. Nippon-keizai shinbunsya, Tokyo, 226 pp. (In Japanese).

Horie, S. (ed.), 1984. Lake Biwa. Monographiae Biologicae 54. Dr W. Junk Publishers, Dordrecht, Boston, Lancaster, 654 pp.

Horie, S. (ed.), 1987. History of Lake Biwa. Inst. Paleolimnol. Paleoenviron. Lake Biwa, Kyoto Univ., Kyoto, 242 pp.

Hosoya, K., 1982. Classification of the cyprinid genus *Sarcocheilichthys* from Japan, with description of a new species. Japan. J. Icthyol. 29: 127–138.

Humphries, J. M., 1984. Genetics of speciation in pupfishes from Laguna Chichancanab, Mexico. In A. A. Echelle &

I. Kornfield (eds.), Evolution of fish species flocks. University of Maine at Orono Press. Orono, Maine: 129–139.

Itihara, M., 1982. The bottom surface of Lake Biwa and its relation to the climax of the Rokko Movements. In Islandarc disturbance. Association for Geological Collaboration in Japan, Monograph, 24: 229–233. (In Japanese)

Kadota, S., 1987. Microfossil organisms identified from Lake Biwa sediments. In S. Horie (ed.), History of Lake Biwa. Inst. Paleolimnol. Paleoenviron. Lake Biwa, Kyoto Univ., Kyoto: 217–223.

Kawanabe, H., 1978. Some biological problems. Verh. Internat. Verein. Limnol. 20: 2674–2677.

Kodera, H., 1985. Paleontological evidence for appearance of a species, gengoro-buna(*Carassius cuvieri* Temminck & Schlegel). A comparative study of the dental tissues between living and fossil species. Earth Sci. 39: 272–281. (In Japanese)

Kozhov, M., 1963. Lake Baikal and its life. Monographiae Biologicae 11. Dr W. Junk Publishers, The Hague, 344 pp.

Lindberg, G. U., 1972. Krupnye kolebaniya urovnya okeana v chetvertichnyi period. Biogeograficheskie obosnovania gipotezy. Nauka, Leningrad, Japanese translation by Shnbori, T. & F. Kanemitsu, 1981. Tokai University Press, Tokyo, 366 pp.

Mc Kaye, K. R., 1980. Seasonality in habitat selection by the gold color morph of *Cichlasoma citrinellum* and its relevance to sympatric speciation in the family Cichlidae. Envir. Biol. Fishes 5: 75–78.

Miura, T., 1970. Biwakodake ni sumu sakana, koyushu. In Biwako-hyakka. NHK Otsu Hosokyoku, Otsu: 97–102. (In Japanese)

Miyadi, D., 1984. Haifu-dobutsu-ki. Iwanamishoten, Tokyo, 200 pp. (In Japanese)

Miyadi, D., H. Kawanabe & N. Mizuno, 1976. Coloured illustrations of the freshwater fishes of Japan. [Genshoku Nippon tansui-gyorui zukan.] Hoikusha Publishing Co., Ltd., Osaka, 462 pp. (In Japanese)

Mori, Sh., 1984. Preliminary report of diatom analysis on 1400 m core sample. In H. Kunishi & S. Horie (ed.) Analytical work on the deep drilling core sample in Lake Biwa for study of natural environmental succession. Special report on various analyses of the 1400 meters core sample obtained in Lake Biwa.: 59 pp.

Mori, Sy., 1984, Molluscs. In S. Horie (ed.), Lake Biwa. Dr W. Junk Publishers, Dordrecht, Boston, Lancaster: 331–337.

Mori, Sy. & T. Miura, 1980. List of plant and animal species living in Lake Biwa. Mem. Fac. Sci., Kyoto Univ., Ser. Biol. 8: 1–33.

Morino, H., 1985. Revisional studies on *Jesogammarus-Annanogammarus* group (Amphipoda: Gammaroidea) with descriptions of four new species from Japan. Publ. Itako Hydrobiol. Stn. 2: 9–55.

Nagoshi, M., 1982. Diel vertical migration of zooplankters and fish larvae in Lake Biwa. Bull. Fac. Fish., Mie Univ. 9: 1–10.

Nakajima, T., 1983. Cyprinid fish in Lake Biwa. Animals and Nature, 13: 16–22. (In Japanese)

Nakajima, T., 1986. Pliocene cyprinid pharyngeal teeth from Japan and east Asia Neogene cyprinid zoogeography. In T. Uyeno, R. Arai, T. Taniuchi & K. Matuura (eds.) Indo-Pacific Fish Biology: Proceedings of the Second International Conference on Indo-Pacific Fishes. Ichthyological Society of Japan, Tokyo: 502–513.

Nakajima, T., 1987. Biwako ni okeru gyoruiso no seiritsu to shubunka. In N. Mizuno & A. Goto (eds.), Nippon no tansuigyorui. Sono bunpu hen-i shubunka o megutte. Tokai University Press, Tokyo: 215–229. (In Japanese)

Nishino, M., 1987. The benthic fauna of Lake Biwa. The natural history of Japan. 1(7): 18–23. (In Japanese)

Research Group for Natural History of Lake Biwa, 1983. Fossil assemblages from the Pleistocene Katata Formation of the Kobiwako Group at Ogi-cho, Otsu City, Central Japan. Bull. Mizunami Fossil Mus. 10: 117–142. (In Japanese)

Sage, R. D., P. V. Loiselle, P. Basasibwaki & A. C. Wilson, 1984. Molecular versus morphological change among Cichlid fishes of Lake Victoria. In A. A. Echelle & I. Kornfield (eds.), Evolution of fish species flocks. University of Maine at Orono Press. Orono, Maine: 185–201.

Sasa, M. & K. Kawai, 1987. Studies on chironomid midges of Lake Biwa (Diptera, Chironomidae). L. Biwa Study Monogr. 3: 1–119.

Smith, G. R. & T. N. Todd, 1984. Evolution of species flocks of fishes in north temperate lakes. In A. A. Echelle & I. Kornfield (eds.), Evolution of fish species flocks. University of Maine at Orono Press. Orono, Maine: 45–68.

Takagi, K., 1952. A critical note on the classification of *Chaenogobius urotaenia* and its two allies. Japan. J. Ichthyol. 2: 14–22. (In Japanese)

Takagi, K., 1966a. Taxonomic and nomenclatural status in chaos of the gobiid fish, *Chaenogobius annularis* Gill, 1858 – I. Review of the original description, with special reference to estimation of the upper jaw relative length as a taxonomic character. J. Tokyo Univ. Fish. 52: 17–27. (In Japanese)

Takagi, K., 1966b. Taxonomic and nomenclatural status in chaos of the gobiid fish, *Chaenogobius annularis* Gill, 1858 – II. Specific heterogeneity of *C. annularis* Gill sensu Tomiyana, with description of the genus *Rhodoniichthys*, gen. nov. J. Tokyo Univ. Fish. 52: 29–46. (In Japanese)

Takahashi, S., 1974. Sexual maturity of Isaza, (*Chaenogobius isaza*) I. The seasonal changes of growth and sexual maturation. Bull. Japan. Soc. Sci. Fish. 40: 847–857. (In Japanese)

Takahashi, S., 1981a. Sexual maturity of the Isaza, *Chaenogobius isaza*. II. Gross morphology and histology of the ovary. Zool. Mag. 90: 54–61.

Takahashi, S., 1981b. Vertical distribution and diel migration of Isaza, *Chaenogobius isaza*, Pisces in Lake Biwa. Zool. Mag. 90: 145–151.

Takahashi, S., 1981c. Sexual maturity of the Isaza, *Chaenogobius isaza*. III. Effects of water temperature on vitellogenesis. Zool. Mag. 90: 265–270.

254

Takahashi, S., 1982. Sexual maturity of the Isaza, *Chaenogobius isaza*. IV. Vitellogenesis and the subjective diel thermoperiod caused by vertical migration. Zool. Mag. 91: 29–38.

Takahashi, S. & T. Hidaka, 1984. The offshore life of Isaza larvae *Chaenogobius isaza* in Lake Biwa. Bull. Japan. Soc. Sci. Fish. 50: 1803–1809.

Tokui, T. & H. Kawanabe, 1984. Fishes. In S. Horie (ed.) Lake Biwa. Monographiae Biologicae 54. Dr W. Junk Publishers, Dordrecht, Boston, Lancaster: 339–360.

Tomoda, Y., 1984. Fauna and flora of Lake Biwa. An approach to the recent geohistory of the Lake Biwa. Bull. Japan Sea Res. Inst., Kanazawa Univ. 16: 59–91. (In Japanese)

Uéno, M., 1973. Tankyakumoku, Amphipoda. In M. Uéno (ed.) Tamiji Kawamura's freshwater biology of Japan. [Nippon tansui seibutsugaku.] Zukan no Hokuryukan, Tokyo: 467–472. (In Japanese)

Uéno, M., 1975. Evolution of life in Lake Biwa. In S. Horie (ed.) Paleolimnology of Lake Biwa and the Japanese Pleistocene. 3: 5–13.

Uéno, M., 1984. Biogeography of Lake Biwa. In S. Horie (ed.), Lake Biwa. Monographiae Biologicae 54. Dr W. Junk Publishers, Dordrecht, Boston, Lancaster: 625–634.

Watanabe, N. C., 1970. Studies on three species of *Semisulcospira* in Lake Biwa. I) Comparative studies of shell form and habitat. Venus. 29: 13–30. (In Japanese)

Watanabe, N. C., 1980. Some factors affecting the distribution and abundance of the two melanian snails, *Semisulcospira decipiens* and *S. reticulata*, in Lake Biwa. Jap J. Limnol. 41: 212–224. (In Japanese)

Watanabe, N. C., 1984. Studies on taxonomy and distribution of the freshwater snails, genus *Semisulcospira* in the three islands inside in Lake Biwa. Jap. J. Limnol. 45: 194–203.

Yamamoto, K., 1967. First report of the regular limnological survey of Lake Biwa (Oct. 1965–Dec. 1966) III. Zooplankton. Mem. Fac. Sci., Kyoto Univ., Ser. Biol. 1: 62–77.

Yamamoto, K., 1968. Second report of the regular limnological survey of Lake Biwa (1967). II. Zooplankton. Mem. Fac. Sci., Kyoto Univ., Ser. Biol. 2: 92–106.

Yasuno, T., 1983. Fossil pharyngeal teeth of sub-family Cultrinae collected from the Miocene Kani Group and Plio-Pleistocene Kobiwako Group in Japan. J. Fossil Res. 16: 41–46. (In Japanese)

Yokoyama, T., 1984. Stratigraphy of the Quaternary system around Lake Biwa and geohistory of the ancient Lake Biwa. In S. Horie (ed.), Lake Biwa. Monographiae Biologicae 54. Dr W. Junk Publishers, Dordrecht, Boston, Lancaster: 43–128.

Index